Magnetic Reconnection

Magnetic Reconnection

A Modern Synthesis of Theory, Experiment, and Observations

Masaaki Yamada

PRINCETON UNIVERSITY PRESS

PRINCETON AND OXFORD

Published by Princeton University Press
41 William Street, Princeton, New Jersey 08540
99 Banbury Road, Oxford OX2 6JX

press.princeton.edu

ISBN (cloth) 978-0-691-202419
ISBN (pbk) 978-0-691-180137
ISBN (e-book) 978-0-691-232980

British Library Cataloging-in-Publication Data is available

Editorial: Ingrid Gnerlich and Whitney Rauenhorst
Cover Design: Wanda España
Production: Jacqueline Poirier
Publicity: Matthew Taylor (US) and Charlotte Coyne (UK)

Cover Image: Graphic representation of the Magnetospheric Multiscale (MMS) mission poster, without text. Courtesy of NASA

Printed on acid-free paper ∞

Printed in the United States of America

10 9 8 7 6 5 4 3 2 1

Contents

Preface

Magnetic reconnection is an important fundamental process at work in laboratory, space, and astrophysical plasmas, in which magnetic field lines change their topology and convert magnetic energy into plasma particles by acceleration and heating. Magnetic reconnection occurs explosively in solar flares, coronal mass ejection (CME), in the interaction of solar winds with the earth's magnetosphere, and during the self-organization process of current-carrying fusion plasmas. An explosive CME with an ejected energy of 5×10^{30} erg, which is equivalent to 100 million "H-bombs" (each equivalent to 1 million tons of TNT explosive), is caused by magnetic reconnection and strongly affects our satellite communications and the welfare of space astronauts. A sudden disruption of plasma current in tokamak reactors is caused by magnetic reconnection, and control of this phenomenon is key to the successful realization of a tokamak fusion reactor. At the Princeton Plasma Physics Laboratory, Princeton University, I have been working for three decades with my team to understand the mechanisms of magnetic reconnection using dedicated laboratory facilities.

The topic of magnetic reconnection physics, which has remained a fascinating and important fundamental area of scientific investigation for many years, encompasses significant seminal contributions, as well as active ongoing research in laboratory and natural space and astrophysical plasma environments. In this monograph, I will present the fundamental physics of magnetic reconnection in laboratory and space plasmas, starting from the basic concept, theory, and observations from space satellites. After a brief review of the well-known early work on its concept, important recent experimental and theoretical works will be compared to validate modern significant findings. In the area of local reconnection physics, many important findings and discoveries have been made in the last two decades based on two-fluid physics, which describes the dynamics of electrons and ions separately. Profiles of the reconnection layer, Hall currents, and the effects of a guide field, collisions, and microturbulence will be discussed to understand the fundamental processes in the reconnection layers in both space and laboratory plasmas. The essential feature of reconnection is that it energizes plasma particles by accelerating and heating them, thus converting the magnetic energy to particle energy. Despite the long history of reconnection research (more than a half century), how this energy conversion occurs has been a major unresolved problem in plasma physics. In this monograph we will address the energy flow processes extensively and discuss the physics of energy conversion and partitioning in the reconnection layer. Collaborative research accomplishments between laboratory experiments and space observations will be extensively discussed.

Recently, several good books have been published on magnetic reconnection. While some of them are excellent collections of multiple contributions from experts who attended specific conferences, they often lack a coherent view on the physics of magnetic reconnection. This book is aimed at senior graduate students, post docs, and general researchers who would like to learn the present status of this field and the essence of magnetic reconnection physics, as well as recent advancements. Magnetic reconnection is a very popular subject in plasma physics, with hundreds of papers being published every year. While I have tried to cover most of the key processes of magnetic reconnection, I have made no attempt to comprehensively review the entire field. The work presented in this monograph centers on work familiar to me and is thus limited due to the space allowed for the book and to my knowledge.

This monograph is made primarily of two parts. The first six chapters are devoted to the historical development of concepts of magnetic reconnection, theory, numerical simulations, and major laboratory experiments. With a brief review of well-known work on its concept together with the most recent results, typical reconnection phenomena observed in space and laboratory plasmas are presented in chapter 2 and important theoretical progress based on magnetohydrodynamics (MHD) is described in chapter 3. A one-dimensional Harris sheet equilibrium with kinetic physics is studied in chapter 4, referring to both early theoretical and early experimental results. Chapter 5 describes the evolution of two-fluid physics and formulation. In chapter 6, the primary laboratory experiments of past and present times are described.

The remaining nine chapters describe the most recent advancements of research on magnetic reconnection, which have benefited from collaboration between laboratory experiments, space observations, and numerical simulations. Chapters 7 and 8 are devoted to observations of magnetic reconnection in astrophysical plasmas, in particular reconnection in solar flares, CME, reconnection in the Crab Nebula or supernova, and the dynamics of the magnetic reconnection layer in the magnetosphere. Chapter 9 is devoted to magnetic self-organization in laboratory plasmas or global reconnection phenomena. In chapter 10 we address extensively the energy flow processes and present the mechanisms of energy conversion and partitioning that have been discovered in the recent few years. Chapter 11 covers the most recent studies of the energy inventory in the reconnection layer, including the author's analytical model of energy conversion from magnetic field to plasma particles. In chapter 12 we directly compare the dynamics and energetics of the asymmetric reconnection layer observed both in the laboratory plasma of the Magnetic Reconnection Experiment (MRX) and in the magnetopause by the Magnetospheric Multiscale Satellite (MMS). In chapter 13, after a brief introduction to MHD dynamo theory, we consider how magnetic field originates in the universe and how magnetic reconnection plays a role in the dynamo, putting a special focus on two-fluid physics dynamo mechanisms. In chapter 14 we consider magnetic reconnection in large systems. In astrophysical plasmas, the ratio of global to kinetic scales is large and the ratio of mean free path to plasma scales is small, thus MHD models are considered to be practical to treat space astrophysical phenomena. The appearance of multiple layers would become dominant particularly in large three-dimensional plasma systems. A short summary of the book is presented in chapter 15.

I am very grateful to the many physicists with whom I have worked over the years on the fascinating research area of magnetic reconnection. I especially thank Russell Kulsrud for his constant encouragement and theoretical support in writing this book and for his invaluable advice and insights throughout the course of my studies of magnetic reconnection. I also thank my long-time colleague Hantao Ji who has provided excellent energy and inspiration that have enabled us to achieve many exciting discoveries since the beginning of the MRX program. Many colleagues and my graduate students have made significant contributions to the research results cited throughout this book, as mentioned at the end of chapter 15. Finally, and most importantly, I express my deepest gratitude and appreciation to my wife, JoAnn, for her sincere, whole-hearted support of my research activities.

Magnetic Reconnection

Chapter One

Introduction

This monograph describes how our understanding of magnetic reconnection, a fundamental process in the universe, has developed from a classical concept based on magnetohydrodynamics (MHD) to a modern concept based on kinetic and two-fluid physics theory, by which many phenomena observed in laboratory and space plasmas are now explained.

1.1 CONCEPT OF MAGNETIC RECONNECTION AND ITS DEVELOPMENT

Magnetic reconnection is a fundamental physical process in which magnetic-field-line configuration changes its topology, leading to a new equilibrium state of lower magnetic energy. During this process, part of the magnetic energy is converted into the kinetic energy of plasma through acceleration or heating of charged particles, which is the most important aspect of magnetic reconnection. In astrophysical and laboratory plasmas, magnetic reconnection occurs ubiquitously, rearranging the configuration of magnetic field lines and simultaneously changing macroscopic quantities of plasmas such as flow and temperature. Magnetic reconnection is seen in the evolution of solar flares, coronal mass ejection, and in the interaction of solar winds with the earth's magnetosphere. It is considered to occur in the formation of stars. It also occurs during the self-organization process of current-carrying fusion plasmas.

In magnetic fusion devices, plasma is confined by the combined forces of internal and external magnetic fields. Thus, the interaction of magnetic field lines with plasma determines the confinement features of hot plasmas. In toroidal fusion devices, toroidal currents are usually induced to heat the plasma and generate magnetic field configurations that effectively confine the hot fusion plasma by compressing pinch forces. There is a remarkable feature common to these configurations: the plasmas constantly tend to relax to a quiescent state through global magnetic self-organization processes in which magnetic reconnection plays a key role. Understanding and controlling magnetic reconnection in fusion devices is essential to creating a reliable fusion reactor core.

Magnetic fields can be found everywhere in the universe at all scales: in the earth's magnetosphere, in the solar corona, and on larger scales from the interstellar medium to galaxy clusters. How are magnetic fields generated in the universe? How do they determine the properties of plasmas? Understanding magnetic reconnection provides

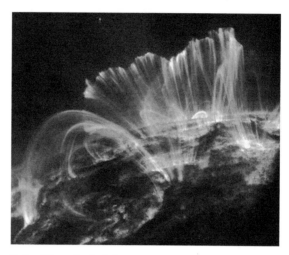

Figure 1.1. See Color Plate 1. Soft-X-ray picture from the TRACE satellite. [https: //www.solar-facts-and-advice.com/solar-flares.html]

a key to these fundamental questions about the universe. When magnetic energy significantly exceeds the plasma's kinetic or thermal energy, the magnetic energy tends to be converted to kinetic energy through magnetic reconnection. When there is abundant kinetic energy in a plasma with respect to magnetic energy, just like in the early universe ($W_p \gg W_B$), magnetic fields are considered to be generated through a converse process, a dynamo mechanism in plasma. Even in this dynamo process, magnetic reconnection often plays an important role.

Solar flares exhibit perhaps the clearest visual examples of magnetic reconnection and have been investigated for many decades. Through soft-X-ray pictures, which are considered to represent magnetic-field-line configurations of the solar atmosphere, we can visualize illuminating examples of the global topology change of plasma configurations (Tsuneta, 1996; Masuda et al., 1994; Gabriel et al., 1997; Golub et al., 1999; Lin et al., 2003). As shown in TRACE satellite data (Golub et al., 1999; figure 1.1), the topologies of soft-X-ray images are seen to change within a timescale of minutes or hours in the solar atmosphere, in which the magnetic diffusion time for a typical flare, based on the classical calculation for collisional diffusion, is estimated to be as long as 1 million years. These observations suggest the presence of fast changes of the global field-line topology, implying the existence of an anomalously fast magnetic reconnection process. Giovanelli (1946) noted that the abundant magnetic field energy in the chromosphere could be converted to electron kinetic energy during this process. Although the theory of MHD was not used in his calculation and the evolution of the sunspot field was treated as though it was a low-frequency wave, satellite measurements later showed that his concept is indeed valid and can be applied to solar corona reconnection.

In the early days of plasma research, a powerful way of describing the plasma dynamics was developed based on MHD, which treat plasma as a one-fluid element.

MHD theory was built upon the foundation of hydrodynamics by implementing the theory of electromagnetism. This MHD was found to be very effective, particularly when the Lundquist number (which is defined as the ratio of the magnetic diffusion time ($= \mu_0 L^2 V_A / \eta$) to the crossing time of the Alfvén waves ($= L / V_A$) in the region) is high ($S \gg 1$): $S \sim 10^{12}$ for a solar flare plasma of 10,000 km and $S > 10^6$ for tokamak plasmas. In this situation, plasma dynamics can be formulated based on the principle of flux freezing, namely that plasma always moves with magnetic field lines (as if it is frozen to them) with no dissipation. We call this principle "ideal MHD" dynamics. In ideal MHD, the plasma resistivity caused by collisions between electrons and ions and the viscosity caused by like-particle collisions are neglected in most cases. On the other hand, it was also realized that ideal MHD breaks down in a region of magnetic reconnection because the flux freezing principle does not hold in reconnecting plasmas. In other words, magnetic reconnection, in which field lines change their topology inducing magnetic energy dissipation at the reconnection layer, cannot be described by ideal MHD.

How do magnetic field lines move around in plasmas and how do they reorganize? Ideal MHD, developed in the early 1950s, describes the dynamics of highly conductive plasmas, where the electric field parallel to the magnetic field line, $E_\parallel$, vanishes (Sweet, 1958; Parker, 1957; Vasyliunas, 1975; Dungey, 1995). In this idealized model, magnetic field lines always move with the plasma and remain intact and never break or tear apart, as we will see in chapter 3 in detail. To consider the physical picture of this situation more precisely, we can represent any magnetic field by a set of lines that fills the system. The lines are tangent to the magnetic field and their density equals the field strength. If the system is time dependent, the features of the lines are different at every instant. If the plasma moving with the field lines is infinitely conducting, a physical identity can be assigned to the lines. If the magnetic field lines move with the plasma, they will continue to represent the magnetic field at any later time. This allows us to picture the magnetic field clearly. The field thus consists of strings embedded in the plasma which are neither created or destroyed. The magnetic force is represented by imagining the strings to have longitudinal tension and transverse pressure. If the strings are sharply bent, the curvature force replicates the magnetic tension force. If the lines are put closer together and bunched in a region, there is a transverse force due to the magnetic pressure force. Any plasma on a given line stays on that line as it moves, and cannot move to another line. This is basically the flux freezing feature associated with ideal MHD. We will revisit this concept later in detail in chapter 3.

Let's consider two magnetic field lines that are approaching each other in a small region of plasma. Outside this region, plasma fluid is frozen to field lines as described by ideal MHD. When the two field lines approach very close at an angle in a narrow region (figure 1.2), the magnetic field gradient becomes large. This interaction of magnetic field lines generates a current sheet due to Ampère's law $\nabla \times \boldsymbol{B} = \mu_0 \boldsymbol{j}$. We note that since the presence of a current sheet requires different motions of electrons and ions, strictly speaking this phenomenon cannot be described by single-fluid ideal MHD theory. The exact treatment of this region requires two-fluid physics as described in chapters 4 and 5. In MHD theory, we called this a diffusion region. In this area, the field lines are not frozen to the plasma, and they lose their identity, break, and reconnect.

Magnetic reconnection

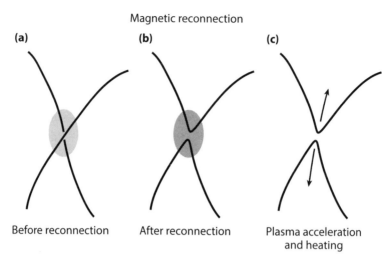

Figure 1.2. Schematic view of magnetic reconnection. After reconnection, plasma heating and acceleration follow.

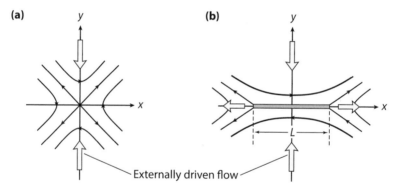

Figure 1.3. Formation of current sheet by externally driven flow. [From Biskamp (2000).]

After reconnection occurs, the two newly connected field lines accelerate plasma fluid due to a tension force generated by the reconnection. This interaction of field lines leads in most cases to a singular sheet of high current density in plasma where $E_\parallel$ becomes sufficiently large ($E_\parallel = E \cdot B/B \neq 0$) to induce nonideal-MHD plasma behavior and to cause the magnetic field lines to lose their connectivity and identity. This is why we call it a diffusion region.

As shown in two-dimensional geometry by figure 1.3, Dungey (1953) showed that such a current sheet can indeed be formed in a plasma by the collapse of the magnetic field near an X-type neutral point, and suggested that "lines of force can be broken and rejoined in the current sheet." We note that if it were not for a plasma, this would

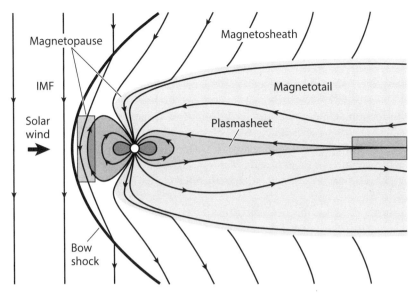

Figure 1.4. Cross-section of the simplest model of the magnetosphere in the day and night meridian. [From https://mms.gsfc.nasa.gov/science.html.]

not happen. Instead, the two opposing field lines would meet with an X-type crossing of angle 90 degrees, satisfying the Maxwell equations in a vacuum, $\nabla \times \boldsymbol{B} = 0$ (because there is no current sheet) and $\nabla \cdot \boldsymbol{B} = 0$. This sheet in a plasma is called a neutral sheet or a current sheet. As previously mentioned, it is often called a diffusion region since magnetic field lines lose their connectivity, diffuse, and reconnect in the sheet. When the field lines are reconnected, the topology of magnetic configuration changes and $\boldsymbol{j} \times \boldsymbol{B}$ forces expel the plasma from the diffusion region and result in the conversion of magnetic energy into kinetic energy. Thus, it is important to note that while the topology of magnetic configurations changes by magnetic reconnection, the conversion of magnetic energy to kinetic energy occurs at the same time and the plasma gains energy. This is a very important aspect of magnetic reconnection, as mentioned before.

An important example of flux freezing and magnetic reconnection in a space plasma is shown in figure 1.4, illustrating a simplified two-dimensional schematic picture of the solar-wind interaction with the earth's magnetosphere. The plasma on the incoming solar wind is embedded on solar-wind magnetic field lines that are separated from the magnetospheric lines. In the ideal MHD picture, there is no way for the solar-wind plasma and energetic particles to penetrate into the earth's magnetosphere, owing to the flux freezing principle. The solar wind is accordingly forced to move around the magnetosphere and is blown downstream.

At the magnetopause, where the solar wind presses against the magnetic pressure of the earth's dipole field, the interacting region becomes very thin, and the motions of plasma particles, ions, and electrons are quite different with respect to the magnetic fields of both sides and in the thin magnetic reconnection region. To describe this type of reconnection layer, a more general theory than MHD is necessary for a proper treatment of the neutral layer, that takes into account the different behaviors of electrons and ions. Reconnection layers, such as those created in the magnetopause (Vasyliunas, 1975; Dungey, 1995; Kivelson and Russell, 1995), have thicknesses that are comparable to the ion skin depth c/ω_{pi} ($\sim 50\,\mathrm{km}$). This situation leads to strong two-fluid effects, especially the Hall effect, in which magnetized electrons flow perpendicular to the magnetic fields in the neutral sheet. This effect induces a large reconnection electric field at the reconnection region and is thus considered to be responsible for speeding up the rate of reconnection, which is larger than the classical MHD rate (to be described in chapter 5).

In such a situation, magnetic reconnection takes place at the front and the tail parts of the magnetosphere, even if the plasma is truly infinitely conducting. Because of magnetic reconnection, some of the solar-wind lines break near the surface separating them and they reconnect to lines in the magnetosphere, which also break. As a result, some of the solar-wind lines end up attached to the magnetosphere, allowing the solar-wind plasma to penetrate the magnetosphere. This process can be regarded as the converse of flux freezing because of flux dissipation. Solar cosmic rays can also get into the magnetosphere because of magnetic topology changes and are often measured.

How such physical processes occur and how fast line breaking takes place have been the subjects of research for more than a half century. Thanks to recent collaborative research using observations, experiments, and numerical and theoretical works, significant progress has been made in understanding magnetic reconnection. Early work based on elementary MHD physics demonstrated the possibility of reconnection, but predicted reconnection rates that are too slow to explain the observations. As a result of the application of more advanced physics that take into account two-fluid physics and the kinetic effects of plasma particles, much insight has been obtained, and the reasons why reconnection is so much faster than first theorized have become clearer. It is essentially a partial breakdown of the remarkable property of flux freezing described by ideal MHD.

Thus, magnetic reconnection in the magnetosphere is treated using the two-fluid theory. It should be noted, however, that the flux freezing concept can still be applied in a modified form to the two-fluid regime in which electrons are still magnetized but ions are not. In this regime, magnetized electrons move with field lines for the most part, as if the flux freezing principle works only for electron fluid. On the other hand, ions are generally not magnetized and the different motions of electrons and ions can generate electric field in the reconnection plane. They also induce a large Hall electric field in the out-of-reconnection plane and as a result cause a fast reconnection as described in chapter 5. The induced electric fields introduce a new strong mechanism of particle acceleration and heating. This regime is sometimes called the electron-MHD regime. The region in the center of the reconnection layer, where even electrons do not move with field lines anymore and diffuse, is called the electron diffusion region. A good part of this monograph is devoted to a description of the key dynamics of this unique reconnection region.

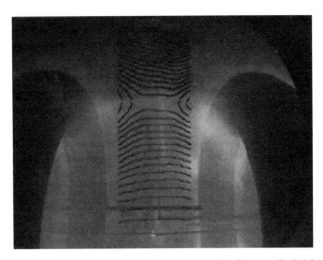

Figure 1.5. See Color Plate 1. Picture (time integrated) of controlled driven reconnection discharges in MRX and a flux plot from magnetic probes from an MRX movie. The flux contours deduced by assuming toroidal symmetry are considered to represent magnetic field lines without guide field. [https://mrx.pppl.gov/mrxmovies/Collisio nal.mov]

1.2 RECENT DEVELOPMENT AND PROGRESS OF UNDERSTANDING MAGNETIC RECONNECTION

Progress in understanding the physics of magnetic reconnection has been made in three research fields of the discipline in the past several decades: space and astrophysical observations, theory and numerical simulations, and laboratory plasma experiments. Space and astrophysical observations have provided much key suggestive evidence that magnetic reconnection plays an important role in natural plasmas and have generated strong motivation for fundamental research. Theory and numerical simulations provide important analysis and insights to help break down the complex reconnection phenomena into a set of fundamental key processes and to gain understanding of each process. Magnetic fusion plasma experiments provide examples of magnetic reconnection through self-organization of the plasma configurations. Laboratory experiments dedicated to the study of the fundamental reconnection physics can measure quantitatively the characteristics of reconnection dynamics by monitoring the essential plasma parameters simultaneously at a large number of points in the reconnection region (Yamada et al., 2010).

Figure 1.5 presents an example of contours of magnetic flux which were deduced from experimentally measured data using internal magnetic probes located at multiple locations in the reconnection region of the Magnetic Reconnection Experiment (MRX; Yamada et al., 1997a; Yamada, 1999). Dedicated laboratory experiments quantitatively cross-check theoretically proposed physics mechanisms and models, and provide a bridge between space observations and theoretical ideas, such as two-dimensional two-fluid reconnection models, by generating a typical reconnection layer. On the other

Figure 1.6. MMS satellite mission. Four satellites measure key components of local plasma parameters to document the electron and ion dynamics. [From https://en.wiki pedia.org/wiki/Magnetospheric_Multiscale_Mission.]

hand, space satellites can provide detailed data at selected points with simultaneous multiple sophisticated diagnostics. Recent significant progress in data acquisition technologies has allowed us to directly compare the observed data from satellites and laboratory experiments recently published (Yamada et al., 2018). In laboratory experiments, even an evolution of magnetic field lines was able to be monitored with respect to time. Remarkably, through this cross-cutting research, a new common picture of the two-fluid magnetic reconnection layer has emerged, aided by numerical simulations mostly performed in two-dimensional geometry. We use a significant part of this monograph to describe the two-fluid physics mechanisms that have become clearer through our cross-discipline studies.

In particular, a new cluster satellite system, called the Magnetospheric Multiscale Satellite (MMS) was launched in March 2015. Their mission goal was to explore the physics of magnetic reconnection in spatial scales extending down to the thin electron skin depth. Figure 1.6 shows a graphic picture of four satellites that measure key components of local plasma parameters to document the electron and ion dynamics. The four spacecraft are placed at times in a tetrahedral configuration with a separation of about 7–10 km, or $\sim$ 3–5 times the expected value of the electron skin depth at the magnetopause. Since the current sheet moves past the spacecraft at speeds of over 100 km/s, resolving these fine-scale structures requires field measurements at a 1 ms cadence and particle distribution function measurements at a 20 ms cadence, which is challenging for a spacecraft mission.

To date, the MMS mission has made many significant findings, identifying the structure and the dynamics of the electron diffusion region both in the magnetopause and the magnetotail reconnection layer. In the first phase of the MMS mission, the dayside magnetosphere reconnection region was investigated. At the subsolar magnetopause, where the solar-wind plasma meets the magnetospheric plasma, reconnection is very asymmetric with an upstream plasma density larger than that of the magnetosphere by a factor of 10–20. Subsequently, the magnetic field strength is smaller by a factor of 2 to 3. This asymmetric reconnection is of much interest and is often very important for real physical situations in both space and astrophysical plasmas (Mozer

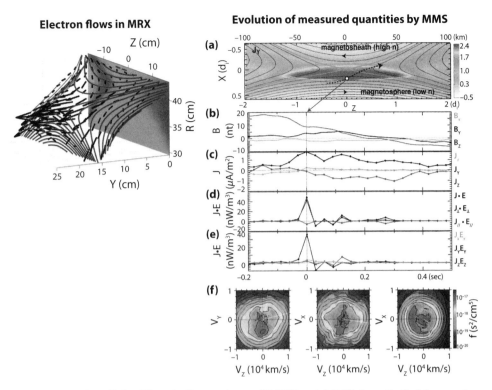

Figure 1.7. See Color Plate 1. Comparison of MRX and MMS data. Left: Measured electron flow vectors in MRX (red arrows). Measured magnetic flux contours are shown by blue lines. Right: (a) Approximate MMS trajectory through the electron diffusion region of the magnetosphere. The trajectory is determined based on a comparative study of MMS data and 2D numerical studies. (b)–(f) Time evolution of key components of local plasma parameters showing that $\boldsymbol{J}_\perp \cdot \boldsymbol{E}_\perp$ becomes maximum at the electron diffusion region (d). The electron velocity distributions in (f) show that they predominantly flow in the Y-direction as shown in the MRX data. The documented MMS data are remarkably consistent with the electron dynamics measured by MRX. [From Yamada et al. (2018).]

and Pritchett, 2011). Recently, through a collaboration between MMS research and MRX, the key physics of asymmetric reconnection have been intensively investigated and illuminated (Yoo et al., 2017; Yamada et al., 2018).

Both in MRX and in the magnetopause plasmas, the length of the reconnection layer was measured to be very similar, about 3 times the ion skin depth, indicating the same physics mechanisms are at play. Taking advantage of this situation, the dynamics and energetics of the magnetic reconnection layer were comparatively studied in the context of two-fluid physics. Despite huge differences between the length scales of the reconnection layers ($2L \sim 30$ cm in MRX versus ~ 250 km in the magnetopause) and the ion skin depths ($d_i \sim 5$–6 cm in MRX versus ~ 50 km in the magnetopause), remarkably similar characteristics are observed regarding the dynamics of electrons and ions, as well as energy deposition profiles and energy partitioning. Let us look into common characteristics observed in MRX and MMS.

Figure 1.7 shows a comparison of MRX and MMS data sets in different formats. Electron flow vectors (red arrows) measured in MRX show strong out-of-reconnection-plane electron flows. Also, a strong energy deposition to electrons is measured to occur through a large value of the $j_e \cdot E$ quantity (Yamada et al., 2014) in MRX. Here it should be noted that the uppercase letter J was used for electron current density in the MMS data set, while lowercase j was used in the MRX data as well as in most of this book. On the right, panels (a)–(f) show the time evolution of key components of local plasma parameters, documenting the electron dynamics in the magnetopause. These MMS data show a strong spike in the quantity $j_e \cdot E$ when the satellite system flies through the region just south of the X-point where reconnecting field lines meet and reconnect. This observation is in remarkable agreement with the profile of electron flow vectors measured on MRX as seen in the left-hand panel of figure 1.7. When the energy deposition rate to electrons, $j_e \cdot E$, is decomposed into $j_{e\perp} \cdot E_\perp + j_{e\parallel} E_\parallel$, i.e., separating the inner product into perpendicular and parallel components with respect to the local magnetic field lines, $j_{e\perp} \cdot E_\perp$ is measured to be significantly larger than $j_{e\parallel} E_\parallel$ as shown in panel (d) of figure 1.7. In addition, the measured electron velocity distributions in the three directions are consistent with the MRX data of electron flow vectors shown on the left.

Further observational verifications of electrons' motion frozen to field lines outside the electron diffusion region were made both in MRX (Yoo et al., 2013) and MMS (Burch et al., 2016b), and excellent agreement was found between the dynamics and energetics of electrons. This agreement demonstrates that the same two-fluid mechanisms in two-dimensional analysis operate well in both systems, despite vastly different scales ($\sim 10^6$), while various three-dimensional phenomena including micro-fluctuations are expected to be involved. This will be discussed in more detail in chapter 12.

1.3 MAJOR QUESTIONS

We address the following major questions, which have been studied intensively for the past 30 years:

(1) Why is the reconnection rate so fast in collisionless conductive plasmas? What is a scaling for the reconnection rate on collisionality?
(2) What are the mechanisms of magnetic reconnection in collisionless plasma? How does two-fluid physics influence the dynamics and speed of local reconnection? What determines the structure of the reconnection layer?
(3) How is magnetic energy converted to the kinetic energy of electrons and ion? In what channel does the energy flow take place?

(4) How do fluctuations and turbulence affect the reconnection dynamics or vice versa? Which fluctuations are most relevant, how are they excited, and how do they determine the reconnection rate and influence the conversion of magnetic energy?

(5) How do reconnection features change as the size of the plasma system increases? How are plasmoid structures formed and how do they influence the reconnection rate?

(6) How is the local physics, which has been studied in great detail, connected to the large global environment around the reconnection layer? How is the reconnection layer generated in a global boundary of different sizes?

(7) Why does reconnection occur impulsively in most cases?

Keeping these questions in mind, we will study most of the significant modern experimental findings and discoveries in magnetic reconnection research and discuss many of the theoretical understandings to which they have led.

To begin, we review magnetic reconnection research and significant studies that have continued up to the present time, beginning with the well-known seminal ideas of Dungey, Sweet and Parker, and Petschek, based on MHD. While theory led the early research progress in this area, more recent research has been dominated by experiments and numerical simulations. Since the early work is fairly well known and presented in textbooks, we focus on recent findings and developments of most significance. There are a number of different views as to which physical processes are most important for reconnection. While the relative importance of two-fluid processes of a laminar current sheet versus three-dimensional fluctuation-induced effects of multiple reconnection sites or plasmoids are still debated, our goal is to provide a broad understanding of different theories and observations.

One of the most important questions has been why reconnection occurs much faster than predicted by classical MHD theory. During the past two dozen years, notable progress in understanding the physics of this fast reconnection has been made. Extensive theoretical and experimental work has established that two-fluid effects, resulting from the fundamentally different behavior of ions and electrons, are important within the critical layer where reconnection takes place. Two-fluid effects are considered to facilitate the fast rate at which reconnection occurs in the magnetosphere, stellar flares, and laboratory plasmas. Dedicated laboratory experiments and magnetospheric satellite measurements show strikingly similar data in the profiles of magnetic fields and electrostatic and magnetic fluctuations. Recent improvements in the understanding of the role of reconnection in magnetic self-organization processes in laboratory and space–terrestrial plasmas will also be covered in this monograph.

Despite the long history of reconnection research, how the conversion of magnetic energy occurs remains a major unresolved problem in plasma physics. A good amount of the recent studies on energy conversion are presented in the present monograph. In the past several years, it has been realized that energy conversion in a laboratory reconnection layer occurs in a much wider region than previously considered. The mechanisms for energizing plasma particles in the reconnection layer are identified, and a quantitative inventory of the converted energy is presented for the first time in a well-defined reconnection layer in a laboratory plasma study (Yamada et al., 2014, 2018). In this monograph, a new analytical study is considered for a key step toward resolving one of the most important problems in reconnection physics.

A special effort is made to cover both the major experimental results and recent space observations that have provided useful information on the physics of magnetic reconnection over the past few decades. This book is quite different in emphasis from recent review papers and books, which have emphasized theoretical aspects or results from numerical simulations. Since the main objective of this book is not a review, many fine works in this field are not covered because of space limitations, because of our primary focus on recent experimental findings, and because of our intention to convey my views to the readers.

Magnetic reconnection is a very popular subject in plasma physics. For some years, the numbers of papers submitted to annual meetings of the Division of Plasma Physics of the American Physical Society (APS) have exceeded 100 (out of 1,500–1,800 total papers). To cover wider aspects of the physics of magnetic reconnection, I would like to refer to the books by Priest and Forbes (2000), Biskamp (2000), Birn and Priest (2007), the reviews Zweibel and Yamada (2009), Yamada et al. (2010), and the collection of edited reviews in Gonzalez and Parker (2016). Magnetic reconnection research covers plasmas of many types, including weakly ionized, electron–positron pairs, and relativistic plasmas. The reader seeking special material should consult additional references including Uzdensky (2011) for reconnection in relativistic or astrophysical environments and Ji et al. (2022) for recent development of reconnection research in large systems.

An important perspective is that magnetic reconnection is influenced and determined both by local plasma dynamics in the reconnection region and global boundary conditions. One major question is how large-scale systems generate local reconnection structures through formation of current sheets—either spontaneously or via imposed boundary conditions. In this regard, we will look into the question of how multiple reconnection layers are formed in a large plasma system. When we consider a large system in which reconnection takes place, we think all classical models do not simply apply, particularly when long global lengths are assumed for the current layers. Recently, more research has been carried out on the formation process of current layers in a larger system and has found that a current sheet often breaks up to form multiple reconnection layers. It would be of great importance to develop and elucidate a general theory of current layer formation in a highly nonsymmetric magnetic equilibrium such as is observed in the magnetopause or the sun. We will address magnetic reconnection in the magnetopause where strong density asymmetry exists across the reconnection layer. There may be mechanisms to generate multiple small-scale current sheets in which field-line reconnection takes place with multiple X-lines. These structures can often be small enough to decouple the motion of electrons from that of ions in collisionless plasmas. These smaller-scale sheets can fluctuate, leading to faster reconnection, and a large number of these layers should lead to a large energy release as seen, for example, in the magnetosphere and the reversed field pinch (RFP) plasmas for fusion research. In RFP plasmas, reconnection in multiple layers of flux surfaces is observed to generate a significant magnetic self-organization of the global plasma, invoking strong ion heating. While we expect that a theory from a first principle can lead us to a breakthrough for solving this problem, we have recently initiated a new experimental effort to address this important issue (see chapter 14).

In this monograph, we describe the fundamental physics of magnetic reconnection at work in laboratory and space plasmas, starting from concept, theory, and observations from space satellites, and also the most important progress in the research fronts. With a brief review of the well-known work on its concept, together with the most recent results in chapter 1, typical reconnection phenomena observed in space and laboratory plasmas are presented in chapter 2, and important theoretical progress based on MHD is described in chapter 3. A one-dimensional Harris sheet equilibrium with kinetic physics is studied in chapter 4, referring to both early theoretical and experimental results. In the area of local reconnection physics, many findings have been made regarding two-fluid physics analysis and are related to the cause of fast reconnection. Chapter 5 describes the evolution of two-fluid physics and formulation. Profiles of the reconnection layer, Hall currents, and the effects of a guide field, collisions, and microturbulence are discussed in chapter 5 to understand the fundamental processes in reconnection layers in both space and laboratory plasmas. In chapter 6, the primary laboratory experiments of past and present times are described.

Chapters 7 and 8 are devoted to observation of magnetic reconnection in astrophysical plasmas, in particular reconnection in solar flares, coronal mass ejection, reconnection in the Crab Nebula or supernova, and the dynamics of the magnetic reconnection layer in the magnetosphere. Some readers may find chapter 8 to be too detailed and hard to follow since I use specific wording and descriptions used in the space physics community. Chapter 9 is devoted to magnetic self-organization in laboratory plasmas or global reconnection phenomena. In chapter 10, we address extensively the energy flow processes and present the mechanisms of energy conversion and partitioning, which have been discovered in the recent few years. Furthermore, more accurate recent satellite observations will be presented regarding magnetic reconnection and its energetics in space astrophysical plasmas and those will also be covered in this book. Chapter 11 covers the most recent studies of the energy inventory in the reconnection layer. In chapter 12, let us directly compare the dynamics and energetics of the asymmetric reconnection layer observed both in the laboratory plasma of MRX and in the magnetopause by MMS and discuss our results in the context of two-fluid physics, aided by simulations. In chapter 13 we consider how magnetic field is generated in the universe and how magnetic reconnection plays a role in the dynamo. Since the focus of this monograph is two-fluid physics mechanisms, we mainly consider here the two-fluid effects of dynamo action in fusion laboratory plasmas, after a brief introduction to MHD dynamo theory. In chapter 14 we consider magnetic reconnection in large systems. In astrophysical plasmas, the ratio of global to kinetic scales is large and the ratio of mean free path to plasma scales is small, thus MHD models are often considered to be practical to treat space astrophysical phenomena. The appearance of multiple layers would become dominant, particularly in large three-dimensional plasma systems. Readers who might find it difficult to follow the detailed technical description of results in some chapters, such as 8 and 9, might be recommended to skip them and move on in order to grasp the whole picture of magnetic reconnection.

Chapter Two

Magnetic reconnection observed in space and laboratory plasmas

2.1   MAGNETIC RECONNECTION IN SOLAR FLARES

Solar flares have been central objects for studying the physical mechanisms of magnetic reconnection. Since the inception of the concept, magnetic reconnection has been thought to play a major role in the evolution of solar coronae as well as in CMEs (coronal mass ejection (Parker, 1979; Priest and Forbes, 2000). The topologies of soft-X-ray pictures of coronae are seen to change within a timescale of minutes or hours, much shorter than magnetic diffusion time and even than classical Sweet–Parker time (Parker, 1957). The study of the dynamics of solar flares has been intensified through detailed pictures of solar coronal activities photographed by Skylab satellites in the 1970s, through more modern satellites such as Yohkoh, SOHO, TRACE, RHESSI, SDO, and more recently Hinode, IRIS, and Parker. Recent satellites have revealed the solar atmosphere with unprecedented spatial and temporal resolution, covering spectral wavelengths from ultraviolet through soft- and hard-X-rays to gamma rays. With many large coronal loops seen actively interacting with themselves, as shown in figure 1.1, their topology is observed to change rapidly on a very short timescale of a few minutes during an eruptive phase.

It was conjectured that conversion of magnetic energy to plasma particles should occur during reconnection in the solar corona, where a much higher plasma temperature than that of the photosphere (~ 0.5 eV) is always observed. Finding the true cause of the heating of the corona to more than 10^6 degrees (100 eV) has been a major objective of solar plasma research (Birn and Priest, 2007). While there are other possibilities, such as wave heating, reconnection is a strong candidate for the coronal heating mechanism since the magnetic field represents the dominant energy source in the corona, exceeding the plasma's thermal pressure.

Sources of the magnetic field at the photosphere are dynamic and highly turbulent. The magnetic flux at the surface in the quiet sun is replaced every 14 hours (Hagenaar, 2001), which indicates the very dynamic nature of the solar magnetic field. Furthermore, Close et al. (2004) investigated the statistical properties of the magnetic field lines of the lower corona (under 2,500 km) by constructing magnetic field lines from magnetograms of the SOHO data, tracking them, and recalculating their connectivity. The tireless motion of these magnetic flux concentrations, along with the continual appearance and disappearance of opposite-polarity pairs of fluxes, releases a substantial amount of energy that may be associated with a whole host of physical processes in the solar corona. Their calculations show that the emergence and cancellation of flux in the photosphere has a profound effect on the corona. They concluded that the time for all the field lines to change their connection is the astonishingly short time of 1.4 hours, even in the quiet phase, since a substantial amount of reconnection occurs in response to emergence and cancellation of flux. This suggests that very fast reconnection is taking place at numerous places on the photosurface.

There are different types of eruptions in the solar atmosphere, such as CMEs, prominence eruptions, and eruptive flares, and they are considered to be related. CMEs are large-scale ejections of mass and magnetic flux from the corona into interplanetary space. They are thought to be produced by a loss of equilibrium in coronal magnetic plasma structures, which induces abrupt changes in magnetic topology. A typical CME carries roughly 10^{15} Wb of flux and 10^{13} kg of plasma into space (Priest and Forbes, 2000). During the active period of the sun, one CME is seen per day. The intermittent emergence of new flux from the convection zone and reshuffling of the footpoints of closed coronal field lines causes coronal field stress to accumulate. When the stress exceeds a certain threshold, the stability of the magnetic field configuration breaks and erupts. This model is called a storage-and-release model. While this plausible explanation is difficult to verify by observation due to the limited measurements of magnetic field topology, many numerical simulations and experimental investigations have been carried out recently, as described in chapters 7 and 9

Almost all active phenomena occurring in the solar atmosphere seem to be related to magnetic reconnection, directly or indirectly. This is probably a consequence of the universal nature of the low-β plasmas in the solar corona, where magnetic energy dominates over kinetic or thermal energy. It is expected that magnetic reconnection will have a significant impact on heating as well as dynamics in the solar corona. In addition, there is evidence that even dynamic phenomena in the chromosphere ($\beta \sim 1$) and photosphere ($\beta \gg 1$) may be related to reconnection. In the reconnection layer, where magnetic field changes drastically, currents are induced and concentrated in thin filaments, so that the magnetic energy density in the filaments is rapidly dissipated. Thus, once reconnection occurs in the filaments, the influence of reconnection can be significant. Also, recent stellar observations have reported many dynamic activities in various stars, such as jets and flares from young stellar objects and binary stars. Even superflares have been discovered on many solar-type stars. These dynamic events are much more energetic than solar flares, but the basic properties of these explosive events appear to be similar to solar flares. Although evidence is still considered "indirect," both theories and observations suggest similarity between solar flares and stellar flares.

For the past half century, numerous theories and numerical simulations have been attempted in order to determine the mechanisms of the CME. The CSHKP model (Carmichael, 1964; Sturrock, 1966; Hirayama, 1974; Kopp and Pneuman, 1976) has been regarded as a standard model, while some modification was made later. Initially, coronal arcades of magnetic field lines are in equilibrium, supporting a high-density filament called a prominence, which resides on top of the arcade lines as shown in

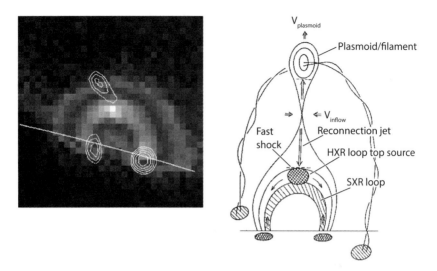

Figure 2.1. Hard-X-ray image of the top of an arcade from the Yohkoh satellite [from Masuda et al. (1994)] and a CME model [from Shibata et al. (1995)].

figure 2.1. When this equilibrium between the flare magnetic field with prominence and its overlaying coronal arcade breaks, a CME occurs.

An intriguing illustration of magnetic reconnection occurs in the wake of a CME. As the mass is ejected from the solar surface it pulls out field lines of a magnetic loop. As the ejected mass moves away from the sun, the opposing magnetic field lines of the loop are drawn out (see figure 2.1), and these field lines begin to reconnect at an X-point, which would form an X-line in three dimensions. The reconnection sends particles down the field lines and when the particles hit the surface they emit radiation that appears as a pair of ribbons at the photosurface. As more lines reconnect, the X-line rises and the ribbons separate outward correspondingly. The correlation of this rise and the separation of the foot ribbons nicely illustrates the reconnection event. (See Pneuman, 1984 for a more detailed description and Harvey and Recely, 1984 for a specific event.)

Some theoretical work has focused on two-dimensional models of the evolution of force-free magnetic arcades, in which field-line footpoints are advected by flows in the solar photosphere. A two-dimensional flux rope model has been proposed by Forbes and Priest (1995) to describe the eruptive processes of solar flares by a sequence of ideal MHD (magnetohydrodynamic) equilibria. They demonstrated that the equilibrium of a flux rope jumps from one state to the other through the formation of a current sheet or reconnection layer in the solar atmosphere. Recent work addresses three-dimensional effects.

Past and recent satellites have provided a wealth of observational evidence of magnetic reconnection. Cusp-shaped flare loops consistent with the classical CSHKP model were observed and a plasmoid ejection model was proposed (Shibata et al., 1995). The profiles of hard-X-ray emissions show evidence of particle acceleration at the top of soft-X-ray flares, concomitantly with the appearance of impulsive flares or CMEs. Masuda et al. (1994) observed that magnetic reconnection occurs above the flare arc as predicted by the classical CSHKP model for long-duration-event (LDE) flares and found that high-speed jets generated by reconnection intersect with the top of the reconnected flare loop to produce a hot region that is represented by strong hard-X-ray emission. Based on this observation, Shibata proposed a model modifying the earlier flare models as shown in figure 2.1. Yokoyama et al. (2001) measured the reconnection speed based on evolution of the soft-X-ray pictures from Yohkoh and concluded that the reconnection speed is in the relatively wide range of 0.001–$0.05 V_A$ (V_A is the Alfvén velocity). This reconnection rate is significantly larger than the classical Sweet–Parker rate, which will be described in chapter 3.

2.1.1 Particle heating and acceleration in solar flares

The features of energetic particles in solar flares are studied using the hard-X-ray and γ-ray imaging system of the Reuven Ramaty High-Energy Solar Spectroscopic Imager (RHESSI) (Lin et al., 2003). Emissions from energetic ions up to gigaelectron volt energies and electrons up to 100 MeV energies are observed during large solar flares (Lin, 2006). Comparative studies of these images with TRACE images show that the locations of these emissions overlap with the arcade footpoints on the photosphere, suggesting that these emissions are due to collisions of energetic particles with the solar surface. These results are consistent with the physical picture shown in figure 2.1. The observed energy spectra are often of a power-law form, and the estimated total energy from these particles can be as large as 50% of the total released energy of solar flares. These results suggest that there exist efficient mechanisms for accelerating nonthermal particles to high energies during magnetic reconnection. This poses a significant challenge to our understanding of magnetic reconnection physics as theoretical investigations on particle acceleration have just begun. It has been suspected that a large amount of change of flux in a short timescale would accelerate plasma particles to an enormously high energy level of multiple megavolts. One other idea proposed is based on Fermi acceleration from contracting magnetic islands with magnetic reconnection (Drake et al., 2006). Some signatures of magnetic islands correlating with energetic electrons have been observed in magnetospheric plasmas (Chen et al., 2008b).

The H_α emission is often accompanied by X-rays with energy levels of tens to hundreds of kiloelectron volts, microwave emission, and, in some flares, γ-ray emission. The hard-X-rays come due to bremsstrahlung from electrons with energies in the tens to hundreds of kiloelectron volts. The microwaves are due to gyro-synchrotron radiation from the same electrons, whereas the H_α is excited by the fast electrons as they slow down in the chromosphere. The γ-rays result from electron–positron collisional annihilation, neutron capture on protons, and the decay of excited nuclear states, and they are evidence that ions are accelerated at least to tens of megaelectron volts. The particle energy spectra are nonthermal and typically fit by broken power laws with spectral indices in the range 4–6. Most of the emission comes from the chromospheric footpoints of the coronal loops, where the high gas density makes the

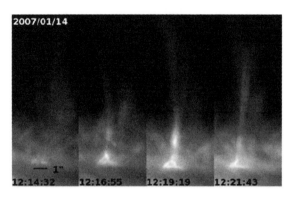

Figure 2.2. Observation of a plasma jet through time evolution of typical X-ray jets observed in the Ca II H broadband filter of Hinode/SOT. Times are shown in UT. [From Shibata et al. (2007).]

interaction time short. However, the presence of microwave hotspots at the loop tops, as well as other morphological evidence, suggests that the particles are accelerated in the corona. The emission is typically sustained for hundreds to thousands of seconds, but varies on timescales as short as several microseconds. This could be due either to an intermittent acceleration mechanism or to propagation effects (Zweibel and Haber, 1983).

2.1.2 Small flares and jets in low altitude coronae

Space-based solar observations revealed that the solar atmosphere is full of small-scale flares, called microflares, nanoflares, and even picoflares, and that these small-scale flares are often associated with jets. One example is the X-ray jet, discovered by Yohkoh/SXT (Shibata et al., 1992). There are many pieces of observational evidence that show that the jets are produced by magnetic reconnection (Shibata et al., 1995). Yokoyama and Shibata (1995) performed MHD simulation of reconnection between an emerging flux and an overlying coronal field and successfully explained the observational characteristics of X-ray jets on the basis of their simulation results.

A second solar satellite from Japan, called Hinode, was launched in 2006 to study the properties of the lower solar atmosphere. Its data are revealing much about the evolution of the chromospheric corona, which exists in the 500–2,000 km range of the solar atmosphere, as well as how plasma waves might transport energy to the corona. Using high-resolution images taken with SOT (Solar Optical Telescope), numerous tiny chromospheric anemone jets (whose apparent footpoint structures are similar to a sea anemone in three-dimensional space) were discovered in the active region of the chromosphere. Shibata et al. (2007) reported the ubiquitous presence of chromospheric jets at inverted-Y-shaped exhausts outside sunspots in active regions as shown in figure 2.2. These jets are typically 2,000 to 5,000 km long and 150 to 300 km wide, and their velocity is 10 to 20 km/s.

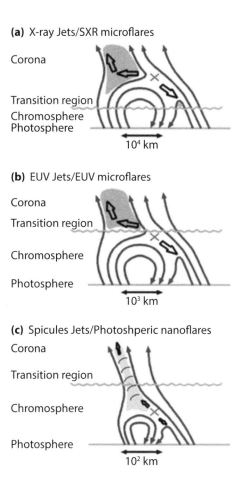

(a) X-ray Jets/SXR microflares

Corona

Transition region
Chromosphere
Photosphere

10^4 km

(b) EUV Jets/EUV microflares

Corona

Transition region

Chromosphere

Photosphere

10^3 km

(c) Spicules Jets/Photoshperic nanoflares

Corona

Transition region

Chromosphere

Photosphere

10^2 km

Figure 2.3. A schematic illustration of magnetic reconnection that occurs at various altitudes in the solar atmosphere. (a) X-ray jets/SXR microflares. (b) EUV jets/EUV microflares. (c) Chromospheric anemone jets/nanoflares. [From Shibata et al. (2007).]

Since the morphology of the chromospheric anemone jets is quite similar to that of the coronal X-ray jets, it was suggested that magnetic reconnection occurs at the feet of these jets (Shibata et al., 1992), although the length and velocity of the jets are much smaller than those of the usual coronal jets (figure 2.3). They suggested that magnetic reconnection similar to that seen in the corona occurs at a much smaller spatial scale throughout the chromosphere and that the heating of the solar chromosphere and corona may be related to small-scale reconnection, as seen in figure 2.3. We will describe the mechanisms of magnetic reconnection in X-ray jet flares in more detail. On the other hand, De Pontieu et al. (2007) found evidence of Alfvén waves propagating with a speed of 10–25 km/s and argue that the waves are energetic enough to accelerate the solar wind and possibly heat the quiet corona. With these two different views from the same satellites, investigations have been continued in order to answer one of the most important questions: How is the solar corona heated to a temperature more than two orders of magnitude higher than that of the photosphere?

2.1.3   Past and recent observations and development

It is important to note that the global magnetic energy of solar coronae is stored through a slow emergence of flux from the photosphere. If reconnection happens quickly all the time, CME does not develop. One physical mechanism that can trigger a storage-and-release eruption, CME, is a loss of equilibrium that occurs when the vertical force balance of the magnetic field breaks down suddenly. This is called torus instability. The basic idea is that an upward perturbation of the flux rope will be unstable if the downward restraining forces acting on the rope decay more quickly with height than do the upward driving forces. For torus instability, the restraining forces are assumed to be generated primarily by the interaction between the flux rope and an ambient "strapping" magnetic field and are generally parallel with the solar surface. Thus, if this strapping field decreases too quickly with height, then its associated restoring force is not strong enough to prevent the flux rope from erupting. While this concept was developed in fusion research, Kliem and Török (2006) derived the condition for torus instability. They used a somewhat more general formulation than the one by Bateman (1978) and treated the time evolution of the instability. They considered two cases: a freely expanding toroidal ring relevant in the laboratory and for CMEs and an expanding ring with constant total current, which captures an important effect of the footpoint anchoring on an expanding partial ring and can be relevant in the initial stage of CMEs. They concluded that torus instability is a possible mechanism for CME by justifiably neglecting the effects of gravity and plasma pressure.

In order to study the mechanisms of solar flare eruptions, several laboratory experiments have been carried out in past decades by simulating an eruption in the laboratory (Hansen and Bellan, 2001; Soltwisch et al., 2010; Tripathi and Gekelman, 2010). These laboratory arched-flux-rope experiments relied on the dynamic injection of either plasma or magnetic flux at the footpoints in just a few Alfvén times, $\tau_A = L/v_A$. Recently, a laboratory experiment was carried out on MRX (Myers et al., 2015). In this new MRX setup, an experimental study of magnetic reconnection beyond the local reconnection layer was carried out, considering the impact of impulsive reconnection phenomena on the global topology of astrophysically relevant laboratory plasmas. The special plasma configuration studied here was that of an arched, line-tied magnetic flux rope. This configuration is of particular interest due to its central role in storing and releasing magnetic energy in the solar corona (Chen, 1989; Titov and Démoulin, 1999; Savcheva et al., 2012). Recent MRX flare experiments enforced a strict separation of timescales between the footpoint driving time $\tau_D \sim 150\,\mu s$ and the dynamic Alfvén time $\tau_A \sim 3\,\mu s$, such that the observed eruptions were driven by storage-and-release mechanisms. It was shown that toroidal magnetic flux generated by magnetic relaxation (reconnection) processes generates a new stabilizing force that prevents plasma eruption. The results lead to the discovery of a new stabilizing force for solar flares

(Myers et al., 2015). A detailed description of this experiment and the analysis will be presented in chapter 9.

2.2 MAGNETIC RECONNECTION IN THE MAGNETOSPHERE

The solar wind travels inside the solar system, carrying magnetic fields with it. When the solar wind interacts with a planetary magnetic field, or with other solar winds with different velocity vector components, magnetic boundary layers develop. In these boundaries, current sheets develop and magnetic reconnection occurs. When the field lines meet nearly antiparallel in these boundaries, a current sheet develops with the magnetic field becoming zero (neutralized) at the sheet center. It is sometimes called a neutral sheet. Current sheets are seen on both the dayside (magnetopause) and the nightside (magnetotail) of the earth's magnetosphere as shown in figure 2.4 (Dungey, 1995; Vasyliunas, 1975; Kivelson and Russell, 1995) at places where interactions occur between the magnetic fields of the solar wind and the earth's dipole field. Such current sheets are expected around all other magnetized planets.

Satellite observations showed that the current sheath thickness is of the order of the ion skin depth or the ion gyroradius. We note that in the magnetopause where the solar winds meet the magnetosphere, the magnetic pressure of the magnetosphere is almost equal to the plasma pressure ($\beta \sim 1$) of solar winds and the ion skin depth is thus about the same as the ion gyroradius. The ion gyrodepth is typically 50–200 km in the magnetosphere, while it is larger by an order of magnitude (500–2,000 km) in the magnetotail. In this situation, the reconnection dynamics cannot be described by the conventional MHD theory of reconnection. This is because ions and electrons behave differently in the reconnection region, requiring two-fluid and kinetic physics. Also, the reconnection could be a very turbulent process, both in time (intermittent) and space (patchy), since the relative drift of electrons with ions can excite electrostatic or electromagnetic fluctuations. A number of researchers have observed electric and magnetic turbulence in the magnetopause as well as in the magnetotail.

The magnetopause is the boundary that separates the geomagnetic field and plasma of terrestrial origin from the solar-wind plasma (Hughes, 1995). Figure 2.4 is a schematic of the magnetosphere showing it on both the dayside and the nightside. On the dayside magnetopause, pressure balance is maintained between the incoming solar winds and the earth's magnetic field. Ampère's law applied across the boundary shows that currents have to flow in the boundary sheet as shown in the figure. On the nightside of the magnetosphere there is a magnetotail in which the lines of force stretch behind the earth in a direction away from the sun. As seen in figure 2.4, a current sheet is formed between the tail lobes, which is occupied by the magnetic field lines that connect to the two polar regions of the earth. The energy and plasma in these magnetotails are intermittently released into the inner magnetosphere during magnetic substorms. It is generally believed that, during a substorm, solar-wind plasma and energy are injected into the magnetosphere and then released from it through magnetic reconnection processes (Hughes, 1995).

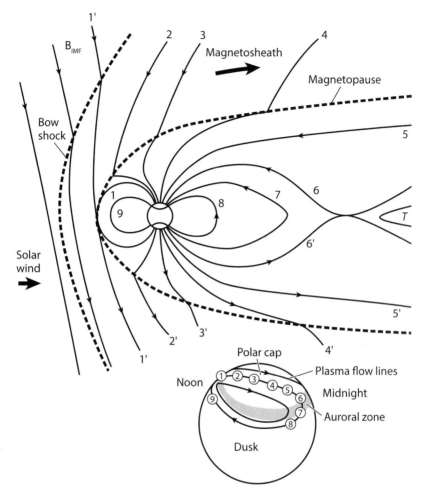

Figure 2.4. Schematics of magnetic reconnection to generate magnetic reconnection in the magnetosphere. IMF field lines (1′) reconnect the earth dipole field line (1) at the magnetopause. Field lines 6 and 6′ reconnect at the second X-line at the tail. [Adapted from Hughes (1995).]

When the magnetic fields on either side of the dayside magnetopause have different tangential components, a current sheet develops and dissipates, as the magnetic fields reconnect. If a southward interplanetary magnetic field (IMF) and the northward earth dipole field meet at the magnetopause, reconnection occurs efficiently. The reconnected field lines, still tied to the polar cap on one end, are embedded in the solar wind on the other side and are blown away to the nightside of the earth. Dungey showed that this motion of the reconnected magnetic field lines would induce the observed pattern of plasma flow in the upper atmosphere of the polar cap, shown in figure 2.4 as lines 1–5. The plasma on the flux tube driven away by the solar-wind flow would sense an electric field of $E = v_{SW} \times B_{SW}$ in the dawn-to-dusk direction. This electric field shows up in the polar cap and drives the flow of electrons through the ionosphere in the noon-to-midnight direction.

If this process were to continue indefinitely, the entire geomagnetic field would become connected to the open field lines of the IMF. Actually, another reconnection at another X-line happens and half of the reconnected flux returns to closed magnetospheric lines in the lobe which is connected to the earth while the other half is blown downstream with the solar wind. The newly connected closed dipole (dipolarized) field line contracts toward the earth, increasing the kinetic pressure of the dipolarized plasma. The stressed dipole field lines flow toroidally around the earth from the nightside to dayside. The convective flow circuit is closed, as shown in figure 2.4 as lines 6–9. This figure illustrates the plasma flow lines caused by the sequences of reconnection processes as lines 1–9. The entire process, described here as a steady process, actually happens intermittently in bursts, which are called substorms. This Dungey picture convincingly describes the fundamental role of magnetic reconnection in substorms.

Recently, using measurements by MMS satellites, in the strongly asymmetric reconnection layers of the magnetosphere, the energy deposition to electrons has been found to occur primarily in the electron diffusion region where electrons are demagnetized and diffuse. A large potential well is observed within the reconnection plane and ions are accelerated by the electric field toward the exhaust region. The present comparative study identifies the robust two-fluid mechanism operating in systems over six orders of magnitude in spatial scales and over a wide range of collisionality. We will describe this important observation in chapter 9.

2.3 MAGNETIC RECONNECTION IN SELF-ORGANIZATION IN FUSION PLASMAS

A large amount of experimental evidence for magnetic reconnection is found in fusion research devices by using measurements of field-line rearrangement through breaking and reconnection of magnetic field lines. Here we observe the evidence for magnetic reconnection through self-organization of magnetic field configuration, which consists of multiple layers of magnetic flux surfaces made of equally pitched magnetic field lines, as shown in figure 2.5. An axisymmetric tokamak plasma consists of nested flux surfaces made of magnetic field lines of an equal global pitch. When magnetic reconnection occurs in a certain flux surface, the pitch of the field lines changes through breaking and reconnection of field lines.

Most fusion laboratory experiments are carried out in toroidal (donut-shaped) plasma systems that satisfy, for the most part, the conditions for an MHD treatment of the plasma. Typical experimental examples of magnetic reconnection are found in "sawtoothing" tokamak fusion plasmas with Lundquist numbers exceeding 10^7 and in magnetic self-organization in spheromak and RFP (reversed field pinch) plasmas. Many experiments have been carried out to investigate magnetic reconnection phenomena in these devices, to control the stability and the confinement features of the

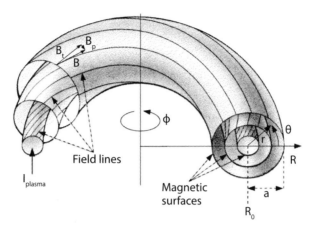

Figure 2.5. Magnetic flux surfaces in a toroidal fusion plasma. An axisymmetric toroidal pinch plasma consists of nested flux surfaces made of magnetic field lines of an equal pitch. On each flux surface, T_e is constant because of high parallel heat conductivity of electrons. The axis of the toroidal plasma core is called the magnetic axis ($r = 0$) and (r, θ) define the poloidal plane. [Figure from Google search: magnetic flux surface.]

current-carrying plasmas. Generally, it is found that magnetic reconnection is determined both by three-dimensional global boundary conditions and by local plasma parameters in the reconnection layer.

In toroidal fusion devices, toroidal currents are often utilized to heat the plasma and produce poloidal magnetic fields that effectively confine the high pressure plasma by a compressing pinch force. Tokamak, RFP, and spheromak configurations belong to this category. While all these configurations generate confining (by self-pinching) poloidal fields, toroidal fields are supplied differently. In tokamaks, a strong toroidal field is supplied externally. The toroidal field of an RFP is created by the combined effects of internal currents and an externally applied toroidal field. A spheromak does not have any externally applied toroidal field, and its toroidal field is entirely created by an internal plasma current. There is a remarkable feature common to all these current-carrying configurations: their plasmas constantly tend to relax to a quiescent state (of lower magnetic energy) through global magnetic self-organization in which magnetic reconnection plays a key role.

2.3.1   Magnetic reconnection in tokamaks

Sawtooth relaxation oscillations were discovered by von Goeler et al. (1974) in tokamak discharges. They are a typical example of global magnetic reconnection in a plasma. Sawtooth oscillation (Kadomtsev, 1975; Wesson, 1987) is observed as a periodic repetition of peaking and sudden flattening of the electron temperature (T_e) profile in the minor cross-section (a cutoff plane in a toroidal position) of a tokamak, as shown in figure 2.6. The conventional diagnostics for this phenomena are soft-X-ray diodes measuring bremsstrahlung emission along different chords across the plasma.

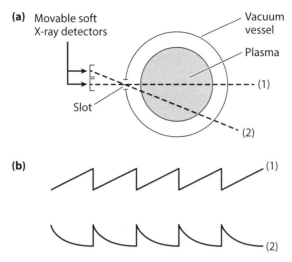

Figure 2.6. Observation of sawtooth oscillation by soft-X-ray diagnostics. (a) Schematic setup to measure the minor cross-section of a tokamak with two (inner and outer) chords. (b) Soft-X-ray signals along the inner (1) and outer (2) chords. [Adapted from Biskamp (2000).]

The observed signals usually have a sawtooth shape, whence the name. Recently, the electron cyclotron emission diagnostics have accurately measured the evolution of the temperature profiles.

An axisymmetric tokamak plasma consists of nested toroidal flux surfaces on each of which T_e is constant because of high parallel heat conduction along magnetic field lines. The MHD stability of a tokamak plasma is determined by the safety factor q, which is expressed (in chapter 9) as 2π times the inverse of the rotational transform of toroidal magnetic field lines (Wesson, 1987). A peaked T_e-profile generally leads to a more highly peaked current profile because of higher electrical conductivity at the center of plasma. The resultant strong peaking of current density makes the plasma unstable to a helical MHD kink mode, which develops near a resonant flux surface. The resulting helically deformed plasma induces magnetic reconnection near the $q = 1$ surface, as shown in figure 2.7. This reconnection produces a topological rearrangement of the internal magnetic field lines of the flux surfaces, relaxing the plasma into a lower energy state. During this reconnection process the q-profiles were measured to change. Kadomtsev (1975) proposed that the reconnection event (crash) should lead to a uniform current-density configuration with $q = 1$ after the crash phase, and a flat electron temperature (T_e) profile. The same cyclic evolutions are repeated afterward. Dozens of experimental studies were carried out to verify his work. While the temperature flattening was always observed after the crush, the measured q-value after the crash was not always 1. However, a small but measurable change of q-value was recognized in all measurements, which shows evidence of the magnetic reconnection process (Soltwisch, 1988; Levinton et al., 1993; Yamada et al., 1994; Park et al., 2006b). During a sawtooth crash in the Tokamak Fusion Test Reactor (TFTR) tokamak plasma, magnetic reconnection was observed to cause only a partial mixing of field lines. This was

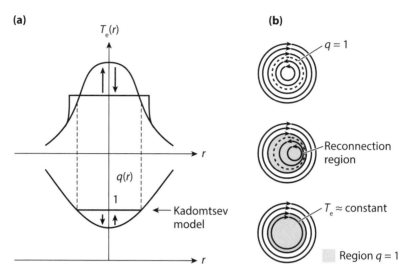

Figure 2.7. (a) Schematic view of changes in T_e- and q-profiles versus minor radius in a minor cross-section (poloidal plane) during sawtooth crash in a tokamak plasma. (b) Description of the Kadomtsev model in a poloidal plane: $m = 1$, $n = 1$ MHD instability develops near the $q = 1$ flux surface and induces magnetic reconnection. [From Kadomtsev (1975).]

evidenced by the small changes in the q-profile, which were documented by Levinton et al. (1993), Yamada et al. (1994), and Nagayama et al. (1996). This change of q-value represents magnetic reconnection. Recent progress in the study of tokamak sawtooth reconnection will be discussed in more detail in chapter 9.

2.3.1.1 *Magnetic relaxation in RFPs*

The RFP (reversed field pinch) is an axisymmetric toroidal pinch in which plasma is pinched (squeezed) and confined by a poloidal magnetic field created by a toroidal plasma current, and by a toroidal field created internally and externally. As postulated by Taylor (1974), the RFP configuration is generated by a process of self-organization in plasma, in which plasma settles into a state of minimum energy for a given helicity.

In an RFP discharge, magnetic reconnection occurs during this self-organization process of a toroidally confined plasma, and can be both continuous and impulsive. The magnetic energy is stored in a force-free magnetic equilibrium configuration via slow adjustment to an external driving force. Then through a sawtooth event the magnetic field suddenly reconnects and the plasma reorganizes itself to a new MHD equilibrium state. In this device, local reconnection on different flux surfaces leads to a global relaxation whose macroscopic properties are studied as shown in figures 2.8 and 2.9. The RFP magnetic field is sheared (figure 2.8) with its pitch changing its direction from a toroidal direction at the center to a nearly poloidal direction at the edge. Because of the shearing of the field lines, reconnection occurs at multiple radii, with each radial

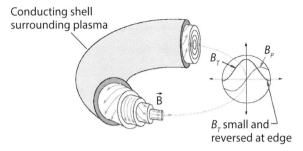

Figure 2.8. Schematic of an RFP plasma configuration showing magnetic field lines, strongly sheared, where B_T is the toroidal field and B_P is the poloidal field. Reconnection can occur at multiple surfaces, such as those indicated in the cutaway view of the toroidal plasma. The radial dependence of the poloidal and toroidal magnetic fields is plotted. [From Sarff et al. (2005).]

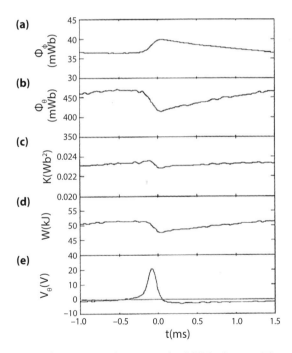

Figure 2.9. A magnetic reconnection event in MST plasma. Time evolution of (a) toroidal magnetic flux, (b) poloidal magnetic flux, (c) magnetic helicity K (explained in chapter 9), (d) magnetic energy W, and (e) one-turn poloidal voltage. [From Ji et al. (1995).]

location corresponding to a rational surface at which the safety factor $q = m/n$ (m and n are poloidal and toroidal mode numbers respectively). Often, the multiple reconnections occur suddenly and simultaneously, leading to a sudden global rearrangement of the entire magnetic field. During these reconnections it is found that global helicity tends to be conserved, while the total magnetic energy is dissipated (Ji et al., 1995).

In a representative RFP device, the Madison Symmetric Torus (MST), simultaneous reconnection at different radii corresponding to different n with $m = 1$, was observed. When the current density profile becomes highly peaked, tearing modes develop reconnecting magnetic field lines, and plasma reorganizes itself rapidly to a new MHD equilibrium state. In this self-organization of magnetic field lines, a conversion of magnetic flux and energy from poloidal to toroidal occurs. Figure 2.9 shows the time evolution of toroidal and poloidal magnetic flux, together with that of magnetic helicity K (discussed in chapter 9) and magnetic energy W, indicating an abrupt conversion of poloidal magnetic flux to toroidal flux and making the total magnetic energy smaller (Ji et al., 1995). The ion temperature increases significantly at the expense of magnetic energy. Multiple reconnection events are often observed and the reconnection is impulsive in time. Recent theoretical and experimental results show that the different reconnections are coupled. Essentially, all effects of magnetic self-organization in MST (dynamo, ion heating, momentum transport) are strongly amplified when multiple, coupled reconnections occur. One of the most important questions for global reconnection is why reconnection occurs impulsively. The study of magnetic reconnection in RFPs is discussed in more detail in chapter 9.

2.4   AN OBSERVATION OF A PROTOTYPICAL RECONNECTION LAYER IN A LABORATORY EXPERIMENT

Recently, more than a half dozen dedicated laboratory devices have been built to study the basic mechanisms of magnetic reconnection. The MRX device is a typical example. Reconnection in MRX is driven by utilizing a flexible toroidal plasma configuration (Yamada et al., 1997a). An environment is created in which the MHD criteria are satisfied on a global scale, with a large Lundquist number and a size much larger than the ion gyroradius. One advantage of the modern dedicated experiments is that the current sheets are made to be toroidally continuous and free from constraints caused by termination by endplates or electrodes.

In MRX, reconnection is driven in a controlled manner with toroidal-shaped flux cores that contain coil windings in both the toroidal and poloidal directions. By pulsing currents in these coils, two annular plasmas are inductively created around each flux core (Yamada et al., 1981, 1997a). After the plasmas are created, the coil currents are programmed to drive magnetic field lines toward a reconnection point (X-point), producing a narrow neutral sheet or current layer. The dynamics of the local reconnection layer are then studied in it. This is an experimental realization of Dungey's concept shown in figure 1.3. When the experiments are carried out in collision-dominated plasmas with no guide magnetic field (antiparallel reconnection), a typical two-dimensional Sweet–Parker diffusion region profile (a rectangular box shape), with two Y-shaped

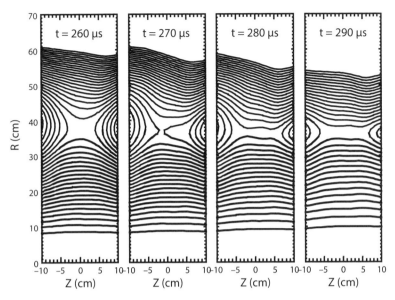

Figure 2.10. Demonstration of magnetic reconnection seen through measured field lines in MRX. In the low-β plasma outside the neutral sheet, poloidal flux contours represent magnetic field lines and are seen to break and reconnect. [From Yamada et al. (1997a). https://mrx.pppl.gov/mrxmovies/Collisional.mov]

ends, is observed. The time evolution of the magnetic field lines is measured and displayed as a movie. (See the MRX website at http://mrx.pppl.gov/mrxmovies.) This movie shows the time evolution of the measured flux contours of the reconnecting field being driven to the reconnection layer, reconnecting and exiting to the exhaust region. By monitoring these contours, the reconnection rate can be documented as a function of plasma parameters (Yamada et al., 1997b,a; Ji et al., 1998, 1999). During the past decades, extensive studies have been carried out in many other laboratories that are dedicated to the study of magnetic reconnection. In later chapters, the results of the MRX and other dedicated laboratory devices are described in detail.

Chapter Three

Development of MHD theories for magnetic reconnection, and key observations in laboratory and space plasmas

In the early days of plasma physics in the 1950s, scientists found a very powerful and useful way of describing the plasma dynamics using MHD (magnetohydrodynamics), treating plasma as one fluid element. Ideal MHD theory, which excludes resistive diffusion due to plasma resistivity, was developed to describe the dynamics of highly conductive plasmas ($\eta = 0$), where magnetic field lines always move with the plasma and remain intact, with $E_\parallel = 0$. This MHD theory was built upon the foundation of hydrodynamics. Thus we start with a description of plasma and associated electromagnetic fields in fluid physics. While it was found to be very effective, it was also realized that it cannot straightforwardly describe the key physics of magnetic reconnection, where a large current density is generated separating the motion of electrons from ions.

3.1   EARLY HISTORY OF MHD THEORY ON MAGNETIC RECONNECTION

The idea of magnetic reconnection first originated in attempts to understand the heating of the solar corona and the origin of the enormous energy observed in solar flares. It was recognized early (Giovanelli, 1946; Hoyle, 1949; Dungey, 1953) that solar flares occur in the neighborhood of sunspots where the strong magnetic field is observed to have magnitudes of several kilogauss. Such fields contain large quantities of energy, and if a mechanism could be found to convert this energy into heat, radiation, and kinetic energy, it would provide an origin for the energy emitted by a solar flare. However, a simple estimate of the resistive decay time of the magnetic field shows that a direct conversion of magnetic energy is too slow to heat coronae. For example, the magnetic diffusion time for a standard-size solar flare ($L \sim 10,000$ km) would be 1 million years ($\sim 10^{14}$ s). On the other hand, a topology change in a solar flare occurs in just a few minutes ($\sim 10^3$ s). Instead, attention moved to the origin of the energetic particles associated with the flares. Giovanelli (1946) showed that the changing field strengths in the sunspot fields would produce large voltages that were capable of accelerating charged particles to high energies. It was suspected that such voltages in the presence of the magnetic fields would lead to local mass motions when the electric field was applied across the magnetic field, and would be shorted out when it was parallel to the field. Keeping these points in mind, they looked into neutral point regions where the magnetic field was zero.

The annihilation of the magnetic field in the neutral layer by MHD forces was emphasized by Dungey (1953) who pointed out that near neutral points the decreasing magnetic field would generate current layers and invoke magnetic reconnection, leading to fast conversion of magnetic energy as mentioned in chapter 1. Sweet (1958) pointed out, at the 6th International Astronomical Union (IAU) conference in 1956, that the electric current density j could be concentrated in thin layers (the current sheets of Dungey, 1953), where its magnitude is enhanced such that Ohmic dissipation, proportional to j^2, could release magnetic energy at a greatly increased rate. In reality, it is not just Ohmic dissipation that leads to the conversion of magnetic energy to other forms. Another energy conversion mechanism can come from a change in the configuration of the magnetic lines so that they develop the strong curvature of field lines in the current layer. The energy conversion is caused by the curved magnetic field lines unfolding and accelerating the plasma out of the ends of the current layer (exhaust), simultaneously lowering the magnetic energy and accelerating the plasma, increasing its kinetic energy. This kinetic energy could sometimes lead to shocks and to viscous dissipation, which turns the energy into radiation and accelerates particles. This sequence of events was termed magnetic reconnection at that time, although it is only the first stage that involves physical reconnection of the magnetic lines.

At an IAU conference, Sweet (1958) introduced a detailed model for reconnection and conversion of energy. The same model was independently discovered by Parker and elaborated by Parker (1963) and it eventually became known as the Sweet–Parker model of magnetic reconnection. The introduction of the Sweet–Parker model based on MHD theory led to a long period of research into the physics of magnetic reconnection. Although the Sweet–Parker model leads to much faster conversion of magnetic energy than is expected from resistive decay in the absence of current sheets, estimates showed that it is still much too slow to account for the fast conversion observed in solar flares. The early history of magnetic reconnection research thus became attempts to modify the simple Sweet–Parker model to explain the fast topology changes of solar flares, and high-speed reconnection in the magnetosphere.

In most of the cases of magnetic reconnection presented in chapters 1 and 2, the Sweet–Parker reconnection rate would be very much smaller than that needed to explain the observed amount of reconnected flux. This serious discrepancy was apparently removed by an important modification to the Sweet–Parker model, introduced by Petschek in 1963 (see Petschek, 1964). Petschek theory introduces slow shocks. If these shocks emerge at a sufficiently small distance from the X-point, they produce much faster reconnection rates and can lead to results compatible with observations. The theory was controversial because the origin of the shocks was unclear. Eventually, the Petschek model was considered to be of limited applicability barring a discovery of the shock origin. Its main virtue is that if it could be made to work, it would explain the desired faster reconnection rates. Details of the Petschek model are given later in this chapter.

3.2 DESCRIPTION OF PLASMA FLUID IN
MAGNETIC FIELDS BY MHD

How do we treat plasma fluid theoretically? Plasma is made of negatively charged electrons and positively charged ions in equal quantities, if it does not have any impurity (minor component of) ions. It is always neutral with 99.999...% accuracy, consisting of equal amounts of positive and negative charges. If we know the position and behavior of each particle composing a plasma—electrons and ions—then we should be able to completely describe the plasma assuming that we have an enormously powerful supercomputer. But each particle's position and velocity change under the influence of electromagnetic fields, and we have to know every action to solve the problem. Since this approach is unpractical and almost impossible, we rather describe plasma dynamics by employing the scheme used in hydrodynamics for nonmagnetic fluid materials such as water and oil (Kulsrud, 2005).

In MHD theory, the basic state of a plasma is specified locally by its mass density ρ, its velocity V, its fluid pressure p, the electric field E, and magnetic field B. They are all functions of position r and time t. If the collision frequency is high, the electron and ion pressure is isotropic and heat flow can be assumed small. It is also important to note that electrons and ions are considered to compose one fluid element even though a current can flow through it: the one-fluid formulation. If we can neglect the effects of plasma resistivity (this is a valid assumption when the magnetic Lundquist number (S) is large, as we describe in a later section), the plasma can be described by the "ideal MHD equations" neglecting resistive dissipation.

First, all electromagnetic dynamics of plasma should be based on the following four Maxwell equations:

- Faraday's induction equation,

$$\frac{\partial B}{\partial t} = -\nabla \times E;$$
(3.1)

- Ampère's law,

$$\nabla \times B = \mu_0 j + \frac{1}{c^2}\frac{\partial E}{\partial t};$$
(3.2)

- the divergence of B,

$$\nabla \cdot B = 0;$$
(3.3)

- the Gaussian equation,

$$\nabla \cdot E = \frac{q}{\varepsilon_0},$$
(3.4)

where q is the electric charge density.

The second term of eq. (3.2) represents a displacement current and is negligible in most MHD dynamics except for very fast phenomena (e.g., the wave phenomena of fast phase velocity) or in nonneutral plasmas.

Once again, in the MHD formulation the plasma parameters (mean density ρ, pressure $p = nT$, and mean velocity V) are used to characterize the plasma state and they are governed by the fluid equations. The equations that describe the MHD dynamics are

- the continuity equation for the total plasma density ρ,

$$\frac{\partial \rho}{\partial t} + \nabla \cdot (\rho V) = 0;$$ (3.5)

- the equation of motion for the velocity V,

$$\rho \left(\frac{\partial V}{\partial t} + V \cdot \nabla V \right) = j \times B - \nabla p + \rho g,$$ (3.6)

where g is the external (gravity) force;
- the equation of energy conservation,

$$\frac{\partial}{\partial t} \left(\frac{p}{\rho^\gamma} \right) + V \cdot \nabla \left(\frac{p}{\rho^\gamma} \right) = 0.$$ (3.7)

Here p represents the plasma pressure and γ is the adiabatic constant.

Particularly when a plasma is sufficiently collisional, namely the collisional mean free path of plasma particles is much shorter than the system size, plasmas can be described macroscopically as a "one-component fluid" interacting with the local electromagnetic field by MHD. This MHD formulation was established in the early stages of development of the theory for plasmas. The approach is generally valid and effective when the collision rate of each species with itself is high compared to the macroscopic rate of change. In this situation, the mean free paths are short compared to the macroscopic system size.

3.3 THE FLUX FREEZING PRINCIPLE AND MAINTAINING PLASMA EQUILIBRIUM

In order to understand the basics of MHD, we introduce here two important properties of a plasma. The first basic property is *flux freezing*. Flux freezing implies that magnetic field lines maintain their physical reality, and any given set of field lines (which represent the magnetic field in strength and direction) continues to represent the magnetic field at later times. Indeed, this is achieved by the field lines being bodily carried with the plasma. As a consequence, if two plasma fluid elements A and B lie on the same magnetic line of force at time t, they will continue to lie on a common line of force at any later time t'. Furthermore, if a given line of force passes through a fluid element

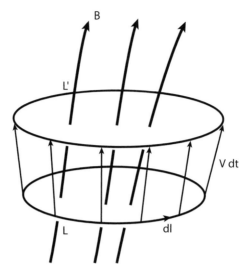

Figure 3.1. Motion of plasma with magnetic flux. Schematic of a lateral move of contour L' floating freely from its initial position at contour L.

A at a particular time, the line that passes through the identical fluid element at a later time is regarded as the same line of force, or magnetic field line.

Let us understand the idea of flux freezing. In a highly conductive plasma, the plasma resistivity is negligible. Generally, the electric field E' in the moving frame of the plasma is written as

$$E' = E + V \times B. \tag{3.8}$$

There is a simple fact that establishes MHD as an appropriate description of the dynamics of a magnetic field in plasma. It is that a plasma moving with velocity v is so conductive that it cannot contain its own significant electric field due to the large mobility of electrons. Thus, in a highly conductive plasma we write Ohm's law as

$$E + V \times B = 0. \tag{3.9}$$

Combining this equation with the induction equation (Faraday's law), we obtain the equation of motion for the field:

$$\frac{\partial B}{\partial t} = \nabla \times (V \times B). \tag{3.10}$$

This equation is called the magnetic differential equation or the induction equation (equation of motion for a magnetic field).

Our mathematical treatment continues with the consideration that a bulk motion of plasma moves with a magnetic flux through a closed contour, as shown in figure 3.1.

Then the plasma moves with magnetic field from contour L to L'. It is clear that $V\,dt$ represents the movement of loop L to L'. Thus we write a change of magnetic flux in a plasma with a specific boundary as

$$\frac{d\psi}{dt} = \int \frac{\partial \boldsymbol{B}}{\partial t} \cdot d\boldsymbol{S} + \oint (\boldsymbol{V} \times d\boldsymbol{l}) \cdot \boldsymbol{B}, \tag{3.11}$$

where $\boldsymbol{V}$ is the global plasma velocity.

The first term represents a pure time change of flux. The second term comes from the motion of plasma from L to L' during the time $t = 0$ to $t = dt$.

We use Faraday's equation

$$\frac{\partial \boldsymbol{B}}{\partial t} = -\nabla \times \boldsymbol{E} \tag{3.12}$$

and Stokes' vector relationship

$$\oint (\boldsymbol{V} \times d\boldsymbol{l}) \cdot \boldsymbol{B} = \oint (\boldsymbol{B} \times \boldsymbol{V}) \cdot d\boldsymbol{l} = \int (\nabla \times (\boldsymbol{B} \times \boldsymbol{V})) \cdot d\boldsymbol{S}. \tag{3.13}$$

As plasma moves with magnetic field lines, we obtain

$$\frac{d\psi}{dt} = -\int \nabla \times (\boldsymbol{E} + \boldsymbol{V} \times \boldsymbol{B}) \cdot d\boldsymbol{S}. \tag{3.14}$$

Thus from eq. (3.9), a clear principle of flux freezing is now established:

$$\frac{d\psi}{dt} = 0. \tag{3.15}$$

This argument can be understood as follows on figure 3.1. With $d\boldsymbol{l}$ representing the length along L and L', $d\boldsymbol{l} \times \boldsymbol{V}\Delta t$ represents the area swept out by $d\boldsymbol{l}$ to L' in time Δt. The magnetic flux through the contour L is the same as that of the area through the contour L' since magnetic flux is conserved.

This geometrical characterization of flux freezing can also be carried out in a more mathematically exact way by making use of Clebsch coordinates to describe the magnetic field as described in Yamada et al. (2010). It is known that an arbitrary divergence-free vector field, such as a magnetic field $\boldsymbol{B}$, can be expressed in terms of two scalar functions of position α and β as

$$\boldsymbol{B} = \nabla\alpha \times \nabla\beta. \tag{3.16}$$

It is well known that this representation is automatically divergence-free, so that two such scalar functions can be found to satisfy eq. (3.16). This formulation is very popular to describe the equilibria and global characteristics of fusion plasmas, such as tokamaks (Shafranov, 1956; Wesson, 1987), reversed field pinches, and spheromaks. We note that

ψ and ϕ are often used instead of α and β to describe the equilibrium of toroidal fusion configurations.

If we assume that α and β are scalar functions of time and space such that

$$\frac{d\alpha}{dt} = \frac{\partial\alpha}{\partial t} + V \cdot \nabla\alpha = 0,$$

$$\frac{d\beta}{dt} = \frac{\partial\beta}{\partial t} + V \cdot \nabla\beta = 0, \tag{3.17}$$

then B as given by eq. (3.16) satisfies eq. (3.10).

It is clear now that a line of force, or magnetic field line, is represented by $\alpha = \text{const.}$ and $\beta = \text{const.}$, and by eq. (3.17) these are constant following a plasma element, so this shows that any line is thus bodily transmitted by the plasma motion.

There is another mathematical way to represent flux freezing. For the solar magnetic field, we have that if the velocity in the solar surface is zero, then α and β are fixed at the solar surface, the field line is frozen at the surface, and the footpoint mappings and the topology are preserved during any motions in the atmosphere. In magnetic confinement fusion plasmas, we can define the flux surfaces of plasma by α and β, replacing α and β by ψ and ϕ in the customary notation. In this case, $\psi = \text{const.}$ represents a poloidal flux surface with a toroidal function of $\phi = \text{const.}$ in a toroidal geometry. In a flux surface of $\psi = \text{const.}$, the plasma pressure is considered to be constant since there is no exchange of field lines between flux surfaces: $p = \text{const.}$

The second important property is that any static low-β plasma equilibrium is largely determined by the topology of the magnetic field. This can be proven by considering the footpoint mapping determined by the solar-magnetic-field topology, neglecting its plasma pressure. Then, of all the magnetic fields B that are divergence-free and have this topology, the magnetic field that minimizes the total magnetic energy is a force-free equilibrium and is the unique force-free equilibrium associated with this topology, as shown by Kulsrud (Yamada et al., 2010).

To see why this theorem holds for the solar magnetic field, consider minimizing the magnetic energy

$$\mathcal{E} = \frac{1}{8\pi} \int B^2 \, d^3x = \frac{1}{8\pi} \int (\nabla\alpha \times \nabla\beta)^2 \, d^3x \tag{3.18}$$

over all possible functions α and β, again with the condition that α and β are fixed on the solar flux surfaces. This is the same condition as the footpoints being held fixed.

Now, if we vary $\mathcal{E}$ by changing α by $\delta\alpha$, integrate by parts, and use Stokes' theorem to get rid of the integrated term, which vanishes on the boundary since $\delta\alpha = 0$ there, then

$$\delta\mathcal{E} = \frac{2}{8\pi} \int \delta\alpha(\nabla\beta \cdot \nabla \times B) \, d^3x = 0, \tag{3.19}$$

so $j \cdot \nabla \beta = 0$ everywhere since $\delta \alpha$ is essentially arbitrary. Similarly, from varying β we get $j \cdot \nabla \alpha = 0$. Thus, expanding the triple product,

$$j \times (\nabla \alpha \times \nabla \beta) = \nabla \alpha (j \cdot \nabla \beta) - \nabla \beta (j \cdot \nabla \alpha) = 0, \tag{3.20}$$

or $j \times B = 0$, which is the condition for a force-free equilibrium.

For the case of nonzero pressure p, the energy is also a minimum when p is properly constrained, but the argument is presented in a more sophisticated way (Kruskal and Kulsrud, 1958). By different arguments, B. Taylor found that a state of $p = \text{const.}$ (and $\lambda = j \cdot B / B^2 = \text{const.}$) would lead to a minimum energy state.

This unique relationship between topology and equilibrium shows that any change in topology by reconnection has a significant impact on the entire equilibrium.

3.4 BREAKDOWN OF FLUX FREEZING AND MAGNETIC RECONNECTION

In the MHD formulation described above, all equilibria are characterized by their topology. Any change in the topology implies a different equilibrium with a different energy. The tendency of plasmas to lower their energy allows a plasma to carry out a magnetic reconnection. The main question is how fast this can happen, and this is perhaps the most important question concerning magnetic reconnection. As we learned earlier, magnetic reconnection cannot happen if eq. (3.9) is exactly satisfied.

In a plasma with finite resistivity, Ohm's law would lead us to a different conclusion. A more exact magnetic differential equation on MHD for the evolution of B that allows reconnection must include the resistivity term. For an analysis of the local reconnection layer using the resistive MHD formulation, the motion of magnetic field lines in a plasma can be described by combining Ohm's law and Maxwell's equations,

$$E + V \times B = \eta j, \tag{3.21}$$

where η is the plasma resistivity and j is current density.

Once again, combining with

$$\frac{\partial B}{\partial t} = \nabla \times E \tag{3.22}$$

yields

$$\frac{\partial B}{\partial t} = \nabla \times (V \times B) + \frac{\eta}{\mu_0} \nabla^2 B. \tag{3.23}$$

The last term gives a diffusion of the magnetic field over a timescale of $\mu_0 L^2 / \eta$. This relation shows that the field lines are not exactly tied to the plasma but can slip in the diffusion region. If this diffusion region is small compared to the scale of the magnetic field then one can consider the field as frozen in most of the region, even if the resistivity is not exactly zero.

As mentioned before, a solar flare is generally preceded by the twisting of field lines by a slow motion of the footpoints in the solar surface, leading to a slow change in the magnetic field in the solar atmosphere. This twisting gradually increases the stored magnetic energy. Many visible examples are observable in the solar coronae shown in figure 1.1. After the energy has increased enough, a sharp shear of field lines occurs at a certain region and magnetic reconnection occurs, breaking flux freezing conditions. While this reconnection occurs, footpoints in the solar surface hardly change but the footpoint mapping and the topology change, throwing the magnetic field out of equilibrium. After reconnection, the plasma then relaxes by global motions of the plasma to a new equilibrium of lower energy, releasing its increased energy that had been produced by the motion in the solar surface. Thus we can say magnetic reconnection is a counter phenomenon to flux freezing. This discussion started with a concept that ideal plasma motions do not break lines or change their topology, based on the concept of flux freezing. In this manner, one sees that a sudden change in topology by a nonideal motion leads to a rapid conversion of magnetic energy into kinetic energy and then a subsequent conversion of this kinetic energy into heat, radiation, or particle acceleration by some viscous process. This abrupt change of topology is a nonideal change that magnetic reconnection can trigger. It is of considerable importance just because it can lead to a rapid conversion of magnetic energy to other forms.

3.5   RESISTIVE MHD THEORIES AND MAGNETIC RECONNECTION

Sweet and Parker addressed magnetic reconnection problems in a situation where solar coronal fields are merging in a two-dimensional reconnection boundary layer in which oppositely directed field lines meet and merge, as shown in figures 3.2 and 3.3. During this merging process, a diffusion of magnetic field lines occurs due to resistivity. Let us consider this model in detail here.

3.5.1   Sweet–Parker model

If we consider magnetic field lines approaching each other in a plasma, magnetic field gradients become locally strong at the meeting point. Plasma flows can lead to singular current density sheets where $E_\parallel$ becomes sufficiently large (finite) to induce non-MHD plasma behavior so that a magnetic field line can lose its original tie (identity) with plasma particles due to diffusion. Dungey (1953) showed that such a current sheet can indeed be formed by the collapse of magnetic field near an X-type neutral point as described in chapter 1, and he suggested that magnetic field lines can be broken and rejoined. When the field lines are reconnected, the topology of magnetic configuration can change and large $j \times B$ MHD forces often result.

In Sweet's original model for magnetic reconnection, he considered what would happen if two pairs of sunspots approached each other as considered in figure 3.2. Although this concept has not been exactly verified by observations (as described in

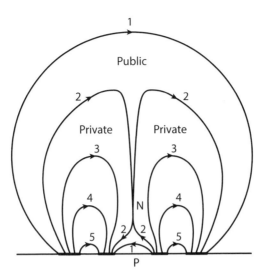

Figure 3.2. Sweet model for analysis of a reconnection layer. Oppositely directed field lines merge in the solar flare. Through the reconnection process at the N point, two private field lines, labeled 2, become a single public line, and after that line 3 goes through the same process. This reconnection layer around the N point continues to divide the same public and private lines and all other private lines go through it. (Although this model has not been verified by recent solar observations, it provided a basic idea for a magnetic reconnection layer.) [Modified from Sweet (1958).]

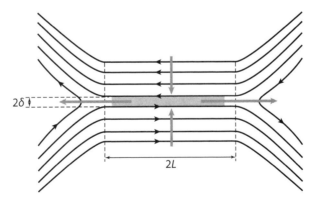

Figure 3.3. Sweet–Parker model for analysis of a reconnection layer. Oppositely directed field lines merge in the diffusion region of width 2δ and length $2L$. [From Yamada et al. (2010).]

more detail in chapter 7), let us follow his thoughts. Each pair has its own magnetic dipole field and as these sunspot pairs approach each other, their dipole fields, frozen in the solar atmosphere, would be crushed together forming a thin layer. He compared this to the case of a vacuum nonconducting solar atmosphere in which the magnetic field would simply add vectorially, and the neutral (N) point of intersection would simply shift from line to line, with the consequence that some lines originally connected to a single dipole would end up connecting one sunspot to the other. Comparing this to the highly conducting solar atmosphere case, he showed that the line passing through the N point would continue to pass through it. Through the reconnection process at the N point, two private field lines, labeled 2, become a single public line, and after that line 3 goes through the same process. This reconnection layer around the N point continues to divide the same public and private lines and all other private lines go through it.

Sweet and Parker addressed magnetic reconnection problems in a situation where solar coronal fields are merging and transforming the reconnection region into a two-dimensional reconnection boundary layer in which oppositely directed field lines merge as shown in figure 3.3. In their model, magnetic fields of opposite polarity approach in the rectangular-shaped reconnection region where the incoming field lines merge, and newly reconnected field lines emerge and move away. During this process, a diffusion of magnetic field lines occurs due to resistivity. This two-dimensional MHD model introduced the important concept that the magnetic reconnection rate can be calculated quantitatively through a magnetic-field-line transfer between two geometrically separated plasma regions, assuming uniformity in the third dimension.

In resistive MHD plasmas, hydromagnetic flows can lead to the formation of a neutral sheet where the plasma flow is constrained to a finite-size resistive region and the electric field E is balanced with ηj. In the rectangular diffusion region shown in figure 3.3, the resistivity term becomes sufficiently large that a magnetic field line can diffuse and lose its original identity and reconnect to another field line. In a steady state, eq. (3.23) can be simplified to

$$V_{in} B = \frac{\eta}{\delta \mu_0} B. \tag{3.24}$$

Using the continuity equations of plasma flows to connect inflow (length L) and outflow channels (width δ),

$$V_{in} L = V_{out} \delta. \tag{3.25}$$

From the pressure balance between the upstream magnetic pressure ($B^2/2\mu_0$) and the downstream ($mv^2/2$) regions ($B=0$), namely $V_{out} = V_A$, a very simple formula is derived for the reconnection speed V_{in}:

$$\frac{V_{in}}{V_A} = \frac{1}{\sqrt{S}}, \tag{3.26}$$

where we repeat that V_A is the Alfvén velocity and $S = \mu_0 L V_A/\eta$ is the Lundquist number, which is defined as the ratio of the Ohmic diffusion time ($= \mu_0 L^2 V_A/\eta$) to the crossing time of the Alfvén waves ($= L/V_A$) in the reconnection region.

In this resistive MHD formulation (Parker, 1957), magnetic fields diffuse and dissipate in the rectangular reconnection region as illustrated in figure 3.3, where the incoming plasma flux is balanced with the outgoing flux, satisfying the continuity equations for plasma fluid and magnetic flux. The reconnection rate depends on the Lundquist number S, which is usually extremely large: S can be 10^4–10^8 in laboratory fusion plasmas, 10^{10}–10^{14} in solar flares, and 10^{15}–10^{20} in the interstellar medium of the Galaxy. The Sweet–Parker reconnection rate derived above is thus far too slow to describe the observed reconnection phenomena. This slowness comes from the (unnatural) assumption that the plasma and magnetic flux have to go through the narrow rectangular neutral sheet of thickness $\delta = L/\sqrt{S}$.

One of the first applications of reconnection was made by Dungey (1961), who showed that the reconnection of solar-wind magnetic lines could account for the entire gross structure of the magnetosphere as discussed in the introduction. There are two important questions concerning the Dungey model: Does this reconnection occur? If it does, how fast does it happen? As solar-wind lines impinge and are compressed in the magnetosphere they have a choice: to go around it much as water goes around a boat moving through water, or to reconnect. The fraction that reconnect, and end up in the earth's magnetotail, is controlled by the reconnection rate. Assuming this is the Sweet–Parker rate, approximately 1 in 100,000 lines reconnect (Kulsrud et al., 2005), as opposed to the measured value of 1 in 10 (Hughes, 1995). The predictions of Sweet–Parker strikingly disagree with the observations. Actually, observations shows that 5–10% of the incoming solar-wind field lines reconnect.

3.5.2   Physical interpretation of the Sweet–Parker model

As the lines in the current layer reconnect, they move out of this layer and go to the exhaust region. As the inflowing field lines steadily pass into the current layer to be reconnected, plasma mass is also carried into the layer, and as the lines leave the layer, mass must also be carried out at a velocity whose magnitude is limited. Thus the time to reconnect the inflowing field lines is of order $t_R = L/v_R \approx \sqrt{S}(L/v_A)$. This time should be compared with the resistive decay time for the region if there were no current layer, which is $L^2/\eta \approx SL/v_A$, a time longer by $\sqrt{S}$ than the time to reconnect. Since the Lundquist number is very large in the corona, often larger than 10^{12}, the reconnection model leads to a very much shorter time to destroy or rearrange the magnetic field lines immersed from the surface of the sun.

The physics that leads to such slow reconnection rates in this model is summarized as follows: First, in the two-dimensional Sweet–Parker model, a very narrow current layer has to be generated for a large-Lundquist-number plasma because the effective resistivity is so small. The thinness of the current layer is constrained: by $\delta = L/\sqrt{S}$ or by the necessity for the plasma on the reconnecting lines to be expelled along this thin rigid current layer. The breaking of the lines occurs over a very narrow region near the center during a topological change. The reconnected lines inside the layer, although the field strength is weak, have large curvature so they can unfold themselves by magnetic tension, as well as the pressure gradient force considered by Sweet and Parker. This tension force is of the same order as the pressure force, so that the Sweet–Parker model still gives the correct order of magnitude for the reconnection rate. This would lead to the same thickness of the layer along the entire exhaust line as at the central neutral point.

The Sweet–Parker model is a beautifully simplified concept in which flows of plasma and magnetic flux go through a two-dimensional uniform reconnection layer in which magnetic energy is dissipated evenly with a uniform out-of-plane reconnection current density profile. This model can be realistic when the Lundquist number S does not exceed unity by a large amount: $S < 10^4$. But it becomes unrealistic when S is significantly larger than unity; for example, S can be very large in a solar flare. For $S > 10^4$, the aspect ratio of a rectangular reconnection layer would be over 100. It is rather difficult to imagine that such a uniform current sheet can be maintained in a stable form in a realistic three-dimensional environment. It should make the current sheet unstable, as we will find in chapter 14. This model would also impose an unrealistic energy conversion process on the reconnection layer; as an example, most of the incoming magnetic energy has to be dissipated uniformly in the rectangular-shaped reconnection layer. In other words, magnetic energy has to be slowly converted to particle energy without breaking or deforming the current layer. This will be readdressed from an energetic point of view in later chapters.

There are important factors to modify the above Sweet–Parker picture if it is applied to reconnection that occurs on the dayside of the magnetosphere:

(1) The thickness of the reconnection layer is generally comparable to the ion gyro-radius or the skin depth of the plasma and the ions become demagnetized. Thus we expect quite different motion of ions with respect to the still magnetized electrons, and we have to take into account two-fluid effects, which will be described in detail in chapter 5. In addition, due to the significant stagnation of the solar wind at the magnetopause reconnection point, the two-fluid picture of magnetic reconnection has to be modified as shown in chapter 9.

(2) The effects of plasma flow should be considered. Although the solar-wind flow is slowed down by the earth's bow shock, it is not reduced to zero except at the exact subsolar point. On either side of this point the solar wind still has a considerable velocity, and when it reaches the magnetosphere it turns tangent to it. The original solar wind is traveling at about 10 times the Alfvén speed. When it turns to flow along the magnetosphere, the velocity is still quite substantial and this flow eases the burden of acceleration of the flow along the magnetopause. This in turn speeds up the reconnection rate at the subsolar point and this effect has to be taken into account to evaluate the rate of the magnetopause reconnection.

3.5.3 Petschek model

Shortly after the Sweet–Parker theory was developed, another model was proposed by Petschek (1964) to resolve the dilemma of the slow reconnection rate through a narrow reconnection channel by introducing shocks that open up the reconnection layer to a wedge shape. The situation was greatly improved by this Petschek model in which

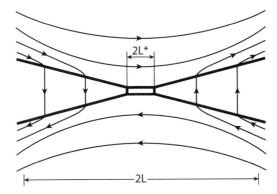

Figure 3.4. Petschek model. [From Yamada et al. (2010).]

inclusion of slow shocks in the outflow region would greatly speed up the mass flow and remove a major hurdle for the Sweet–Parker theory, where a large amount of mass must flow through the very narrow current channel of constant width.

To increase the flow over a wider channel, extra forces are needed. These forces can be provided by slow shocks (see figure 3.4). In the Petschek model, these shocks start at the ends of a short Sweet–Parker layer, and fan out with an angle v_R/v_A. As the external plasma crosses the shocks it is accelerated in the downstream direction up to the Alfvén speed by the intense currents in the shocks. A simple analysis shows that the reconnection is then given by a reconnection rate corresponding to the much shorter length L^* of the Sweet–Parker layer, where L^* is the distance from the X-point at which the shocks start. (From now on we refer to Sweet's N point as the more conventional X-point.)

In the downstream region, the magnetic field in the channel is essentially a $B_\perp$ field perpendicular to the current layer. The shock speed is that of a slow shock supported by this transverse $B_\perp$ field and is $B_\perp/\sqrt{\mu_0\rho}$. This velocity is equal to v_R, the transverse velocity of the incoming flow, so that the shock is stationary. Since the shock takes care of the downstream flow, the only question that remains is what the distance L^* from the X-point at which the shocks begin is (Kulsrud, 2001).

The reconnection velocity is the Sweet–Parker velocity modified by replacing L by L^*, i.e.,

$$v_R = V_A\sqrt{\frac{L}{SL^*}}, \qquad (3.27)$$

faster by $\sqrt{L/L^*}$ than the Sweet–Parker velocity.

Petschek showed that all the MHD relations were satisfied independent of the choice for L^*, so it appears that L^* could be arbitrarily small. Petschek found there is a limit on the shortness of this length, $L^* > L(\log S)^2/S$, at which length the current in the shocks seriously perturbs the upstream flow (Kulsrud, 2001).

Petschek proposed this as the correct length, leading to

$$V_R(\text{Petschek}) = V_A \frac{\pi}{8 \log S}. \tag{3.28}$$

We note here that this result actually differs by a factor of 2 from the original Petschek result. There is a minor error in the Petschek paper, a correction for which was pointed out by Vasyliunas (1975) and Priest and Forbes (2000).

Since $\log S$ is of order 10 or 20, the Petschek model predicts a very fast reconnection velocity, a finite fraction of the Alfvén velocity. But since in the two main applications S is very large, of order 10^{12}–10^{14}, the length L^* must be extremely short, namely L/S. In most astrophysical and space applications this is a microscopic length. This extremely short length for the Sweet–Parker layer in the Petschek model was not commented on at the time his model was proposed.

It was recognized that the Petschek formula involved the magnetic field strength B_i just outside the layer. Because of the perturbations produced by the shock currents, this could be considerably larger or smaller than the global field strength B_e. Petschek chose his limit on L^* qualitatively so that the upstream magnetic field was not seriously perturbed. The relation between B_i and B_e in the Petschek model was made quantitative by Priest and Forbes (1986), who showed that, for various solutions of the external field, B_i could be considerably stronger than B_e. For some global flows their theory predicts that the reconnection rate can reach the Alfvén speed based on the global field B_e.

The validity of the Petschek model was not challenged until the Biskamp simulation which showed that, for constant resistivity, the Sweet–Parker model was the correct one. In these simulations Biskamp did find shocks but they only emerged at a distance L^* comparable to the global scale L, much larger than would have been predicted by Petschek. The scaled boundary layer numerical simulation of Uzdensky and Kulsrud (2000) confirmed these results.

While the Petschek reconnection rate is consistent with the observed fast reconnection rates in space and has become very frequently cited, it has not been rigorously established because it is not compatible with either resistive MHD characteristics or two-fluid physics mechanisms (Kulsrud, 2001; Yamada et al., 2010). In past decades, a further analysis of this model has been made, employing a locally enhanced resistivity $\eta_{\text{eff}}(r)$, which is consistent with the notion that a high electron current should induce an anomalous resistivity due to the waves generated by high electron current density at the reconnection region (Sato and Hayashi, 1979; Ugai and Tsuda, 1977). The locally enhanced resistivity would increase dissipation near the X-point, accelerating electron flows toward the X-point. This would generate a wedge-shaped reconnection region with "slow shocks" formed near the central high resistivity region, as shown in figure 3.5 (Uzdensky and Kulsrud, 2000). This configuration is free from the constraint of the thin reconnection layer of the Sweet–Parker model and allows for a fast reconnection rate. However, the shock-like structures do not extend all the way to the system size and are thus not consistent with the original wedge-shaped structure of Petschek. We also note that there has been no conclusive experimental evidence to date of shocks observed in association with magnetic reconnection layers in laboratory plasmas.

(a)　　　　　　　　　　　　　　　　**(b)**

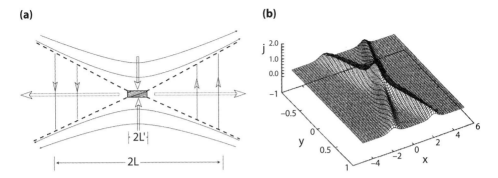

Figure 3.5. (a) Petschek reconnection model and (b) current profile from a numerical simulation by Uzdensky and Kulsrud (2000).

Although many other mechanisms have been found to speed up reconnection on the larger scales found in astrophysics, the simple Petschek mechanism, which produces fast enough reconnection to agree with observations, has been very popular. However, the exact physical mechanism of the Petschek model was unclear for a long time. Some understanding of the shocks emerged in a paper by Ugai and Tsuda (1977). They carried out a detailed numerical simulation in a two-dimensional model for the current layer with appropriate boundary conditions. Their reconnection model had a resistivity that was a specific nonconstant function of space. The resistivity was strongly enhanced in a sizable region about the central X-point of the current layer where the field was zero. (The resistivity was larger at the X-point than in the region exterior to the current layer, by a factor of 100.) Their solution differed from that of Sweet–Parker and indicated the presence of Petschek-like shocks. These shocks emerged from the position in the current layer where the resistivity was most rapidly changing.

A similar numerical simulation was carried out by Sato and Hayashi (1979). These authors allowed the resistivity to be a function of the current density. They triggered the reconnection by imposing a space-dependent inflow into the current layer of flux lines and plasma. Because the resistivity depended on the space-dependent current density, their resistivity was also space dependent as in the Ugai–Tsuda paper. Shocks were again found to emerge at the place where the resistivity was most rapidly changing.

The relation between the resistive scale and the location of the shocks was explored over a wide range of resistivity scales by Scholer (1989). He showed that the shocks always start at the place where the plasma resistivity was most rapidly varying. This is approximately where the resistivity has decreased by a factor of 2 from its maximum value at the current layer center.

Biskamp (1986) was the first person to study the Petschek theory by numerical calculation with a constant resistivity. His calculation was carried out on a global scale with only a small part of his simulation volume occupied by the reconnecting current layer. (This layer emerged naturally during the simulations.) He made multiple simulations, determining the rate of reconnection in each case. Expressing the theoretical rates in terms of magnetic field strengths just outside the current layer and using the length of the current layer as the global length in the two theories, he found that the Sweet–Parker formula for the reconnection rate agreed with all of the eight computational rates. The Petschek formula did not agree with any. He concluded that the Sweet–Parker model was the correct one. However, he did not emphasize that the resistivity was a constant in space in his calculations. Here it should be noted that numerical calculations in the 1990s were limited to studies at a relatively low Lundquist number ($S < 10^5$) because of large numerical noise. With much more powerful modern computers many new findings are being made, and we will discuss magnetic reconnection in large systems in chapter 14.

The situation for magnetic reconnection at the magnetosphere is clearer than that of a solar flare because the case for a current layer is more compelling. The solar wind possesses a magnetic field, and when it encounters the surface of the magnetosphere one expects that if the solar wind were an ideal plasma, it would be deflected around the earth leaving a cavity, called the magnetosphere. However, on the surface of the cavity, the magnetopause, the solar-wind field is not aligned with the earth's magnetic dipole field, and there is automatically a narrow current layer separating them.

As discussed in the introduction, the reconnected solar-wind lines in the Dungey model of the magnetosphere have a part in the solar wind and a part in the magnetosphere. One end is dragged along with the solar-wind velocity v_S, while the other end is anchored in the electrons in the earth's ionosphere. This part of the line will be dragged through the ionosphere at a much slower rate. Since the ions in the ionosphere are immobilized by collisions with neutrals, the electron motion produces an electrical current in the ionosphere that is measured from the ground, enabling one to actually count the number of reconnected lines flowing across the polar cap of the earth per unit time. It is found that on average 1 line in 10 of the incoming solar-wind field lines is reconnected; see Hughes (1995) which suggests a fast reconnection rate. The Sweet–Parker theory predicts that the fractional number of reconnection lines is $(v_A/v_S)/\sqrt{S}$, where $S = v_A L/\Lambda$. A simple estimate gives a value of about 10^{-5} (Kulsrud, 2005).

3.6 EXPERIMENTAL ANALYSIS OF THE MAGNETIC RECONNECTION LAYER BASED ON MHD MODELS

Important progress has been made using MHD models in analyzing the reconnection processes observed in laboratory experiments and in situ measurements by space satellites. Such analyses quantitatively test the validity of the primary MHD models, and provide insight into non-MHD effects. This section presents recent MHD analyses of magnetic reconnection layers in laboratory and space plasmas.

3.6.1 Experimental test of the Sweet–Parker model in a laboratory plasma

In controlled driven experiments in the MRX (Magnetic Reconnection Experiment), the basic physics of magnetic reconnection was quantitatively studied by measuring the evolution of the measured flux contours of the reconnecting field (Yamada et al., 1997a,b). Experiments were carried out in the double annular plasma setup in which two toroidal plasmas with annular cross-section are formed independently around two flux cores, as shown in figure 3.6. Each flux core contains poloidal and toroidal field coils to generate plasma discharges in both the private and common flux regions around the cores and to create a variety of magnetic-field-line merging situations. After an initial setup period, poloidal field currents in the two flux cores are decreased to generate a reconnection layer where common flux lines are pulled back toward the X-point. In this way, the contours of both poloidal and toroidal magnetic flux are driven toward the central reconnection region. Typical plasma parameters are $n_e = (0.1–1) \times 10^{14} \, \text{cm}^{-3}$, $T_e = 5–15 \, \text{eV}$, $B = 0.2–1 \, \text{kG}$, and $S = 500–1,000$.

The experiments were carried out first in high-density plasma that satisfies a condition for collisional plasma: $\lambda_{\text{mfp}} < c/\omega_{pi}$. The "rectangular" profiles of the reconnection region were verified to be very close to the shape Parker predicted, as seen in figure 3.6 (Yamada et al., 1997a). The reconnection rate was measured by monitoring the time motion of the poloidal flux contours shown in figure 3.6 as a function of plasma parameters and compared with the Sweet–Parker model. When the collision frequency was high, the classical Sweet–Parker reconnection rate was measured based on the resistivity calculated by Spitzer resistivity (Trintchouk et al., 2003; Kuritsyn et al., 2006). It should be noted that this rectangular shape is consistent with the uniform influx of radial inflow with respect to Z in figure 3.6 (Biskamp, 2000). When the collisionality was reduced by operating the experiment in a low-density condition, the shape of the reconnection region was observed to change with higher reconnection rates.

3.6.2   MHD analysis of the current sheet with an effective resistivity

To cope with the significant enhancement of the resistivity for the collisionless regime ($\lambda_{\text{mfp}} > c/\omega_{pi}$), an anomalous resistivity theory was used initially by employing an ad hoc enhanced value of the resistivity, η_{eff}, in eq. (3.21). It should be noted that this formulation is sometimes useful in describing the fast reconnection rate by MHD, although there was no theoretical basis developed at that time for imposing uniformly enhanced resistivity from turbulence or other mechanisms. A generalized Sweet–Parker model was developed by employing the measured effective resistivity η_{eff} to quantitatively explain the reconnection rates observed in MRX (Ji et al., 1998, 1999).

The first question is whether classical MHD theory can quantitatively describe the reconnection process in highly collisional plasmas in which the one-fluid assumption holds. A neutral sheet experiment (Syrovatskii, 1971; Frank, 1974) was carried out in a collisional plasma in which the electron mean free path is much shorter than the plasma size. A quantitative analysis of the reconnection rate was not made. The reconnection speeds, inferred from the evolution of the reconnected flux, were reported from collisional experiments on TS-3 (see table 6.1) (Yamada et al., 1991; Ono et al., 1993) with and without a guide field. The reconnection speed was much faster without a guide field (see section 6.2.1).

The first quantitative tests of the classical Sweet–Parker model were performed on MRX (Yamada et al., 1997a; Ji et al., 1998, 1999), where all important quantities were measured or inferred. Figure 3.7 shows an example of the magnetic profile in MRX at high collisionality. The prototypical rectangular shape of the diffusion region is seen.

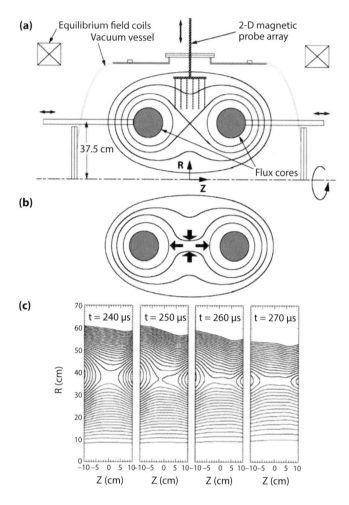

Figure 3.6. (a) Illustration of reconnection driven by inductive flux core coils in the MRX apparatus. Internal magnetic probes document the time evolution of the magnetic profile during reconnection. (b) By programming induction coil currents, two types of reconnection can be driven near the X-point: push or pull reconnection. (c) Time evolution of flux contours during a driven pull reconnection in which common field flux lines (which wrap around both flux cores) are pulled toward the X-point. Sweet–Parker-type magnetic reconnection is demonstrated through measured flux contours. In the low-β plasma outside the neutral sheet, poloidal flux contours represent magnetic field lines. [From Yamada et al. (1997a,b). https://mrx.pppl.gov/mrxmovies/Collisio nal.mov]

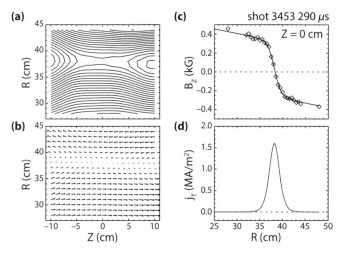

Figure 3.7. An example of a magnetic profile measured in MRX. (a) Vector plot of the poloidal field. (b) Evolution of poloidal flux contours, which represent field lines (measured in MRX by Ji et al., 1999).

The reconnection speed V_R was determined by E_θ / B_Z, where the reconnecting electric field $E_\theta \equiv -(\partial \Psi / \partial t)/2\pi R$, Ψ is poloidal flux, and B_Z is the upstream reconnecting magnetic field. The measured reconnection rate V_R / V_A did not agree with the predicted rate of $S^{-1/2}$ from the classical Sweet–Parker model. The causes of the discrepancies were found by examining the validity of the assumptions made in each step of the derivation of the Sweet–Parker model.

Examining the continuity equation revealed effects due to plasma compressibility. The relation $V_R = (\delta/L)V_Z$ is replaced by

$$V_R = \frac{\delta}{L}\left(V_Z + \frac{L}{n}\frac{\partial n}{\partial t}\right) \tag{3.29}$$

when the density within the current sheet increases. This effect accelerates reconnection during the density buildup phase.

Downstream plasma pressure also plays a role. From the equation of motion, the outflow is reduced from the usual $V_Z = V_A$ to

$$V_Z^2 = V_A^2(1+\kappa) - 2\frac{p_{\text{down}} - p_{\text{up}}}{\rho}, \tag{3.30}$$

where $\kappa \equiv (2/B_Z^2) \int_0^L B_R(\partial B_Z/\partial R)\,dZ = 0.2\text{--}0.3$ represents the relative importance of the downstream tension force, which is omitted in the Sweet–Parker model. A higher downstream pressure ($p_{\text{down}} \gg p_{\text{up}}$) substantially reduces the outflow to 10–20% of V_A. This reduction indicates the importance of boundary conditions in determining local reconnection rates.

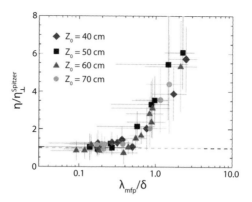

Figure 3.8. Effective plasma resistivity normalized to the transverse Spitzer resistivity as a function of inverse collisionality $\lambda_{\mathrm{mfp}}/\delta$ for different flux core separations Z_0 in no-guide-field cases. [From Kuritsyn et al. (2006).]

Ohm's law along the current (toroidal) direction was examined. Outside the current sheet, $(V \times B)_\theta$ balances with the reconnecting electric field E_θ, but has to be balanced by other terms within the current sheet. In MHD models, the balancing term is the resistive term, and thus an effective resistivity can be determined by $\eta^* = E_\theta/j_\theta$ as shown by Ji et al. (1998). When the plasma is collisional, i.e., the electron mean free path is much shorter than the current sheet thickness, the effective resistivity η^* agrees well with the transverse Spitzer resistivity ($\eta_\perp = 1.96\eta_{\mathrm{parallel}}$; Spitzer, 1962) within 30% error and it varies as $T_e^{-3/2}$ (Trintchouk et al., 2003). In the relatively collisionless regime where the mean free path is much larger than the current sheet thickness, a significant resistivity enhancement over the classical values was measured, as shown in figure 3.8.

The Sweet–Parker model was generalized by Ji et al. (1998, 1999) to incorporate the above three modifications, i.e., plasma compressibility, higher downstream pressure than upstream, and effective resistivity larger than the Spitzer resistivity,

$$\frac{V_R}{V_A} = \frac{1}{\sqrt{S^*}}\sqrt{\left(1 + \frac{\partial n}{\partial t}\right)\frac{V_Z}{V_A}}, \tag{3.31}$$

where S^* is the Lundquist number calculated from the effective resistivity. Figure 3.9 shows good agreement between the observed reconnection rate and that predicted by the generalized model. This result shows that the reconnection process can be described by the Sweet–Parker model, with generalizations, in a stable two-dimensional reconnection neutral sheet with axisymmetric geometry. This generalized Sweet–Parker model applies to cases both with and without a guide field (co-helicity and null-helicity respectively).

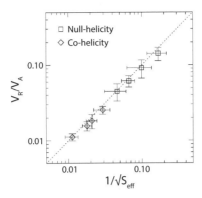

Figure 3.9. Comparisons between experimentally measured reconnection rates and predictions by the generalized Sweet–Parker model at MRX. In co-helicity merging, field lines meet at an angle with a guide field, while in null-helicity merging, field lines meet in an antiparallel manner. [From Ji et al. (1999).]

3.6.3 Non-antiparallel reconnection: Effects of a guide field

Magnetic reconnection usually occurs when two parts of magnetic field lines meet with an angle, but not necessarily in an antiparallel manner. We treat this geometry as an antiparallel reconnection with an additional uniform field. The third vector component of the magnetic field is called a "guide field" and it plays an important role in the reconnection process (Kivelson and Russell, 1995). In the dayside reconnection region of the terrestrial magnetosphere, the southward IMF (interstellar magnetic field), which merges antiparallel to the earth's northward dipole field, reconnects very fast at the meridian plane. On the other hand, the northward-oriented IMF, which merges near parallel to the earth field, reconnects much slower.

In MRX the reconnection resistivity was measured both with and without a guide field (Kuritsyn et al., 2006). In antiparallel reconnection without a guide field, the transverse Spitzer resistivity was measured. With a sizable guide field the measured resistivity was a factor of 2 smaller, consistent with parallel Spitzer resistivity. Spitzer calculated that $\eta_{\perp} = 1.96\eta_{\parallel}$. This was the most conclusive experimental verification of Spitzer resistivity to the author's knowledge. Also in the relatively collisionless operation regime, a significant enhancement over the Spitzer values was measured in both cases.

Merging experiments showed that magnetic reconnection is influenced by the merging angle of the field lines (Ono et al., 1993; Yamada et al., 1997a,b; Brown, 1999; Cothran et al., 2003). To determine the dependence of the reconnection speed on the merging angles of reconnecting lines, the magnitude of the external guide field was varied in TS-3 and MRX while the reconnecting field was kept roughly constant (Yamada et al., 1990; Ono et al., 1993; Yamada et al., 1997b,a). When the guide field is near zero (the reconnecting angle is near 180 degrees), the reconnection speed is maximized. As the reconnecting angle is reduced with increasing guide field, the reconnection speed decreases substantially. In MRX it was observed that the presence of a guide field broadens the neutral sheet, substantially changing the two-dimensional profile from the double-Y shape to an O-shape (Yamada et al., 1997b,a). This transition of the neutral sheet was first recorded by a UCLA group in the electron MHD regime (see chapter 5) (Gekelman et al., 1982).

Generally, the reconnection rates in guide field reconnection are notably smaller than no-guide-field cases. In the context of resistive MHD theory, the observed slower rates are attributed to (1) smaller resistivity for a neutral sheet current parallel to the guide field, (2) suppression of plasma flow by the guide field, and (3) less compressibility of the plasma due to the presence of a guide field. The first factor could be due to the current flow along field lines that causes less microturbulence and reduced Hall effects. The second and third factors can be due to the guide field confining the plasma locally, increasing downstream pressure and reducing plasma compressibility. More work has been carried out to assess the physics of guide field effects, particularly in the two-fluid regime discussed in chapter 5.

An important question raised here is why reconnection occurs so fast in tokamak sawtooth crash, where the guide field is very strong. This may be due to three-dimensional global MHD instabilities that drive fast magnetic reconnection in a localized region, as will be described in chapter 9.

3.6.4   Observation of accelerated plasmas in the exhaust of a reconnection site in space: Consistent with the Petschek model

It was claimed that the recent findings of reconnection exhausts in the solar wind provide a good test bed to verify the Petschek picture of a neutral sheet in which the magnetic field is reversed, together with acceleration of ions in the exhaust region

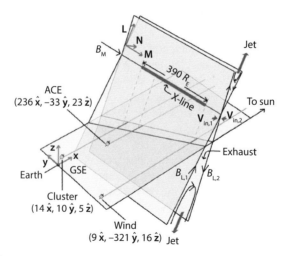

Figure 3.10. Diagram of the encounters of three satellites with regard to the contemplated reconnection X-lines. [From Phan et al. (2006).]

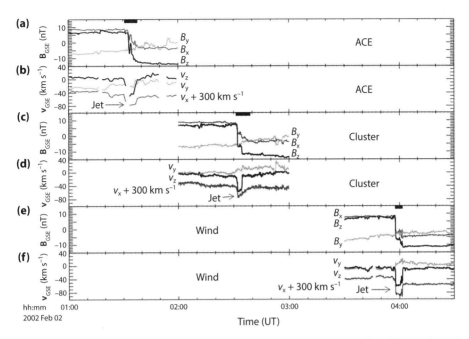

Figure 3.11. Detections of magnetic reconnection exhaust by the ACE, Cluster-3, and Wind spacecraft on 2 February 2002. (a), (b) The magnetic field and plasma velocity in GSE coordinates measured by ACE. (c), (d) The magnetic field and velocity measured by Cluster-3. (e), (f) The magnetic field and velocity measured by Wind. The x-components of the velocity in (b), (d), (f) have been shifted by $+300$ km/s. The bold horizontal bars in (a), (c), (e) indicate the durations of the encounters by the three spacecraft. The magnetic field is rotated 140 degrees across the exhaust. The plasma flow in the exhaust was enhanced by 50 km/s relative to the ambient solar-wind flow speed. The velocity components were correlated (anticorrelated) with the components of the magnetic field at the leading (trailing) edge of the exhaust, as expected from reconnection sunward and northward of the spacecraft. It is concluded that all three (widely separated) spacecraft detected essentially the same current sheet signature. The abrupt changes in the magnetic field B_z at the two edges and a plateau in the B_z-profile in the middle of the current sheet indicate that the current sheet is bifurcated. [From Phan et al. (2006).]

(Gosling et al., 2005; Phan et al., 2006). Solar-wind reconnection is usually generated by the interplay of the two solar winds originating from coronal mass ejection. The magnetic field orientations of the two merging plasmas are well defined. The presence of accelerated ions in the reconnection X-line has been observed by coordinated measurements from the three satellites ACE, Cluster, and Wind: see figure 3.10.

Reconnection in the current sheet (shown in blue in figure 3.10) is considered to occur at the X-line between magnetic field lines with large antiparallel components $B_{L,1}$ and $B_{L,2}$; the resulting bidirectional plasma jets (confined to the reconnection exhausts) are observed far from the X-line. The three spacecraft positions are shown in units of R_E and in geocentric solar ecliptic (GSE) coordinates with the x-axis pointing from the earth to the sun, the y-axis pointing toward dusk, and the z-axis parallel to the ecliptic pole. All three spacecraft were relatively close to the ecliptic plane (in yellow). ACE was $222R_E$ upstream of Cluster while Wind was $331R_E$ "dawnward" of Cluster. Also shown is the LMN current sheet coordinate system, with N along the overall current sheet normal, M along the X-line direction, and L along the antiparallel magnetic field direction. The current sheet is tilted 45 degrees relative to the sun–earth line. The thick solid red line is the hypothesized ($390R_E$) portion of the X-line whose effect is observed by the three spacecraft. The solid orange lines denote the spacecraft trajectory relative to the solar wind, with the red line portion marking the intersection of the exhaust with the spacecraft.

All three satellites detected typical signatures, which would support the Petschek model, in the profiles of magnetic field and ion velocity vectors as seen in the passage of the same bifurcated current sheet; see figure 3.11. Although a direct measurement of a shock structure in the reconnecting region (near an X-line) was not made, the presence of strong current sheets was suggested so that the data were consistent with the Petschek MHD model (Petschek, 1964) in which plasma acceleration is generated by the tension force of the reconnecting field at the exhaust. This implies that the X-line extended at least $390R_E$ (or 4×10^4 ion inertial lengths). If reconnection were patchy, one or more spacecraft would most probably not have encountered accelerated flow. Another fact that is consistent with a coherent and extended X-line is that the reconnection jets detected by all three spacecraft were directed in the same direction, implying that the X-line was north of all spacecraft. Patchy and random reconnection could result in different spacecraft detecting jets directed in different directions.

Most of the observations mentioned in this section have been analyzed in the context of MHD, but it is quite obvious that two-fluid physics analysis is needed to accurately describe the results since the sizes of current sheets and the reconnection regions, for the cases observed, were comparable to the ion skin depth or the ion gyroradii. In the next section, we focus our discussions on two-fluid physics.

Chapter Four

Kinetic description of the reconnection layer: One-dimensional Harris equilibrium and an experimental study

As presented in chapter 3, magnetic reconnection was first described primarily by MHD (magnetohydrodynamic) theory, which was developed in the early stages of plasma research, treating the plasma as a single fluid (Parker and Krook, 1956; Parker, 1957; Sweet, 1958). The MHD framework is based on the assumption that electrons and ions move together as a single fluid, even in the presence of internal currents. When a conductive plasma is described by MHD, it has a large Lundquist number and is approximated well by ideal MHD theory. However, it was realized that the MHD models had to be modified to analyze magnetic reconnection, since the MHD condition does not hold in a thin reconnection layer. The reconnection rate calculated by the Sweet–Parker model, based on resistive MHD theory, is too slow to explain the reconnection rate observed in both space and laboratory plasmas. Reconnection layers in the magnetopause (Vasyliunas, 1975; Dungey, 1995; Kivelson and Russell, 1995) have thicknesses that are comparable to the ion skin depth c/ω_{pi} ($\sim 50\,\mathrm{km}$). This situation leads to strong two-fluid effects. To describe this type of reconnection layer properly, a more general theory than MHD is necessary, taking into account the different behaviors of electrons and ions. Precise measurements of the neutral sheet profile would provide important clues to help understand the physical mechanisms of reconnection. To find the fundamental picture we examine the dynamics of such thin reconnection layers to learn about the fundamental physics of two-fluid dynamics. Here we start with a simple one-dimensional analysis.

4.1 ONE-DIMENSIONAL HARRIS FORMULATION AND SOLUTIONS

We have learned that in the collisionless reconnection layer in the magnetosphere, ions become less magnetized while electrons are fully magnetized, and the relative drift velocity between electrons and ions can be large. In this two-fluid regime, non-MHD phenomena, including induced local electric fields and wave phenomena, are likely to be generated. So the history of magnetic reconnection research has been about how we describe this two-fluid physics in the reconnection layer.

In the neutral sheet in a steady-state equilibrium, the plasma thermal pressure should be in a force balance with the outer magnetic pressure of the opposing reconnecting magnetic fields. This assumption was used in the Sweet–Parker model for MHD. The question is, for a plasma system, can we find such a solution in which the motion of electrons is decoupled from that of ions? Harris (1962) found analytical one-dimensional solutions for a steady-state equilibrium for the magnetic field, plasma pressure, and current density profiles of a current sheet by solving the Vlasov and Maxwell equations with three important assumptions: (1) the presence of no electric field, (2) electrons and ions drifting in opposite directions at the diamagnetic speed, and (3) equal and spatially uniform electron and ion temperatures. The Harris solution is a unique, elegant description of a neutral sheet in a highly conductive plasma and, therefore, is referenced very often; researchers carrying out numerical simulations frequently set their initial conditions to Harris equilibrium. In the early days, numerical simulations and space observations yielded favorable comparisons with the Harris sheet, showing that the sheet thickness is roughly equal to the ion skin depth. Two decades ago, the Harris sheet was investigated experimentally in a laboratory plasma (Yamada et al., 2000) in which the profile of the reconnecting magnetic field was precisely measured, in a steady state, to allow a detailed study of the one-dimensional current sheet profile as a function of the relevant plasma parameters. This is the subject of this chapter. Here, let us first study the kinetic mechanisms in the reconnection layer in a simple one-dimensional analysis.

4.2 THEORY OF THE GENERALIZED HARRIS SHEET

How do we treat plasma in kinetic theory? If we know the position and velocity of each particle composing a plasma, of electrons and ions, then we should be able to completely describe the plasma, assuming that we have an enormously powerful supercomputer, for example using the Klimontovich equations. Since this approach is very difficult, we often describe plasma dynamics by employing a statistical formulation using distribution functions for ions and electrons (and sometimes with additional different species) in six-dimensional phase space $f_{e,i}(\boldsymbol{r}, \boldsymbol{v}, t)$, namely

$$\frac{\partial f_{e,i}}{\partial t} + \boldsymbol{v} \cdot \frac{\partial f_{e,i}}{\partial \boldsymbol{r}} + \frac{q}{m}(\boldsymbol{E} + \boldsymbol{v} \times \boldsymbol{B}) \cdot \frac{\partial f_{e,i}}{\partial \boldsymbol{v}} = 0. \tag{4.1}$$

For a steady-state condition, a collisionless current sheet can be analyzed using the steady-state Vlasov–Maxwell system of equations,

$$\boldsymbol{v} \cdot \frac{\partial f_{e,i}}{\partial \boldsymbol{r}} + \frac{q}{m}(\boldsymbol{E} + \boldsymbol{v} \times \boldsymbol{B}) \cdot \frac{\partial f_{e,i}}{\partial \boldsymbol{v}} = 0, \tag{4.2}$$

$$\nabla \times \boldsymbol{E} = 0, \tag{4.3}$$

$$\nabla \cdot \boldsymbol{E} = \frac{e}{\epsilon_0}\left(\int f_i d\boldsymbol{v} - \int f_e \, d\boldsymbol{v}\right), \tag{4.4}$$

$$\nabla \times \boldsymbol{B} = e\mu_0 \left(\int \boldsymbol{v} f_i d\boldsymbol{v} - \int \boldsymbol{v} f_e d\boldsymbol{v} \right), \tag{4.5}$$

$$\nabla \cdot \boldsymbol{B} = 0, \tag{4.6}$$

where $\boldsymbol{E}$ is the electric field, $\boldsymbol{B}$ the magnetic field, $\boldsymbol{v}$ the particle velocity, and f_e (f_i) the electron (ion) distribution function. Also, q_e and q_i are electron and ion charges ($q_e = -q_i$). Here we neglect the displacement current in the Maxwell equations. In general, the equations are nonlinear and can only be solved numerically. However, Harris (1962) found a one-dimensional steady-state analytical solution with certain assumptions. Let us rederive the Harris solution, relaxing a few of his original assumptions.

Assuming that magnetic field is in the z-direction and a function of x, all solutions of eq. (4.2) must be functions of the constants of the motion, i.e., the total energy $W \equiv mv^2/2 \pm e\phi$ and the canonical momentum in the y- and z-directions, $p_y \equiv mv_y \pm eA_y$ and $p_z \equiv mv_z$ respectively. Here, ϕ is the electrostatic potential and the vector potential $\boldsymbol{A}$ ($\nabla \times \boldsymbol{A} = \boldsymbol{B}$) is assumed to have only a y-component A_y. Consider the distribution function

$$f_{e,i} = n_0 \left(\frac{m}{2\pi T} \right)^{3/2} \exp \left\{ -\frac{m\left(v_x^2 + (v_y - V)^2 + v_z^2\right)}{2T} \pm \frac{e(VA_y - \phi)}{T} \right\}, \tag{4.7}$$

where the constant $T = T_e$ (T_i) is the electron (ion) temperature (spatially constant) and $V = V_e$ (V_i) is the electron (ion) drift (flow) speed in the y-direction. Because the argument of the exponential can be written as $(-W + p_y V - mV^2/2)/T$, $f_{e,i}$ is a function of the constants of the motion and therefore a solution of eq. (4.2). Harris realized that this shifted Maxwellian is the most natural solution.

Here, let us expand the Harris solution by relaxing the original Harris assumptions of $T_e = T_i$ and $V_i = -V_e$ to derive a general Harris solution. First, we again use a coordinate system where the magnetic field is in the z-direction, varying with respect to x. In our one-dimensional model, all variables are assumed to vary only in x, except for T_i, T_e, V_i, and V_e which are all assumed to be constant. The y-direction is the direction of the neutral sheet current. The system of equations can be simplified significantly by assuming that $\boldsymbol{E}$ has only an x-component $E_x = -\partial\phi/\partial x$. And since A_y is assumed to be the only nonzero component of $\boldsymbol{A}$, the magnetic field $\boldsymbol{B}$ has only a z-component $B_z = \partial A_y/\partial x$. It should be noted that $E_y = 0$ and the collisionless assumptions mean that there is no dissipation and hence no reconnection considered in this model.

Substitution of eq. (4.7) into eqs. (4.4) and (4.5) yields two coupled nonlinear differential equations for ϕ and A_y:

$$\frac{\partial^2 \phi}{\partial x^2} = -\frac{en_0}{\epsilon_0} \left\{ \exp \left(\frac{e(V_i A_y - \phi)}{T_i} \right) - \exp \left(\frac{-e(V_e A_y - \phi)}{T_e} \right) \right\}, \tag{4.8}$$

$$\frac{\partial^2 A_y}{\partial x^2} = -en_0\mu_0 \left\{ V_i \exp \left(\frac{e(V_i A_y - \phi)}{T_i} \right) - V_e \exp \left(\frac{-e(V_e A_y - \phi)}{T_e} \right) \right\}. \tag{4.9}$$

By using normalized variables $\hat{\phi} \equiv e\phi/T_e$ and $\hat{x} \equiv x/(c/\omega_{pi})$ (where $\omega_{pi} \equiv \sqrt{n_0 e^2/\epsilon_0 m_i}$ is the ion plasma frequency), eq. (4.8) is converted to the dimensionless form

$$\frac{\partial^2 \hat{\phi}}{\partial \hat{x}^2} = -\left(\frac{c/\omega_{pi}}{\lambda_D}\right)^2 \left\{\exp\left(\frac{e(V_i A_y - \phi)}{T_i}\right) - \exp\left(\frac{-e(V_e A_y - \phi)}{T_e}\right)\right\}, \tag{4.10}$$

where the left-hand side is of order unity but $(c/\omega_{pi}/\lambda_D)^2$ is of order 10^6 (λ_D is the Debye length). Therefore, quasi-neutrality $\left\{\exp(\frac{e(V_i A_y - \phi)}{T_i}) - \exp(\frac{-e(V_e A_y - \phi)}{T_e})\right\} \simeq 0$ must be satisfied, leading to an ambipolar potential

$$\phi = \frac{T_e V_i + T_i V_e}{T_e + T_i} A_y. \tag{4.11}$$

Interestingly, ϕ is proportional to A_y. Substituting eq. (4.11) into eq. (4.9) gives a nonlinear equation in only A_y,

$$\frac{\partial^2 A_y}{\partial x^2} = -e n_0 \mu_0 (V_i - V_e) \exp\left(\frac{e(V_i - V_e)}{T_e + T_i} A_y\right). \tag{4.12}$$

By employing a similar analytical scheme and elaboration to those used by Harris, this equation can be solved analytically using appropriate boundary conditions.

Now we have derived our "generalized" Harris solutions,

$$A_y = -\delta B_0 \log \cosh\left(\frac{x}{\delta}\right), \tag{4.13}$$

$$B_z = -B_0 \tanh\left(\frac{x}{\delta}\right), \tag{4.14}$$

$$j_y = \frac{B_0}{\mu_0 \delta} \text{sech}^2\left(\frac{x}{\delta}\right), \tag{4.15}$$

$$E_x = \frac{T_e V_i + T_i V_e}{T_e + T_i} B_0 \tanh\left(\frac{x}{\delta}\right), \tag{4.16}$$

$$p = n_0(T_e + T_i) \text{sech}^2\left(\frac{x}{\delta}\right), \tag{4.17}$$

where $B_0^2/(2\mu_0) = n_0(T_e + T_i)$. The current sheet thickness δ is given by

$$\delta = \frac{c}{\omega_{pi}} \frac{\sqrt{2(T_e + T_i)/m_i}}{V_i - V_e} = \frac{c}{\omega_{pi}} \frac{\sqrt{2} V_s}{V_{\text{drift}}}, \tag{4.18}$$

where $V_s \equiv \sqrt{(T_e + T_i)/m_i}$ and $V_{\text{drift}} \equiv V_i - V_e$ is the relative drift between ions and electrons. It should be noted that the above solution is more general than the original Harris solution, which is limited to $E_x = \phi = 0$. The original Harris solution can be recovered by setting $T_e = T_i = T$ and $V_i = -V_e = V$ in eq. (4.18) to yield $\delta = (c/\omega_{pi}) (\sqrt{T/m_i}/V)$.

Since the Harris solution is only one of many that might be obtained by choosing f as a function of the constants of the motion (albeit the one most easily solved analytically), one might ask why the experiment "selects" the Harris profile. A possible explanation might be the following. In this experiment, the reconnection proceeds much slower than the Alfvén transit time in which a force balance between the magnetic field pressure and plasma thermal pressure is maintained. Thus reconnection dynamics can come into a play in the slower timescale. Besides, in this collisional layer the ion crossing time of the layer, which is in the range of the ion cyclotron time since the thickness of the layer is approximately one ion gyroradius, and the ion–ion collision time are shorter than the time the plasma takes to flow through the layer. As a result, the ions should be regarded as being in an equilibrium state, neglecting the reconnection velocity V_x. Furthermore, as confirmed by experiment, they are isothermal across the layer, satisfying the full Vlasov–Fokker–Planck equation. We also note that wave phenomena are observed in the less collisional regime and they can also play a similar role to relax the states of ions and electrons in relaxed equilibrium states.

Now, the Harris solution satisfies the Vlasov equation as discussed above. However, it is a shifted Maxwellian solution, eq. (4.7), which makes the ion–ion Fokker–Planck term vanish, and it is probably the only solution of the Vlasov equation which does this. As a result, the ion current is proportional to an exponential in A_y, the first term in eq. (4.9).

The "Harris sheet" solution is a simple analytical equilibrium solution for a plasma confined between oppositely directed magnetic fields. The convenient analytical expressions for the B_z-, j_y-, and p-profiles and the sheet thickness δ lend themselves to direct comparisons with both computer simulations and experiments. However, it is important to keep in mind the limitations and assumptions of the Harris model, including the fact that it is one-dimensional and does not include the effects of reconnection and associated electric fields. And it is necessary to find out how reconnection and two-dimensional effects modify the Harris equilibrium. The Harris model was generalized further by Mahajan (1989) to include time dependence, cylindrical geometry, and various density and velocity profiles.

4.3   EXPERIMENTAL INVESTIGATION OF THE HARRIS SHEET

The Harris theory was experimentally checked and assessed for its application and validity in the MRX (Magnetic Reconnection Experiment) device (Yamada et al., 2000; Ji et al., 2001). As the MRX experimental setup is presented in chapters 2 and 3, we can generate a prototypical reconnection layer in both collisional and collisionless plasmas and can systematically study its characteristics. The hyperbolic tangent shape of a reconnecting magnetic field was observed in the collisional neutral sheets on MRX where the electron mean free path is shorter than the reconnection layer: $\lambda_{\mathrm{mfp}} < L_p$ (see figure 3.7). Also, based on the two-dimensional profiles observed in the MRX reconnection layer, it would be appropriate to the validate Harris sheet theory when a steady-state force balance appears to hold in the plasma along the inflow (x-) direction.

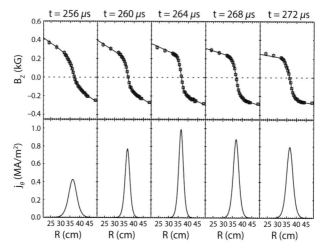

Figure 4.1. Typical time evolution of measured magnetic field profiles and current density profiles during driven magnetic reconnection in MRX. Time evolution of 1D profiles of measured magnetic field (top row) measured by internal magnetic probes and deduced current density (bottom row) along the inflow (R-) direction. In this graph, the R-coordinate was chosen in MRX for the inflow (x-) direction used for our theoretical analysis. The errors in magnetic measurements are typically less than 5%. Note that B_z and j_y are proportional to $\tanh(x/\delta)$ and $\mathrm{sech}^2(x/\delta)$ respectively. [From Yamada et al. (2000).]

Figure 4.1 shows a measurement and its fit to the form

$$B_Z(x) = -B_0 \tanh\left(\frac{x - x_0}{\delta}\right) + b_1 x + b_2. \tag{4.19}$$

The factors b_1 and b_2 are determined by the background quadrupole and equilibrium fields. The latter is an applied equilibrium field necessary to keep the plasma in a desirable position. The factor b_1 does not appear in j_θ because it is canceled by $\partial B_x/\partial Z$ of the background quadrupole field. In Harris equilibrium, the static force balance $\boldsymbol{j} \times \boldsymbol{B} = j_y B_z = \nabla p$ should be maintained between the incoming magnetic field and the plasma pressure during the quasi-steady-state phase of reconnection, since the inflow speed is much slower than the Alfvén speed. The measured magnetic field profiles agree well with the prediction by a generalized Harris theory for nonequal temperatures and drift speeds of ions and electrons (Yamada et al., 2000).

One interesting experimental observation is that the measured reconnecting magnetic field has a shape close to the hyperbolic tangent, despite the many simplifications used to derive this functional shape. The local Maxwellian shape of the distribution function is justified by collisions between like particles, while uniform temperatures can be justified by rapid heat transport (Yamada et al., 2000). It was shown (Ji et al., 2001) that the field profile is insensitive to the normalized drift velocities of charged particles by investigating a nonlinear equation derived from the force balance $\partial B / \partial x = -V(1 - B^2)/\sqrt{2}$, where x and B are normalized by c/ω_{pi} and $\sqrt{2\mu_0 n_0 (T_e + T_i)}$ respectively, and $V \equiv V_d / V_s$. When $T_e + T_i$ is a constant, the magnetic profile is

$$B_z = B_0 \tanh \left(\int_0^x \frac{V}{\sqrt{2}} \, dx \right). \tag{4.20}$$

As long as V is a reasonably smooth function of x, the magnetic profile will be close to the hyperbolic tangent shape (Ji et al., 2001).

4.3.1   Main summaries of MRX results on the Harris sheet

The original Harris theory did not take into account magnetic reconnection: the very good fit of MRX magnetic data to the Harris profile, as well as the agreement between measured and predicted δ, is very remarkable since the MRX plasma is undergoing reconnection. However, based on the discussions in section 4.2 regarding the effects of particle collisions and possibly wave turbulence, the excellent agreement indicates that the Harris profile is the most natural one for a quasi-steady-state reconnecting plasma sheet where collisions keep the plasma in Maxwellian distributions in the velocity space. Phenomenologically, since dissipation is related to an effective resistivity through Ohm's law along the current sheet, $\eta j_y = E_y - V_x B_z$, one expects that static equilibrium in the x-direction (R in MRX) can be maintained during the reconnection process provided $V_x \ll V_A$, which is satisfied in MRX ($V_x / V_A \lesssim 0.1$).

The precise determination of the magnetic field profile in MRX has enabled a detailed study of the neutral sheet thickness δ as a function of other relevant parameters. It is found that $\delta \sim 0.35 c/\omega_{pi}$. In antiparallel reconnection cases in MRX, the ion gyroradius ρ_i is of order c/ω_{pi}. This indicates the importance of pressure balance (between p and $B_0^2/2\mu_0$) and ion gyromotion in determining the structure of the reconnection region. This result is in rough agreement with numerical simulations (Biskamp, 1986; Horiuchi and Sato, 1994; Uzdensky and Kulsrud, 1998; Shay et al., 1998) and observations in the geotail and the magnetopause (Kivelson and Russell, 1995). However, in collisionless reconnection, a new two-dimensional structure appears near the X-point of the reconnection layer and this simplified one-dimensional force balance does not hold anymore with the presence of an in-plane electric field. The detailed features of the neutral sheet profile provide a good indicator for the nature of magnetic reconnection and we will visit this issue in a later chapter.

Figure 4.2 shows the measured thickness of the current sheet with respect to the ion skin depth and theoretically derived Harris sheet thicknesses. An important finding in laboratory neutral sheets is that the measured δ scales with the ion skin depth (Yamada et al., 2000). These results were consistent with other experimental data (Ono et al., 1997; Kornack et al., 1998) where δ was measured to be of order the ion gyroradius, as well as the ion skin depth. If we assume that the pressure balance should hold between magnetic field pressure and the plasma's kinetic pressure, the ion skin depth becomes roughly equal to the average ion gyroradius. A careful comparison of MRX data with the theoretical values expressed by eq. (4.18) regarding the sheet thickness as shown

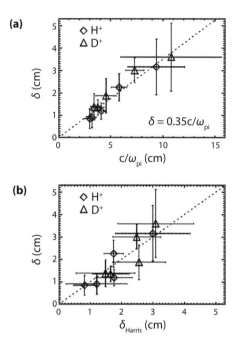

Figure 4.2. (a) Measured neutral sheet thickness δ versus ion skin depth c/ω_{pi}. (b) δ versus theoretical Harris value given by eq. (4.18). [From Yamada et al. (2000).]

in figure 4.2(b) shows that the relative drift velocity of electrons against ions is (or is limited to) 4–5 times the ion sound speed, or $V_{drift} \sim 4$–$5V_S$. This implies that there must be some mechanisms for limiting the relative velocity to this value, which could be a consequence of waves or micro-instabilities induced by the strong relative velocity of electron drift against ions. While they do not appear to affect the force balance of the system as mentioned above, they can accelerate the reconnection rate in the layer. Generally, it is expected that the lower hybrid drift waves are excited by a strong drift velocity of electrons against ions. While the major effects of this micro-instability have not been assessed, we expect that they could exist in the MRX reconnection layer (Ji et al., 2004; Yamada et al., 2010).

Neutral sheets or reconnection layers are observed in magnetospheric plasmas both in the magnetopause and magnetotail sides. The reconnecting magnetic field profile measured by the Polar satellite matches well with the hyperbolic tangent form (Mozer et al., 2002; Bale et al., 2002) but generally deviates quite a lot from the smooth function described here. The reason for this difference appears to be a lack of collisions in the space plasma, which makes a shifted Maxwellian distribution.

In the magnetotail, thin current sheets were observed by the IMP spacecraft (Fairfield et al., 1981), which provided information on the pitch angle distribution function of protons. Detailed observations were made by two ISEE spacecraft (McComas et al., 1986; Sergeev et al., 1993), and further observations were reported by the Cluster spacecraft, which detected for the first time the substructures within the current sheets. Current sheets in a bifurcated state were observed (Runov et al., 2003), sometimes accompanied by flapping motions (Sergeev et al., 2003). These fine structures and fast dynamics could be causes and/or consequences of magnetic reconnection activity in the magnetospheric tail.

4.4 ADDITIONAL COMMENTS AND DISCUSSION

The neutral sheet profile during magnetic reconnection was experimentally studied to identify the basics of two-fluid dynamics of the magnetic reconnection layer. In the relatively collisional condition $\lambda \sim 0.35c/\omega_{pi}$, the measured magnetic field profile is seen to agree remarkably well with the Harris sheet profile even though Harris theory does not take into account reconnection and associated dissipation. These results imply that one can include the effects of quasi-steady-state reconnection within the basic formulation of the Harris equilibrium sheet. Because the Harris shifted Maxwellian velocity distribution function satisfies the full Fokker–Planck equation (leaving out the small electron–ion collision term), it appears also to be the natural state for a reconnecting plasma. In a collisionless reconnection layer there is the possibility that the electron drift velocity becomes very large near the X-point, breaking one of the one-dimensional Harris conditions. The measured neutral sheet thickness δ is seen to scale with the ion skin depth c/ω_{pi} according to $\delta \approx 0.35c/\omega_{pi}$ over a wide range of discharge conditions in collisionality. However, the deviation from a simple one-dimensional Harris sheet profile became apparent as the collisionality was reduced. Also, we note here that shot-to-shot errors become large in the lower-density regime, or large c/ω_{pi}. In the next chapter, we will discuss the feature of reconnection layers in the collisionless regime in which the uniform one-dimensional Harris sheet disappears.

In a collisional plasma it is found that a static force balance is maintained between incoming magnetic field and the neutral sheet plasma pressure during the quasi-steady-state phase of reconnection. An important implication here is that a uniform two-dimensional reconnection neutral sheet in the collisional regime with axisymmetric geometry is consistent with both the Sweet–Parker and the Harris models with generalizations. Resistivity enhancement is observed during reconnection in the collisionless regime. Experimental measurements of the neutral sheet thickness δ agree well with a generalized Harris theory for a nonisothermal plasma ($T_e \neq T_i$). The relationship $\delta \approx 0.4c/\omega_{pi} \sim \rho_i$ suggests that, as mentioned before, a current-driven instability might be excited to limit the current or enhanced resistivity. Among the possible waves, lower hybrid waves are likely candidates since both electrons and ions are involved to excite them and to generate an enhanced momentum transport between them.

In this solution of the Harris model, the ion current is expressed as a diamagnetic current proportional to $T_i(\partial n/\partial x)/nB$ for constant T_i. But if ions are not well magnetized, this component should be small. Thus the Harris condition is not always supported. If the electron temperature is also constant, then its current is proportional to $T_e(\partial n/\partial x)/nB$ and thus also proportional to an exponential in A_y. Hence, under the assumption of constant T_e and constant T_i in the layer, as measured in MRX, one could expect that the magnetic field profile is approximated by the Harris profile.

We note that it can be argued that the electron distribution is not *necessarily* a shifted Maxwellian (or even a solution of the Vlasov equation) since frictional electron–ion collisions would distort such a shifted Maxwellian. However, we assume that, with a moderate collision frequency in the plasma condition used there, this distortion is small and should not modify the electron current from being proportional to the exponential in A_y. Therefore, we can use the Harris solution, i.e., eqs. (4.8) and (4.9) hold. At the same time, the electron–ion collision integral gives rise to the resistivity term ηj_y. If the resistivity is anomalous, the remarks above still apply with the resistivity replaced by the anomalous resistivity.

Chapter Five

Development of two-fluid theory for reconnection coordinated with key observations

As we learned in chapter 4, one can formulate a one-dimensional analytical description of the reconnection layer using the Harris formulation. But it is not straightforward to extend it to a two-dimensional theory to describe the exact features of the reconnection layer. In this chapter, we study first the recent development of numerical simulations of the reconnection layer, compare the results with experimental findings and discoveries, and then study an analytical theory of the reconnection layer.

It has been difficult to develop a two-dimensional analytical theory for a reconnection layer in the two-fluid regime. In the past three decades, significant progress has been made to describe the reconnection layer by effectively utilizing numerical simulations. In addition, thanks to the progress in plasma diagnostics in the past two decades, simulation results can be cross-examined by the detailed magnetic field structure of the reconnection layers measured in laboratory plasmas, as well as in space plasmas. Extensive data have been accumulated in highly conductive plasmas with large Lundquist numbers, which is the ratio of the magnetic diffusion time to the Alfvén transit time, as already defined in chapter 3. The Lundquist numbers of laboratory plasmas are in the range of 10^2–10^7, while the number exceeds 10^{12} in the magnetosphere. When the collisionality is reduced, waves (particularly microturbulence) can be excited due to the different motion of electrons compared to ions. Given the recent progress toward understanding reconnection in collisionless plasmas, this chapter addresses key issues for collisionless reconnection, providing a comprehensive view of two-fluid reconnection dynamics in a collisionless reconnection layer.

5.1 RECONNECTION IN THE MAGNETOSPHERE AND TWO-FLUID DYNAMICS

Magnetic reconnection plays a central role in the interaction between the magnetic field of the solar wind and the earth's dipole field (see figure 1.4). While they often happen impulsively, magnetic reconnection events in the magnetosphere create geomagnetic substorms, which are one important type of magnetic self-organization phenomenon. The magnetospheric plasma is basically collision-free, with the mean free path of both ions and electrons being much larger than the size of the magnetosphere, and thus is governed by typical collisionless two-fluid dynamics. Many satellite missions have been launched to investigate the features of magnetic reconnection in the earth's magnetosphere (Mozer et al., 2002; Øieroset et al., 2002; Cattell et al., 2005). Satellite observations show that the thickness of the current sheath is of order the ion skin depth, which is almost equal to the ion gyroradius in this type of high-β ($\beta \sim 1$) plasma. The ion skin depth (c/ω_{pi}) is typically 50–100 km in the magnetopause while it is larger by an order of magnitude, 500–1,000 km, in the magnetotail. In this situation, the reconnection dynamics cannot be described by the conventional MHD (magnetohydrodynamic) theory of reconnection. This is because ions and electrons behave differently in the reconnection region, requiring two-fluid or kinetic physics. Generally, electrons are magnetized in the current layer except in the electron diffusion layer of a few kilometers in width, while ions are not magnetized in the layer. MHD models that rely on the assumption that electrons and ions move together as a single fluid cannot correctly describe phenomena in the magnetosphere.

The different motion of magnetized electrons compared to demagnetized ions leads to strong two-fluid effects, especially the Hall effect in the neutral sheet. The Hall effect, in which an electric field appears due to the motion of magnetized electrons against nonmagnetized ions, becomes dominant in semiconductors as well as magnetized plasmas (Yamada et al., 2010). This Hall effect, which was not included in the standard MHD model, is considered here to facilitate a large reconnection electric field in the reconnection region and is thus considered to be responsible for speeding up the rate of reconnection over the Sweet–Parker rate.

5.2 RELATIONSHIP BETWEEN THE TWO-FLUID FORMULATION AND MHD

Here, let us discuss the dynamics of electron and ion fluids with respect to electromagnetic fields. We consider two individual fluid equations for electrons and ions, which are moving with fluid velocities of V_e and V_i respectively. We assume that charge neutrality holds with the electron and ion densities being equal ($n_e = n_i$), since we treat phenomena significantly larger than the Debye length (Spitzer, 1962). Here we denote the electron and ion masses by m_e and m_i respectively. MHD treats the plasma as a single fluid, with mass density ρ written as

$$\rho = n_i m_i + n_e m_e = n_e(m_i + m_e) \sim n_e m_i. \tag{5.1}$$

Mass flow density is

$$V = \frac{(n_i m_i V_i + n_e m_e V_e)}{\rho} \sim \frac{(m_i V_i + m_e V_e)}{(m_i + m_e)} \sim V_i + \frac{m_e}{m_i} V_e \tag{5.2}$$

and the current density is

$$j = e(n_i V_i - n_e V_e) \sim e n_e(V_i - V_e). \tag{5.3}$$

These equations are transformed to expressions for the average flow velocity of ions and electrons, V_i and V_e:

$$V_i = V + \frac{m_e}{m_i} \frac{j}{en_e}, \tag{5.4}$$

$$V_e = V - \frac{j}{en_e}. \tag{5.5}$$

The single-fluid MHD basic equations can be obtained by taking combinations of the equations for ions and electrons. The continuity equations for electrons and ions are

$$\frac{\partial n_{i,e}}{\partial t} + \nabla \cdot (n_{i,e} V_{i,e}) = 0. \tag{5.6}$$

These equations can be multiplied by electron and ion masses m_i and m_e respectively, and added together to produce a "mass continuity equation" for an MHD fluid:

$$\frac{\partial \rho}{\partial t} + \nabla \cdot (\rho V) = 0. \tag{5.7}$$

The individual continuity equations can be used (by subtracting one from the other) to generate "charge continuity equation,"

$$\frac{\partial \sigma}{\partial t} + \nabla \cdot j = 0, \tag{5.8}$$

where σ is the charge density.

In a similar way, the two individual momentum balance equations or "equations of motion" for electrons and ions are written as

$$n_e m_e \frac{dV_e}{dt} = -en_e(E + V_e \times B) - \nabla \cdot P_e + R_{ei}, \tag{5.9}$$

$$n_i m_i \frac{dV_i}{dt} = en_i(E + V_i \times B) - \nabla \cdot P_i + R_{ie}, \tag{5.10}$$

where P_e and P_i are the electron and ion pressure tensors. The generalized form of Ohm's law is also equivalent to the equation of motion for electrons. The terms R_{ei} and R_{ie} describe collisional momentum transfer between the two species with the property $R_{ei} = -R_{ie}$. When we add these two equations together, we obtain a "single fluid" equation of motion for MHD through cancellation of charges between electrons and ions:

$$\rho \frac{dV}{dt} = \sigma E + j \times B - \nabla \cdot P, \tag{5.11}$$

where we have used $j = ne(V_i - V_e)$, $V = V_i$, and $P = P_e + P_i$.

This is equivalent to the MHD equation of motion shown in chapter 2. Here we neglect a gravity force. We note that in MHD the pressure tensor becomes isotropic since we expect frequent collisions between plasma particles.

5.3   DEVELOPMENT OF PARTICLE-IN-CELL SIMULATIONS

Thanks to the enormous advancement of computer technology and speed, a new method of calculation has been developed for the past half century to describe plasma dynamics kinetically. In this section, let us look into the recent development of particle-in-cell simulations to describe the kinetic behavior of plasma particles in a collisionless plasma. There are a number of approaches for modeling the kinetic properties of plasma particles. For specific plasma species as described in chapter 4, kinetic phenomena are usually described by the Vlasov equation, for example by a simple one-dimensional analysis. The electric and magnetic fields are determined by Maxwell's equations. The Vlasov equation represents a partial differential equation in a six-dimensional phase space plus time.

The usual approach to kinetic modeling is to represent the distribution function f_j by a number of macroparticles and to compute the particle orbits in self-consistent electric and magnetic fields. The early models (Dawson, 1962) treated the particles as discrete points and computed the electric force acting on each particle by explicitly summing the Coulomb interaction with each of the other $N - 1$ particles. The number of operations then scales as $N(N - 1)/2$. This large number of particle interactions gives a very restrictive limitation on the number of particles that can be employed. The solution that was developed and has now become standard is to introduce a spatial grid on which the particles' charge and current densities are accumulated using an interpolation scheme. The field equations are then solved on this grid, and the forces acting on the particles are obtained by interpolating back to the particles. This is the "particle-in-cell" (PIC) simulation technique. This procedure eliminates fluctuations at scales smaller than the grid spacing and also reduces the number of operations per time step to $N \log N$.

Good introductory guides to such models are given by Winske and Omidi (1993) and Pritchett (2003). Since the earlier work by Dawson (1962) and Birdsall and Langdon (1985), this PIC simulation has been used effectively. The basic idea is quite simple. In a collisionless plasma, particles interact only through electric and magnetic fields, so the point character of particles can be ignored by considering spatial scales that exceed the minimum wavelength of the collective modes. Fields and their sources, charge and current densities, are discretized on a grid mesh size δx. Since a PIC simulation particle represents many plasma particles as a group, it is sometimes called a macroparticle simulation.

In the plasma research community, systems of different species (electrons, ions, neutrals, molecules, dust particles, etc.) can be investigated. The set of equations associated with PIC codes is therefore composed of the Lorentz force as the equation of motion, solved in the so-called pusher or particle mover of the code, and Maxwell's equations determining the electric and magnetic fields, calculated in the (field) solver.

The real systems studied are often extremely large in terms of the numbers of particles they contain. In order to make simulations efficient, or at all possible, so-called superparticles are used. A superparticle (or macroparticle) is a computational particle that represents many real particles; it may be millions of electrons or ions in the case of a plasma simulation or, for instance, a vortex element in a fluid simulation. It is allowed to rescale the number of particles because the Lorentz force depends only on the charge-to-mass ratio, so a superparticle will follow the same trajectory as a real particle would. The number of real particles corresponding to a superparticle must be chosen such that sufficient statistics can be collected on the particle motion.

With the understanding that more than MHD theory is needed to describe the dynamics of the magnetosphere, an instructive comparison of self-consistent kinetics with MHD simulations was made by Chapman and Mouikis (1996). In their comparative study, two simulations were performed, starting from the same initial conditions and employing the same time step: one had a grid spacing equal to one-tenth of the ion gyroradius (called the "hybrid" case) and the other had a grid spacing equal to twice the ion gyroradius (the "MHD" case). The results showed that the evolution of the electromagnetic fields and ion pressure tensor were markedly different in the two cases, demonstrating the importance of kinetic plasma effects in determining the structure of electromagnetic fields and particle distribution functions.

In this monograph we describe the recent PIC simulation results, without getting into their methods, to discuss a number of key issues that bear on the two-fluid or kinetic physics mechanisms of the reconnection layer.

5.4   RESULTS FROM TWO-DIMENSIONAL NUMERICAL
SIMULATIONS FOR COLLISIONLESS RECONNECTION

In recent decades, numerous two-dimensional numerical simulations (Mandt et al., 1994; Ma and Bhattacharjee, 1996; Biskamp et al., 1997; Horiuchi and Sato, 1999; Birn et al., 2001) of the collisionless reconnection layer have demonstrated the importance of the Hall term ($j_e \times B$) based on two-fluid or kinetic codes. In the generalized Ohm equation, it allows a steady (laminar) cross-field current of electrons, which contributes to a large apparent resistivity and generates fast reconnection. Extensive numerical work has been done by Shay and Drake (1998), Shay et al. (1998), Pritchett (2001), Horiuchi and Sato (1999), Daughton et al. (2006), and by many others, with periodic and/or open boundary conditions. In particular, under a collaboration entitled "The Geospace Environmental Modeling (GEM) Magnetic Reconnection Challenge," a concerted effort was made to determine the physical mechanisms and rates of two-fluid reconnection (Birn et al., 2001) and to apply them to the earth's magnetosphere. For this purpose, antiparallel (without a guide field) reconnection was extensively studied in collisionless plasmas.

A common picture has emerged from numerical calculations, which utilized benchmarking studies of reconnection. Figure 5.1 shows a schematic diagram of the field structure and the dynamics of ion and electron flows in a typical reconnection layer (Drake and Shay, 2007), together with results from the PIC simulation of Pritchett

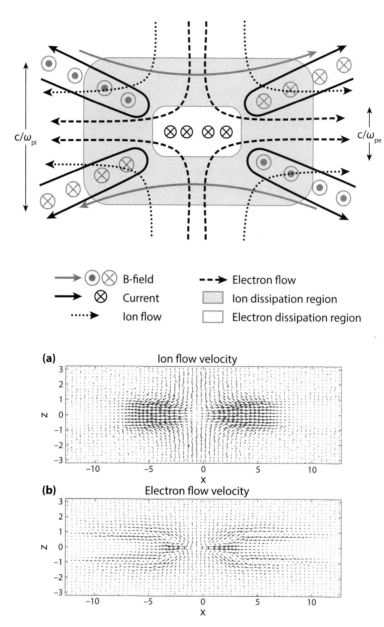

Figure 5.1. (Top) Schematic diagram of the neutral sheet [from Drake and Shay (2007)].
(Bottom) Patterns of ion and electron flows in the neutral sheet [from Pritchett (2001)].
Scale sizes in (x, Z) are in unit of (c/ω_{pi}).

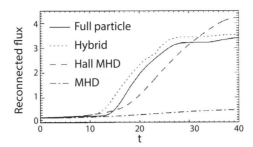

Figure 5.2. Reconnected magnetic flux versus time for four different codes for GEM projects MHD, Hall MHD, hybrid, and full particle codes. The rate of reconnection is the slope of the rising reconnected flux. All models but the MHD model show indistinguishable rates of reconnection and are significantly faster than the MHD model (Sweet–Parker model). [From Birn et al. (2001).]

(2001). As seen in figure 5.1(b), ions become demagnetized as they enter the neutral sheet, turn by 90 degrees in the reconnection plane x–z of their coordinate system, and then flow outward to the exit direction. In contrast, the magnetized electrons mainly flow inward toward the X-point, together with the magnetic field lines. However, near the separatrices, the electrons flow parallel to B field lines toward the X-point. As the electrons' $E \times B$ motion makes them migrate toward the X-point, the magnetic field weakens. The electron drift (E_y/B_x) due to the reconnection electric field E_y becomes larger near the X-point and electrons are ejected to the exit in the reconnection plane. The electron flow patterns generate net circular currents in the reconnection plane shown in figure 5.1 (top) and create an out-of-plane magnetic field with a quadrupole profile, a clear signature of the Hall effect. Similar results were obtained by a simpler two-fluid MHD code that did not describe the exact particle dynamics of electrons and ions (Breslau and Jardin, 2003). The increased electric field derived from the strong Hall term ($j \times B/en$), producing a steady laminar cross-field current of electrons, represents a fast motion of flux lines ($E = -d\Psi/dt$) in the reconnection plane, a fast rate of magnetic reconnection.

A question is how the reconnection rate depends on the dissipation mechanism. An important conclusion of the GEM Challenge (Birn et al., 2001) is that the reconnection speed does not depend on the dissipation mechanism, and it is much larger than the resistive MHD reconnection rate. In figure 5.2, the reconnected flux is shown as a function of time for different simulations using an MHD code, a Hall MHD code (including the $j \times B$ and $\nabla \cdot P$ terms in Ohm's law), a hybrid code (massless electrons and particle ions), and a PIC code. While these data have a similar rate of reconnection during transition phases, all runs were carried out with the same Harris equilibrium with finite initial perturbations. The rate of reconnection is described by the slope of the rising reconnected flux. As seen in figure 5.2, all models but the MHD model show indistinguishable rates of reconnection and are significantly faster than that of the MHD model (Sweet–Parker model). However, there remains a major question concerning how the energy conversion occurs near the X-point or the electron diffusion region since Hall effects do not generate any energy dissipation of the magnetic field. This problem will be discussed again more comprehensively, together with the mechanisms of energy conversion from magnetic field to plasma particles, in chapter 10.

Since the mechanism for breaking field lines in various models differs (electron inertia in PIC and hyper-resistivity in the hybrid models), their results support the idea that the reconnection rate is insensitive to the dissipation mechanism. It can be argued (Drake and Shay, 2007) that because of the dispersion relationship of whistler waves, $\omega/k \propto k$, the total outflow flux of electrons from the dissipation region, $n_e v_x \delta$, is constant, since $v_x \approx \omega/k \propto k$ and $\delta \approx 1/k$. It was then concluded that the reconnection rate is primarily determined by the Hall term, and is insensitive to the dissipation mechanisms. It is considered that the dissipation of magnetic energy in their simulations occurs only in the vicinity of the X-point, within a distance of a few electron skin depths. There still remains the question of whether the GEM Challenge properly addressed the general problem of reconnection, particularly the dissipation mechanism that causes field-line breaking and the conversion of magnetic energy to plasma energy. Further efforts have been made using PIC numerical codes to investigate the effects of boundary conditions (periodic versus open; Daughton et al., 2006). We will also revisit this problem in chapter 10.

A group of PIC numerical calculations have been carried out to assess the Hall effect in the presence of collisions or resistivity. Ma and Bhattacharjee (1996) reported that the neutral sheet profile changes from a double-Y shape to an X shape with impulsive reconnection features as two-fluid effects were turned on with a constant resistivity. When the resistivity was set to be uniform in space and sufficiently large, the familiar rectangular-shaped Sweet–Parker layer was obtained; see figure 5.3. When the resistivity is reduced, characteristic features of the two-fluid dynamics appear with the double-wedge-shaped neutral sheet (figure 5.3). This result is in good agreement with recent observation in MRX (Magnetic Reconnection Experiment), as described later in this chapter.

Impulsive reconnection was observed (Cassak et al., 2005) when the Hall effect was turned on, on top of slow resistive reconnection. For a given set of plasma parameters they observed two stable reconnection solutions: a slow (Sweet–Parker) solution and a fast Hall reconnection solution. Below a certain critical resistivity, the slow solution disappears and fast reconnection occurs suddenly and dominates.

As mentioned before, the GEM Challenge program (Birn et al., 2001) found that reconnection proceeds much faster than resistive MHD reconnection, and the reconnection rate is determined primarily by the Hall term, not by dissipation. Some argued that this results in a separation of the dissipation region of line breaking from the global region. The dissipation region is shorter than the global length and so the problem of transporting plasma a long distance, faced by the Sweet–Parker and Petschek models, is eased. Figure 5.4 shows that there are long separatrices attached to the dissipation region, much as pictured in Petschek model.

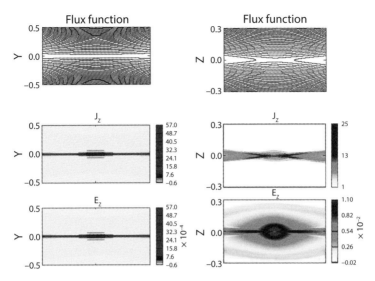

Figure 5.3. Comparison of numerical simulation with resistive MHD (left) and Hall MHD (right) codes in which the electron pressure gradient is included. Without a guide field, the measured profile of the MRX neutral sheet is in remarkable agreement with these results of numerical simulations. [From Ma and Bhattacharjee (1996).]

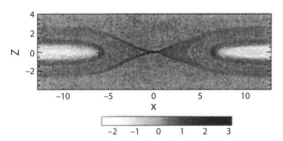

Figure 5.4. Out-of-plane electron current from a hybrid simulation. [From Shay et al. (2001).]

This argument is borne out by the simulations of Daughton et al. (2006), who greatly extended the simulation box and allowed open boundary conditions. They found that the reconnection rate first followed that of the periodic box simulations, but later slowed down as the plasma reached the installed wall. This simulation was applied to the MRX experiment, and it demonstrated that back pressure from a wall at the exhaust region slows down reconnection flows.

While the GEM simulations work well to describe a local reconnection layer in the magnetosphere, where the change of topology is the prime reason for reconnection, they are not well adapted for solar reconnection problems where the conversion of a large amount of magnetic energy is the main concern.

5.4.1 Analytical description of the two-fluid dynamics for the reconnection layer with the Hall field

In the MHD equations it is assumed that electrons and ions move together in a one-fluid approximation, thus the electron and ion velocities are equal, so that we need to keep only one of their velocities, usually the velocity of the heavier ions. This approximation is allowed if the scale size of spatial variations is much larger than the ion skin depth $d_i = c/\omega_{pi}$, where $\omega_{pi} \equiv \sqrt{n_0 e^2/\epsilon_0 m_i}$ is the ion plasma frequency. If the current density $\boldsymbol{j}$ is constant over a current layer thickness δ, the magnetic field change is expressed by

$$\Delta B \approx \mu_0 j \delta. \tag{5.12}$$

Thus the thickness of the current layer is directly related to the change of reconnection magnetic field. Using the relationship expressed by eq. (4.18), we can write

$$\delta = \frac{c}{\omega_{pi}} \frac{\sqrt{2(T_e + T_i)/m_i}}{V_e - V_i} = d_i \frac{\sqrt{2}V_s}{V_{\text{drift}}}. \tag{5.13}$$

When the relative drift velocity $V_e - V_i$ is less than $\sqrt{2T/m_i}$, the ion sound velocity, δ is larger than the ion skin depth, and MHD conditions can be satisfied for $V_R = V_A/\sqrt{S}$. But if the ion and electron velocities were to differ by more than the ion sound speed, non-MHD phenomena could easily be invoked, as discussed in chapter 4.

In many space and astrophysics cases, the calculated thickness of the Sweet–Parker layer δ_{SP} is less than the ion skin depth δ_i. Hence, there is no guarantee that the two velocities are close, or that Sweet–Parker theory is applicable. The ions and electrons can move independently of each other, and the reconnection physics in the layer will differ from that given by the Sweet–Parker model, allowing the layer thickness to be the thicker ion skin depth. For example, the ions could flow in the thicker layer, while the electrons flow in a thinner layer. The ion mass flow can be larger than the flow in the Sweet–Parker layer, while the thinner electron layer can allow the lines to break fast enough to accommodate this faster downstream mass flow. We only need the continuity equation in the more general Sweet–Parker model, while the undetermined thinness of the electron layer is determined by Ohm's law. The resulting reconnection velocity under this simplified model is

$$v_R \approx \frac{\delta_i}{L} V_A, \tag{5.14}$$

which is faster than the corresponding Sweet–Parker reconnection velocity $(\delta_{\text{SP}}/L) V_A$.

The mathematical difference between one-fluid and two-fluid theories appears in the different Ohm's laws. The one-fluid Ohm's law, eq. (3.21), differs from the two-fluid Ohm's law,

$$\boldsymbol{E} + \boldsymbol{V} \times \boldsymbol{B} - \frac{\boldsymbol{j} \times \boldsymbol{B}}{ne} + \frac{1}{ne} \nabla \cdot \boldsymbol{P}_e + \frac{m}{e} \frac{d\boldsymbol{V}_e}{dt} = \eta \boldsymbol{j}, \tag{5.15}$$

where P_e is the electron pressure tensor and $n_e = n$. This equation is called the generalized Ohm's law. Often, the last three terms of the left-hand side are moved to the right-hand side. Equation (5.15) is correct even for one fluid MHD, and is reduced to the ordinary Ohm's law by setting $V_e = V_i = V$ and neglecting the electron inertia and the pressure tensor terms.

The generalized form of Ohm's law is also equivalent to the equation of motion for electrons as described in eq. (5.9),

$$nm\frac{dV_e}{dt} = -\nabla \cdot P_e - ne\left(E + V_e \times B\right),\qquad(5.16)$$

because $j = ne(V_i - V_e)$ and $V = V_i$.

It is necessary to apply two-fluid dynamics to magnetic reconnection when the Sweet–Parker layer is thinner than the ion skin depth δ_i. The ratio of the Sweet–Parker layer thickness to the ion skin depth is $\approx 0.2\sqrt{L/\lambda}$, where λ is the mean free path and L is the global length of the current layer (Yamada et al., 2006). The two-fluid regime is closely related to the collisionless regime. The Hall effect becomes stronger when the mean free path is longer than only a fraction of the global length L.

The two-fluid effect is brought out by the example of a two-dimensional reconnection problem in the x–z reconnection plane, where the reconnection field is along the z-direction. If the initial B_y is zero (no guide field), there can be no B_y field because of a symmetry in MHD theory that can separate the out-of-plane, y-component from the x-, z-components. In two-fluid theory this symmetry is broken by the $j \times B$ Hall term and an out-of-plane B_y component is produced. This was noticed by Sonnerup (1979) in an early discussion of two-fluid theory applied to magnetic reconnection. The same B_y field was found by Terasawa (1983) in two-fluid investigations of the tearing mode.

A physical interpretation of the origin of an out-of-plane field was given by Mandt et al. (1994), who ascribed it to the out-of-plane motion of electrons which by flux freezing "pull" reconnecting magnetic field lines into the y-direction. This interpretation needs some clarification. Inspection of eq. (5.15) shows that, in the absence of the pressure term, electrons are indeed frozen in the electron fluid. When the Hall effects bend the field lines toward the y-direction, the electrons are also accelerated in the y-direction. As the field strength weakens as they approach X-line regions, the electrons pull the field lines further to the y-direction, breaking field lines in this very high β region. As the field lines reconnect and move away to the exhaust, the electrons are pushed out to the y- and z-directions with them.

A second interpretation is that the out-of-plane field would arise from the motion of the electrons in the reconnection plane and from Ampère's law. This picture is closer to the equations that Sonnerup and Terasawa used to show the B_y field's existence. This second interpretation is elaborated by Uzdensky and Kulsrud (2006) as described below (see figure 5.5).

If the magnetic field is purely in the reconnection (poloidal) plane with no externally applied B_y guide field, the ion gyration radius is comparable with the current layer thickness δ, and the ions are essentially unmagnetized. The electron gyration radius is

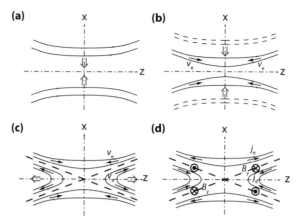

Figure 5.5. The basic idea of out-of-plane field generation. [From Uzdensky and Kulsrud (2006).]

much smaller, so that the electrons are tied to the lines everywhere except in the electron diffusion region where the magnetic field is very small near $x = 0$.

As the reconnection proceeds, the magnetic field lines move into the reconnection current layer, with the electrons tied to them. Their transverse $E \times B$ velocity brings them into the region where the ions are unmagnetized and the ion density is unaffected by the reconnection processes. The electron and ion motions are not entirely decoupled because, by charge neutrality, their densities must be almost equal to avoid large poloidal electric fields. This *quasi-neutrality* condition cannot be accomplished by the transverse electron velocities alone. The electrons must develop velocities parallel to the lines of force and these velocities are strongly constrained by charge neutrality.

In figure 5.5(b), the volume per flux of the field-line tubes increases strongly as the lines approach the separatrix with most of the volume concentrated near the midplane $z = 0$. Without the parallel electron flow, the density near the midplane would drop because the $E \times B/B^2$ flow diverges, so there must be a parallel electron current that produces the out-of-plane (toroidal) magnetic field. (The ion current is assumed small.) The figure shows that this out-of-plane field has a quadrupole character reversing across the axes. Downstream from the separatrix the electrons flow away from the $x = 0$ plane because the flux tubes contract as they move, so the parallel electron current reverses sign.

The electrons are also accelerated to the y-direction along the magnetic field lines bent toward the y-direction due to Hall effects. Consequently, the toroidal inertial force becomes comparable to the out-of-plane reconnection electric field force in Ohm's law. In this situation the electrons are no longer tied to the magnetic lines, or they can slip off the lines where their electron magnetization fails. This leads to a detachment of field lines, which move faster than the electrons in the reconnection plane, and the resulting breakups of field lines.

If we assume the ion density is constant throughout the current layer, then the electron behavior can be treated quantitatively based on the principle of charge neutrality.

Based on this assumption, Uzdensky and Kulsrud (2006) estimated semi-quantitatively the thickness of the region.

5.5 PROFILE AND CHARACTERISTICS OF THE TWO-FLUID RECONNECTION LAYER

As explained earlier in this chapter, electrons and ions move quite differently in a collisionless magnetic reconnection layer. Differential motion between the strongly magnetized electrons and the unmagnetized ions generates strong Hall currents in the reconnection plane as shown in figure 5.1. As magnetic reconnection is induced with oppositely directed field lines being driven toward the X-point ($\boldsymbol{B} = 0$ at the center of the layer), ions and electrons also flow into the reconnection layer. The ions become demagnetized at a distance of the ion skin depth ($d_i = c/\omega_{pi}$, where ω_{pi} is the ion plasma frequency) from the X-point, where they enter the so-called ion diffusion region, and they change their trajectories and are diverted into the reconnection exhaust as seen figure 5.6. The electrons, on the other hand, remain magnetized throughout the ion diffusion region and continue to flow toward the X-point. They become demagnetized only when they reach the much narrower electron diffusion region, as seen in figure 5.6. In this two-fluid model, the expanding exhaust region becomes triangular in shape and the outgoing magnetic flux through this region is expected to be sizable, while the incoming magnetic energy is converted much faster to particle energy in this X-shaped reconnection layer.

In the two-fluid formulation, Ohm's law of MHD should be replaced by the generalized Ohm's law in order to describe the force balance of an electron flow, namely

$$E + V \times B = \eta j + \frac{j \times B}{e n_e} - \frac{\nabla \cdot P_e}{e n_e} - \frac{m_e}{e} \frac{dV_e}{dt}. \qquad (5.17)$$

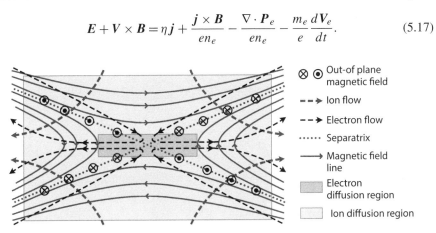

Figure 5.6. Schematic diagram of two-fluid dynamics in the reconnection layer. Electrons and ions move quite differently, generating an out-of-plane quadrupole field (shown on the diagonal separatrix lines). Flows of electrons (dark broken lines) and ions (gray lines) in the reconnection plane, together with reconnecting field-line components projected in the reconnection plane, are shown. [From Yamada et al. (2015).]

Here, conventional notation is used, with E being the electric field, B the reconnecting magnetic field, V_e the electron flow velocity, j the current density, n_e the electron density, P_e the spatially dependent electron pressure tensor, and $V_e + V_i = V$. A large out-of-plane electric field derived from the Hall currents at the reconnection layer ($J_{Hall} \times B$) causes an increase in the reconnection rate by inducing rapid movement of the reconnecting field lines,

$$|V_e \times B_{rec}| \approx E_{rec}, \tag{5.18}$$

where B_{rec} is the reconnecting magnetic field and E_{rec} is the reconnection electric field. This explains why the reconnection rate in collisionless plasmas can be much larger than the classical Sweet–Parker rate. Also, quite different flow patterns of ions and electrons create circular currents that should generate an out-of-plane quadrupole magnetic field, which we call the Hall magnetic field. This "Hall effect" is a very important signature of two-fluid physics.

In the generalized Ohm's law of eq. (5.17), the first term on the right-hand side is negligible in collisionless reconnection, and the second term represents the Hall term. As mentioned before, eq. (5.17) can be reduced to the ordinary Ohm's law by setting $V_e = V_i = V$, and by neglecting the electron inertia and pressure tensor terms. Most of the region shown in figure 5.6, where ions are demagnetized, is called the "ion diffusion region" with $E + V_i \times B \neq 0$. The motion of magnetized electrons is still described by $E + V_e \times B = 0$, until they are near the X-point, where electrons are demagnetized. This central region near the X-point is called the "electron diffusion region." The inertia term and pressure tensor term become large in the electron diffusion region. Generally in eq. (5.17), all vectors should include fluctuation components and $\eta = \eta_0$ denotes the classical Spitzer resistivity based on Coulomb collisions.

5.5.1   Flux freezing by electron fluid

If we describe an equation of motion for electrons in a quasi-steady-state reconnection layer without collisions, neglecting an electron inertial term (fast electron displacement current ~ 0), we obtain

$$E + V_e \times B = -\frac{\nabla \cdot P_e}{e n_e}. \tag{5.19}$$

Since the electron pressure tensor term becomes very small outside the electron diffusion region, we acquire

$$E + V_e \times B = 0. \tag{5.20}$$

Using an analogous argument to chapter 3, we can conclude that eq. (5.20) indicates that *electrons are frozen to magnetic field lines* in the reconnection layer, except inside the electron diffusion region. As described by the flux freezing principle shown for ideal MHD plasmas (section 3.3), we can now say that the flux freezing principle works well for magnetized electrons in the reconnection layer or inside the ion diffusion region. One can also say that electrons do not feel any force from the moving frame of V_e. Near the center of the ion diffusion region, electrons primarily flow in the out-of-plane direction perpendicular to B, and we expect a sizable electric field toward the X-point. This situation would generate a strong potential well around the electron diffusion region with respect to the inflow direction (x). This prediction was verified experimentally in MRX (Yoo et al., 2013), as well as in the magnetosphere by the MMS (Magnetospheric Multiscale Satellite) (Burch et al., 2016b). This will be described in more detail in chapter 10.

5.6 EXPERIMENTAL OBSERVATIONS OF TWO-FLUID EFFECTS IN THE RECONNECTION LAYER

In the past two decades, a number of laboratory experiments have provided important data contributing to the understanding of the local two-fluid physics of reconnection. Since the 1970s, the profile of the reconnection layer has been studied in many laboratory plasmas, by generating it in a controlled manner (Stenzel and Gekelman, 1981; Ono et al., 1993; Yamada et al., 1997a, 2000). In driven reconnection in MRX, profiles of the reconnection layer were extensively investigated by changing plasma parameters such as density and temperature (Yamada et al., 2006; Yamada, 2007). In this section, we study important observations from some of the dedicated laboratory studies that have lead to improved understanding of two-fluid physics in the reconnection layer. Recent major observations from space satellites are also described. The observations are compared with the numerical simulation results mentioned earlier.

5.6.1 Experimental study of dynamics of the two-fluid reconnection layer

In the MRX laboratory experiment, a well-defined reconnection layer is generated in a controlled manner in the two-fluid regime, and the dynamics of the reconnection layer are studied extensively, including features of both the electron diffusion layer and the ion diffusion layer. Figure 5.7 shows a schematic of the MRX apparatus, together with the measured flow of electrons and ions in the reconnection layer, wherein two oppositely directed field lines merge and reconnect.

It should be noted that this formation of a current sheet is similar to the situation in figure 1.3. Each flux core (the darkened section in figure 5.7(a)) contains both toroidal field (TF) and poloidal field (PF) coils. By pulsing both PF and TF coil currents in a controlled manner, a prototypical reconnection layer is generated and the study of particle dynamics and a detailed energy inventory were carried out (Yamada et al., 2014, 2015). For standard conditions of $n_e = 2$–6×10^{13} cm^{-3}, $T_e = 5$–15 eV, $B = 0.1$–0.3 kG, $S \approx 500$, the electrons are well magnetized (gyroradius $\ll L$) while the ions are not. The mean free path for electron–ion Coulomb collisions is in the range 5–20 cm ($>$ the layer thickness) and, as a result, the reconnection dynamics are dominated by

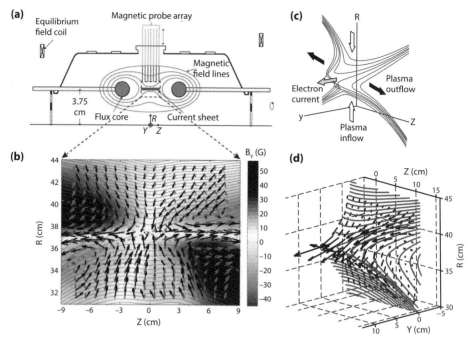

Figure 5.7. See Color Plate 2. (a) MRX apparatus and reconnection drive. (b) Measured flow vectors (the length represents velocity) of electrons (red arrows) and ions (blue arrows) in the full reconnection plane, together with poloidal flux contours as thin blue lines (which represent reconnecting field-line components projected in the reconnection plane) and out-of-plane field contours. A 1 cm vector length stands for 2×10^6 cm/s, color contours represent out-of-plane field strength, and broken green lines depict (experimentally identified) separatrix lines. Toroidal symmetry is assumed. (c) Conjectured 3D view of magnetic field lines moving together with plasma flows. (d) 3D depiction of measured magnetic field lines together with electron fluid flow vectors. [From Yamada et al. (2015). https://mrx.pppl.gov/mrxmovies/Hall.mov]

two-fluid and kinetic physics (Yamada, 2007; Yamada et al., 2010). We employ a geometry (R, Y, Z), where B_Z is the reconnecting field component and Y is the symmetric, out-of-plane axis. Local flow vectors for electrons and ions are measured in the reconnection layer, and completely different flow patterns of ions and electrons are found, as expected. The two-fluid plasma dynamics are described by the generalized Ohm's law, which is derived by multiplying the velocity vector by the Vlasov equation for electrons. When we discuss collisionless reconnection in the two-fluid (or kinetic) physics formulation, the definition of the diffusion region becomes quite different from that of MHD. In a prototypical two-dimensional two-fluid reconnection layer, there are two separate diffusion regions for electrons and ions respectively. Actually, the electron diffusion region resides (near the X-point) inside the broader ion diffusion region as described in figure 5.6.

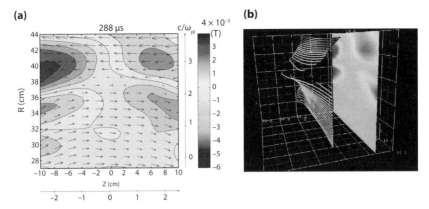

Figure 5.8. See Color Plate 2. (a) The magnetic field profiles of the reconnection region for deuterium plasmas. The arrows depict the measured magnetic field vectors in the $R-Z$ plane. The size of the arrows is normalized to the maximum magnetic field strength 300 G. The color-coded contour plot shows the out-of-plane magnetic field strength B_T. (b) 3D profiles of reconnection magnetic field lines measured in MRX; B_T is also in color-coded contours similar to (a) in the projected back screen. [From Yamada et al. (2006). https://mrx.pppl.gov/mrxmovies/Hall.mov]

5.6.2   Measurements of Hall magnetic fields in laboratory experiments

A comprehensive quantitative study of Hall effects has been carried out in MRX by comparing the results of a two-fluid simulation for the MRX geometry with experimental results (Ren et al., 2005; Yamada et al., 2006). Aided by the numerical work, a deeper understanding of two-fluid reconnection dynamics has been obtained. The results from this study are shown in figure 5.8. Using the three components of the magnetic field vectors measured by a two-dimensional probe array, precise and conclusive measurements of Hall effects in the neutral sheets were carried out in MRX. Figure 5.8(a) shows contours of the measured out-of-plane quadrupole magnetic field in the diffusion region during magnetic reconnection, together with vectors of the reconnecting magnetic field in the $R-Z$ plane. The spatial resolution is 4 cm in the Z-direction and is improved to 1 cm in the R-direction by scanning the probe radially and averaging several shots at each position. The quadrupole configuration of the out-of-plane magnetic field B can be clearly seen. The measured amplitude of this quadrupole magnetic field is of order 30–50 G compared with 100–120 G for the reconnecting field strength.

During reconnection, the reconnecting field lines move into the neutral sheet (reconnection layer) of width comparable to the ion skin depth. As they approach the X-point, ions become demagnetized. The ion flows gradually change direction by 90 degrees, from the R- to the Z-direction in the reconnection $R-Z$ plane (blue lines; in figure 5.7 (b).). It is shown that magnetized electrons flow quite differently (red vectors in 5.7 (b)), still following magnetic field lines until they approach the X-point or separatrix surfaces.

The MRX data in figure 5.9(a) show that as electrons flow through the separatrix regions of the reconnection sheet, they are first accelerated toward the X-point. After

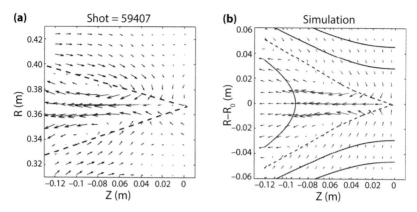

Figure 5.9. (a) The electron drift velocity arrows in a half-reconnection-plane of MRX, deduced from the measured out-of-plane magnetic field components by $V_d = \nabla \times B / e n_e$ and $V_d = V_e - V_i$ with Mach probe measurement of V_i. Separatrices are inferred by broken black lines in a hydrogen plasma, fill pressure = 2 mTorr. (b) Simulation results, where in-plane electron flows are shown by arrows, flux lines by solid lines, and separatrices by broken lines, from the numerical simulation by Breslau and Jardin (2003). [Figure is adapted from Ren (2007), Yamada (2011).]

making a sharp turn at the separator lines, they then flow outward in the Z-direction. When one compares the corresponding flow patterns from the experimental data and from the numerical simulation, one finds excellent agreement and that the data illustrate the essence of the Hall effects. The vectors of electron flow in the MRX data illustrate that, after the initial acceleration, electrons are further accelerated as they pass through the narrow channel section around the central separatrix. The initial acceleration may be due to a larger $E \times B$ ($\sim E_y / B_z$) velocity as the reconnection magnetic field diminishes near the origin ($B_z \sim 0$) with uniform reconnection electric field E_y. To date, these MRX data provide the most quantitative data for Hall currents in a real plasma.

The measured electron flow pattern generates a circular net current pattern in the reconnection plane and thus creates an out-of-plane magnetic field with a quadrupole profile. A two-dimensional profile of the out-of-plane quadrupole magnetic field was measured in MRX by scanning the 90-channel probe array. Figure 5.8(a) shows the color contours of this out-of plane quadrupole field in the diffusion region during collisionless reconnection, together with vectors of the reconnection magnetic field in the reconnection R–Z plane. This has been regarded as a hallmark of the Hall effect (Shay et al., 2001; Birn et al., 2001). This process can be interpreted as a mechanism by which the electrons, which are flowing in the neutral sheet current, tend to pull magnetic field lines toward the 'y'-direction of the electron sheet current. The spatial resolution of this figure is 4 cm in the Z-direction (grid size) and 1 cm in the R-direction, which is obtained by radially scanning the probe array and averaging several discharges at each position. The amount of Hall magnetic field is consistent with results from the numerical simulation (Ren et al., 2005; Yamada et al., 2006).

Color Plate 1

Magnetic reconnection in space and in laboratory plasmas

Figure 1.1

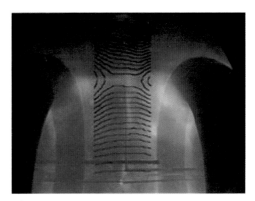

Figure 1.5

Electron flows in MRX

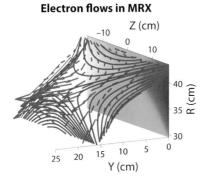

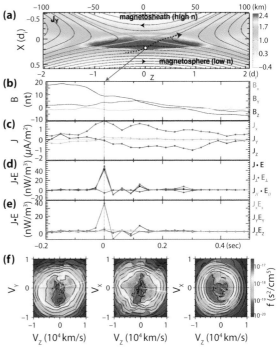

Figure 1.6

Color Plate 2

Two-fluid dynamics of magnetic reconnection measured in MRX

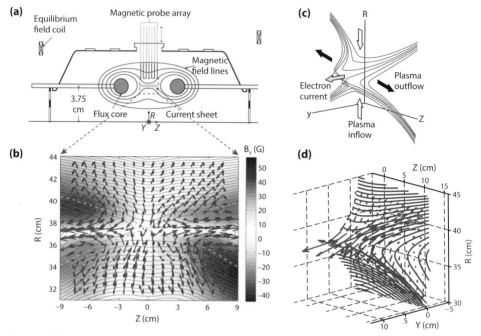

(a)
Equilibrium field coil
Magnetic probe array
Magnetic field lines
3.75 cm
Flux core
Current sheet
R
Y Z

(b)
B_Y (G)
R (cm)
Z (cm)

(c)
R
Electron current
Plasma outflow
Plasma inflow
y
Z

(d)
Z (cm)
R (cm)
Y (cm)

Figures 5.7 and 10.1

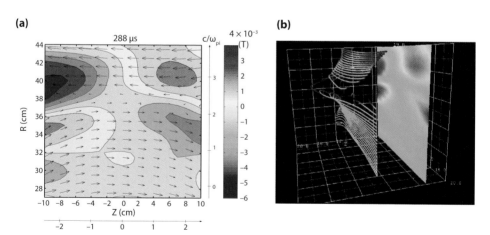

(a)
288 µs
c/ω_{pi}
(T)
R (cm)
Z (cm)

(b)

Figure 5.8

Color Plate 3

Laboratory data on magnetic reconnection

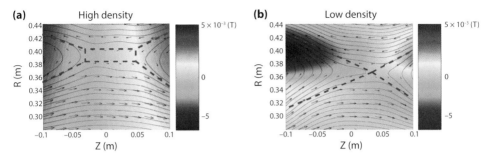

Figure 5.10

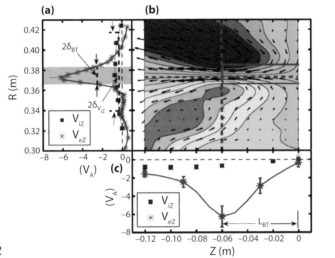

Figure 5.12

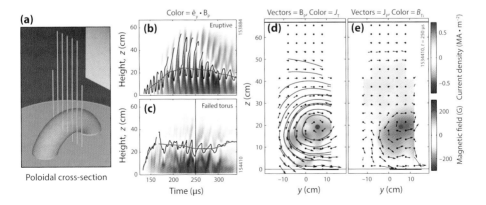

Figure 9.16

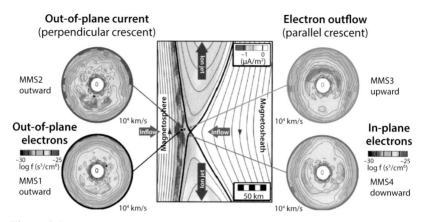

Out-of-plane current
(perpendicular crescent)

Electron outflow
(parallel crescent)

MMS2
outward

MMS3
upward

**Out-of-plane
electrons**

**In-plane
electrons**

MMS1
outward

MMS4
downward

Figure 8.4

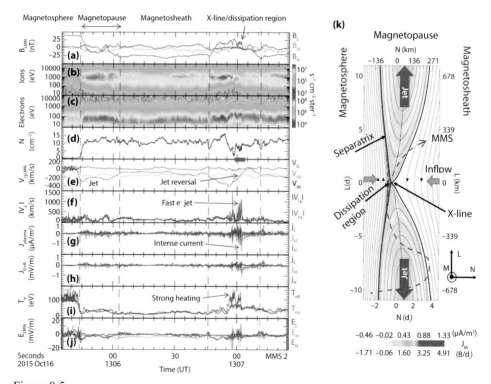

Figure 8.5

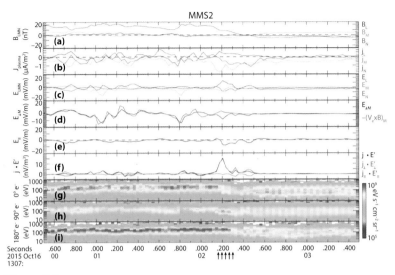

Figure 8.6

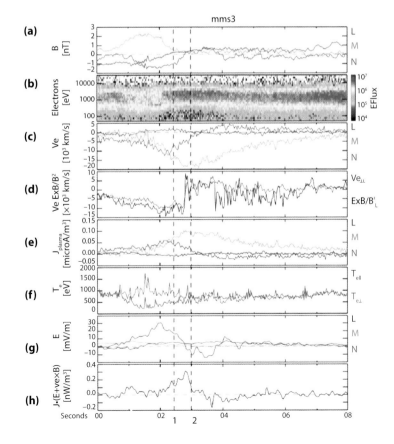

Figure 8.8

Magnetic reconnection in tokamak plasmas

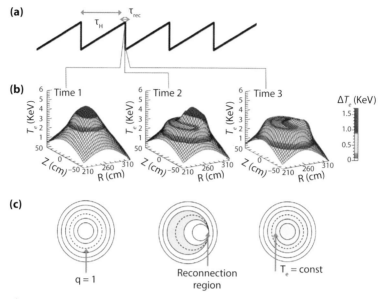

Figure 9.3

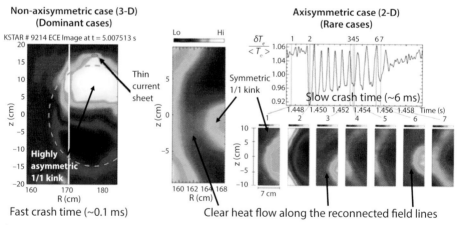

Figure 9.4

Color Plate 7

Dynamics and energetics of reconnection on MRX

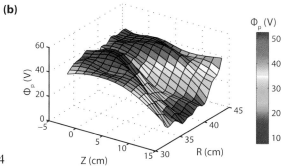

(b)

Figure 10.4

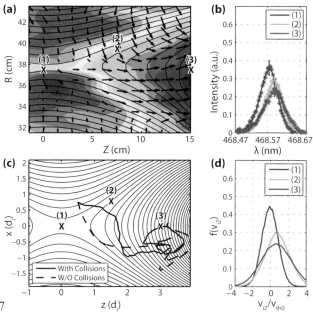

Figure 10.7

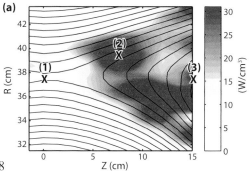

Figure 10.8

Color Plate 8

Cases for asymmetric reconnection and a simulation of a large current layer

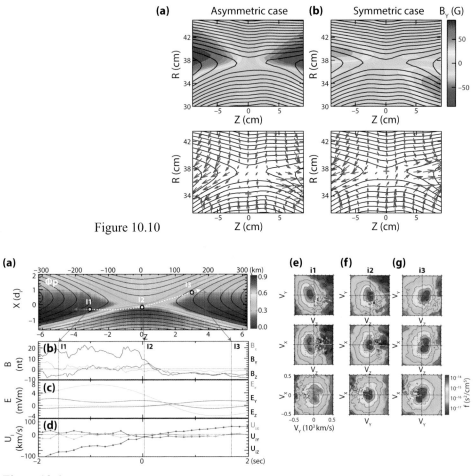

Figure 10.10

Figure 12.4

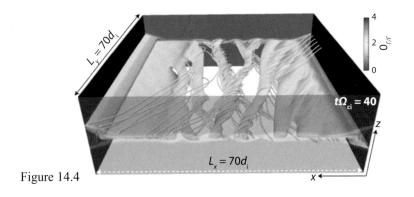

Figure 14.4

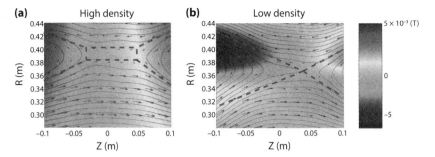

Figure 5.10. See Color Plate 3. Comparison of the experimentally measured reconnection layer profile for two cases. (a) Collisional regime ($\lambda_{mfp} \ll \delta_{sheath}$) and (b) nearly collisionless regime ($\lambda_{mfp} > \delta_{sheath}$). The in-plane magnetic field is shown by black arrows and the out-of-plane field component by the color codes ranging from $-50\,\mathrm{G}$ to $50\,\mathrm{G}$. Dashed pink lines show that the magnetic configuration changes from a rectangular-shaped current sheet (Sweet–Parker type in (a)) to a double-wedge shape (Petschek-like) as collisionality is reduced. The predicted quadrupole structure of the out-of-plane magnetic component, a signature of Hall effects, is observed in (b). [From Yamada et al. (2006).]

It should be noted, however, that the Hall term alone does not create the energy dissipation necessary for conversion of magnetic energy to particle kinetic energy. Instead, it is considered that the electron pressure tensor term in eq. (5.17) and fluctuations can generate energy dissipation particularly at the X-point. As the reconnection proceeds, the lines of force move into the reconnection current layer, with electrons tied to them. When the electrons reach the X-point region, they become demagnetized and diffuse off the field lines, causing the lines to break. Then the diffused electrons are accelerated away from the X-point in both the z- and y-directions, taking energy with them. It has been observed that these electron flows fluctuate on a longer timescale, causing impulsive and turbulent reconnection (Ren et al., 2008). More detailed discussion of electron motions is presented in Yamada et al. (2010). Also, the laminar flows of electrons are analytically described in the calculation for the Hall effects of Uzdensky and Kulsrud (2006), as mentioned earlier.

5.6.3 Profile of reconnection layer changes drastically with collision rates

It is observed that the two-dimensional profile of the neutral sheet changes significantly from a rectangular shape in the collisional regime ($\lambda_{mfp} \ll \delta_{sheath}$) to a double-wedge shape in the collision-free regime ($\lambda_{mfp} > \delta_{sheath}$).

Simultaneously, the reconnection rate is seen to increase as the collisionality is reduced. Figure 5.10 shows how the profile of the MRX neutral sheet depicted by the measured magnetic field vectors and flux contours changes with respect to the collisionality condition. In the high plasma density case, where the mean free path is much shorter than the sheet thickness, the rectangular-shaped profile of the Sweet–Parker model, of the type in figure 3.3, is identified and the classical reconnection rate is measured. In the case of low plasma density, where the electron mean free path is longer than the sheet thickness, an X-shaped profile appears as shown in figure 5.10(b) and the Hall MHD effects become dominant, as indicated by the notable out-of-plane quadrupole field depicted by the color code. There is no recognizable out-of-plane Hall field in the collisional case of figure 5.10(a), where the weak dipole toroidal fields profile is only a remnant of the field created by initial poloidal discharges around the two flux cores. The X-shaped profile of Petschek type, seen in (b), differs significantly from that of the Sweet–Parker model, seen in (a), and a fast reconnection rate is measured in this low collisionality regime. This result is an experimental verification of how collisionality changes the shape of the reconnection layer, simultaneously affecting the reconnection rate. A slow shock, which is a key signature of the Petschek model, is not identified in this regime. This observation is consistent with the numerical results mentioned earlier, which included both two-fluid effects and resistivity (Ma and Bhattacharjee, 1996; Bhattacharjee et al., 2001). Without a guide field, the measured profile of the MRX neutral sheet exhibits remarkable agreement with their numerical simulation results.

While it is difficult to directly measure the two-dimensional spatial profiles of the reconnection region in the magnetosphere, because of the limited number of measuring locations by satellites, the same out-of-reconnection-plane field pattern was measured by Mozer et al. (2002). In the solar atmosphere, two-dimensional neutral-sheet-like patterns have sometimes been recognized through soft-X-ray satellite images of solar flares, but their exact magnetic profiles are unknown. It appears that a reconnection process is underway throughout this area, based on the sequence of high-energy electron flux hitting the footpoints at the photosphere. In order to describe the observed reconnection rate (Yokoyama et al., 2001) by the Sweet–Parker model, and to explain the apparent fast flux transfer, the plasma resistivity or energy dissipation has to be anomalously large throughout a wide region.

5.6.4   Measurements of Hall effects in the reconnection layer of the magnetopause

The two-fluid dynamics of reconnection, which are illustrated in figure 5.1, predict the presence of strong Hall effects due to the decoupling of electron flow from ion flow. In a collisionless neutral sheet such as seen in the magnetosphere, this situation is equivalent to magnetized electrons pulling magnetic field lines in the direction of the electron current and thus generating an out-of-plane quadrupole field.

In the magnetosphere, the two-fluid physics of magnetic field reconnection was recently analyzed in terms of the ion diffusion region of scale size $c/\omega_{pi} \sim 100$ km in the subsolar magnetopause (Mozer et al., 2002).

A symbolic hyperbolic-tangent in-plane field and a sinusoidal out-of-plane Hall magnetic field were observed near the separatrices of the current sheet. Signatures of Hall MHD and the ion diffusion region (the lightly tinted region in figure 5.11 (top)) were seen in the y-component of the magnetic field (into and out of the paper), in the x-component of the electric field (red horizontal arrows), and in the

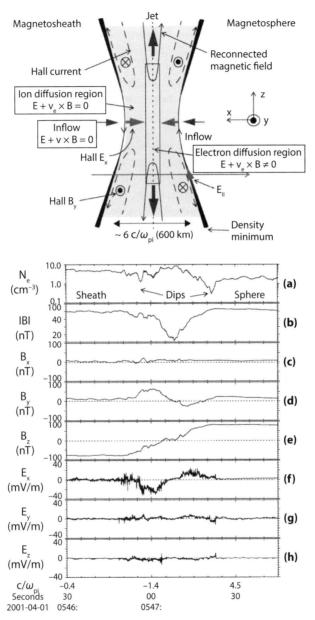

Figure 5.11. Conjectured flight path of the Polar satellite in the modeled diffusion region (top) and data collected on 1 April 2001 (bottom). Panel (a) shows the plasma density; panels (b)–(e) give, respectively, the magnitude and three components of the measured magnetic field at a rate of 8 samples/s. Panels (f)–(h) give the three components of the electric field in a frame fixed to the magnetopause. The electric field data rate was 40 samples/s. Note that the time axis can be translated to a space axis. [From Mozer et al. (2002).]

disagreement between the measured perpendicular ion flow and MHD-based $E \times B/B^2$. The detailed data are shown in figure 5.11 (bottom). The amplitude of the Hall field B_y was 45 nT, or $\sim 0.55 B_{x0}$, where B_{x0} is the asymptotic magnetic field in the magnetosheath.

The maximum normal electric field E_x was ~ 30 mV/m or $\sim 0.5 V_A \times B_{x0}$, which is consistent with recent numerical simulations that demonstrated a large negative potential well around the X-point. The ion diffusion region had a width of about 6 magnetosheath ion skin depths (or ~ 3 magnetospheric ion skin depths) at the location of the spacecraft crossing. At about the same time as their report, evidence of the Hall effect was reported through the detection of a quadrupole B_y field after analyzing the data from Geotail skimming in January 1997 along the dayside magnetopause (Deng and Matsumoto, 2001). Another report of the out-of-plane quadrupole field was made from data from the Wind satellite, which traveled in the reconnection sheet of the magnetotail (Øieroset et al., 2001). More recently, when a satellite flew through the ion diffusion region a reconnection electric field was carefully studied to deduce the reconnection speed (Mozer and Retinò, 2007).

5.6.4.1   *An episode on observations of Hall effects in the magnetopause*

Mozer et al. (2002) observed that the quadrupole Hall field such as seen here was rather rare. Indeed, they found that a prototypical Hall field such as shown in figure 5.11 was seen only once in over 100 passages of the satellite through the reconnection area. Mozer wondered why this was the case, and continuously studied the cases of magnetopause reconnection, concluding that it should not necessarily generate a prototypical quadrupole Hall magnetic field in the out-of-plane direction because of the presence of a strong asymmetry in the inflowing plasma density. In the magnetopause, the stagnated plasma density of the solar winds at the magnetosheath is generally 10–20 times larger than that of the magnetosphere and, because of a strong diamagnetic current, the quadrupole Hall magnetic field is changed to a dipole shape. The data shown in figure 5.11 are a case for a smaller asymmetry. This continuous pursuit by Mozer lead to another important discovery and progress on asymmetric reconnection, which will be described in chapter 10 (section 10.7).

5.7   OBSERVATION OF A TWO-SCALE RECONNECTION LAYER WITH IDENTIFICATION OF THE ELECTRON DIFFUSION LAYER IN A LABORATORY PLASMA

An important result from a recent comparative study of the reconnection layer between experiments and two-dimensional numerical simulations is a verification of a two-scale diffusion layer, in which an electron diffusion layer resides inside the ion diffusion layer whose width is the ion skin depth (Pritchett, 2001). In the reconnection layer of MRX, the electron diffusion region was identified as shown in figure 5.12 and it was found that demagnetized electrons are accelerated to a value that significantly exceeds V_A in the outflow direction in the reconnection plane (Ren et al., 2008).

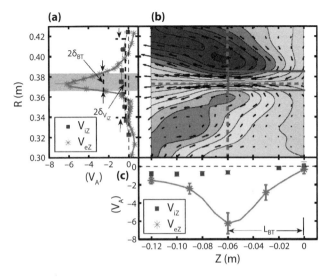

Figure 5.12. See Color Plate 3. (a) Radial profiles of the electron outflow velocity V_{eZ} (magenta asterisks) and ion outflow velocity V_{iZ} (blue squares), measured in a helium plasma. (b) 2D profile of the out-of-plane field B_T (color-coded contours in a similar way to figure 5.8) and the in-plane electron flow velocity V_e (black arrows). (c) V_{eZ} and V_{iZ} as functions of Z. The dashed magenta lines in (b) represent the cuts at $Z = -6$ cm and at $R = 37.5$ cm, along which the profiles in (a) and (c) are taken. [From Ren et al. (2008).]

The width of the outer ion diffusion layer is of order the ion skin depth, ~ 5–6 cm in MRX. The ion outflow channel is shown to be much broader than the electron channel, also consistent with numerical simulations. The width of the electron diffusion region, which is identified by the profile of the electron outflow, scales with the electron skin depth as equal to 5.5–$7.5(c/\omega_{pe})$. The electron outflow velocity scales with the local electron Alfvén velocity ($= 1.2$–$1.6V_{eA}$). However, the measured thickness of the electron diffusion layer is 3–5 times larger than the values calculated by two-dimensional numerical simulations.

Since the outflow velocity affects the reconnection rate, we can plot the maximum electron outflow velocity V_{eZ} against the electron Alfvén velocity V_{eA}, as shown in figure 5.13 for plasmas of three different ion species (the electron Alfvén velocity is calculated using the reconnecting magnetic field evaluated at the edge of the electron diffusion region and the central density). Note that the data show no ion-mass dependence within error bars, since the points come together on a single line despite the variation in the ion species. The measured values scale with the electron Alfvén velocity, namely $V_{eZ} \sim 0.11V_{eA}$, indicated by the linear best fit shown in the figure. This result is quite different from numerical calculation, which shows $V_{eZ} \sim V_{eA}$. This MRX result has been confirmed by recent MMS data (Burch et al., 2020), which reported that the outflow velocity is 10–20% of V_{eA} with $V_{eY} \gg V_{eZ}$, where the Y-axis is out of the reconnection plane. A careful check of the effects of collisions has been made

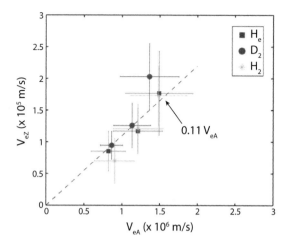

Figure 5.13. Measured peak electron outflow velocity V_{eZ} as a function of V_{eA}. Discharges with three different ion species are shown: helium (solid squares), deuterium (solid circles), and hydrogen (asterisks). The dashed line ($V_{eZ} \sim 0.11 V_{eA}$) is the linear best fit to the data. [From Ren et al. (2008).]

to determine how much of the enhancement of the thickness should be attributed to collisions and other effects (Ren et al., 2008; Ji et al., 2008). While the electron outflow seems to slow down by dissipation in the electron diffusion region, the total electron outflow flux remains independent of the width of the electron diffusion region. We note that, even with the presence of the sharp electron diffusion region, the reconnection rate is still primarily facilitated by the Hall electric field, as concluded by the GEM project. While some three-dimensional wave activities are suspected, it is not yet determined what causes the broadening of the electron outflow channel in both z- and y- (in the direction of the out-of-reconnection-plane current) directions. Also, we note that this electron outflow often occurs impulsively as the collisionality of the plasma is reduced.

Thus, the MRX experiment identified a two-scale diffusion layer in which the electron diffusion layer resides within the outer ion diffusion layer as shown in figure 5.6, with the width of the ion skin depth (Yamada et al., 2016a) as mentioned. In this situation we define the ion diffusion layer as the regime of $E + V_i \times B \neq 0$ and the electron diffusion layer as the regime of $E + V_e \times B \neq 0$. In the electron diffusion region, electrons are demagnetized with $E + V_e \times B \neq 0$ (Yamada et al., 2015), while in the ion diffusion region, electrons are still magnetized by the relation $E + V_e \times B = 0$ (Ren et al., 2008), and $E + V_i \times B \neq 0$ is satisfied with $V_i \neq V_e$. It was also concluded that Hall effects determine the reconnection rate in the broad ion diffusion region; in eq. (5.18), $E_Y \sim (V_e \times B)_Y$. This subject will be revisited in chapter 10.

Recently, it has been reported that a very thin electron current layer was measured in a laser-produced high-β plasma (Fox et al., 2020). In their short pulse experiment (1 ns), a long thin electron current layer of aspect ratio of over 100 with $L = 4$ mm and a half-width of $28\,\mu$m was experimentally identified. It was tentatively concluded that the reconnection rate (which is yet to be conclusively measured) could be explained by the pressure tensor term in the generalized Ohm's law (eq. (5.15)) using a formula developed by Hesse et al. (1999): $E_{rec} \sim (1/e)\sqrt{2m_e T_e}(\partial V_e/\partial x)$, where V_e is the electron inflow velocity.

5.8   WAVES IN THE RECONNECTION LAYER AND ENHANCED RESISTIVITY

5.8.1   Waves in the reconnection layer and their effects

As the collisionality is reduced, waves (particularly microturbulence) can be excited in the two-fluid reconnection layer due to the different motions of electrons compared to ions. When the relative motion of electrons against ions exceeds the ion acoustic velocity current, driven micro-instability can be excited and strong momentum transfer occurs from electrons to ions, leading to enhanced resistivity. The relative electron drift velocity reaches this key value when the thickness of the reconnection layer becomes less than the ion skin depth (Yamada et al., 2006, 2010), and it should significantly increase the reconnection rate through the resistivity enhancement.

This enhanced resistivity in the generalized Sweet–Parker model discussed in chapter 3 (Ji et al., 1999) should have several important effects that would aid the magnetic reconnection process. It should keep the reconnection layer as thick as the ion skin depth. A thinner layer would increase the resistivity by a larger amount and the increased electron and magnetic diffusion would force the layer thickness back to the ion skin depth. When the layer is thicker than the classical Sweet–Parker thickness in collisionless reconnection, it would make a faster outflow of the plasma in the generalized Sweet–Parker model.

The enhanced resistivity allows the electrons to flow across the field lines and weakens the argument for quadrupole Hall magnetic fields and interferes with other two-fluid collisional effects. It is sometimes observed that when magnetic fluctuations occur in magnetic reconnection experiments, indicating the presence of instabilities, the quadrupole magnetic fields indeed weaken. This makes sense since we expect that enhanced fluctuation should perturb the current paths of electrons to make laminar Hall magnetic fields (such as shown in figure 5.11). When Hall effects are seen, the fluctuations are sometimes absent or weak. In the latter case, fluctuations may be too weak to produce much anomalous resistivity.

A large number of candidate waves and instabilities have been suggested during collisionless reconnection. But in laboratory experiments and space observations, no specific wave has been verified as a convincing source for anomalous resistivity. In the appendices of this book, typical waves in plasma are presented, together with a simplified derivation of their dispersion relations, to show the propagation properties of waves. Table A.1 in the appendix presents the dispersion relations for waves based on the propagation direction with respect to the stationary background magnetic field, fluctuating vector components with respect to the propagation vector k, and polarization of waves. While many of the listed waves in table A.1 can be involved in the reconnection layer, electron whistler waves and lower hybrid waves have been the most investigated because the former is expected to cause electron heating and acceleration and the latter generates strong dissipation, as well as anomalous resistivity in the reconnection layer. Ion acoustic waves are basically sound waves in plasma whose dispersion characteristics can also be derived by $\omega/k = \sqrt{p/\rho} = \sqrt{(T_i + T_i)/m_i}$; they are strongly damped modes in the high-β reconnection layer plasma because of the strong effects of Landau damping. Another common wave is the Alfvén wave, whose dispersion can also be derived by $\omega/k = \sqrt{p/\rho} = \sqrt{B^2/\mu\rho}$. In any case, in order to generate an enhanced resistivity for the electron current flow in the reconnection layer, waves have to involve ion dynamics so that the electron momentum flows are resisted by heavier ions. Thus, lower hybrid waves, which are usually represented by magnetized electrons and demagnetized ions, have been considered as the most likely candidates for generating the enhanced resistivity.

In the case of reconnection without a guide field, the plasma beta in the reconnection layer is generally very large, compared to unity, in the center of the current sheet or reconnection layer. This means that in the center of the current layer any instability must be electromagnetic. This rules out the simpler electrostatic instabilities. The application of theories of local instabilities does not generally work since they are often convective, and propagate out of the instability region of the layer before they can grow. This occurs because the current layers are found to be even thinner than the ion skin depth, by a factor of as much as 3.

The bulk of research on plasma instabilities has been devoted to instabilities in collisionless shocks, and other discontinuous regions, rather than those in reconnection layers. A survey of the literature on such instabilities finds that many are not applicable to reconnection. In reconnection layers one finds that the relative electron–ion drift velocity can be much larger than the ion acoustic speed, so we should expand the spectrum of the appropriate parameter regime.

Historically, the most widely quoted instability is that of Krall and Liewer (1971). Although this is a purely electrostatic instability, it is important as the first instability that brings out features that a micro-instability-driving anomalous resistivity should have. It has been detected in MRX in the lower-β outer boundaries of the reconnection layer (Carter et al., 2002a). The original treatment of the instability is fully kinetic, and because its wavelength is of order the electron gyroradius, the treatment is complicated.

Guided by the difficulties found in earlier theories, Wang et al. (2008) discovered a local instability that has the appropriate property of a very small group velocity across the layer. It does not propagate out of the instability region before growing enough to generate appreciable wave-induced anomalous resistivity. The instability turns out to be a normal mode, but their quasi-mode treatment brings out the physics more clearly and is much easier both linearly and nonlinearly. The instability itself is not a strong generator of enhanced resistivity but it can nonlinearly drive a magnetoacoustic mode. These nonlinear coupled modes will lead to a solution to the anomalous resistivity problem. These modes have properties consistent with the experimentally observed fluctuation in MRX and with Daughton's numerical simulations. A similar instability, called the modified two stream instability, was reported in the literature (McBride et al., 1972; Lemons and Gary, 1977). This instability is driven by a local electron drift velocity against ions, a situation that can occur in collisionless shocks, but needs to be modified in order to apply to the reconnection current sheets of the high-β regime.

Global eigenmode analyses in a Harris sheet (Harris, 1962) (described in chapter 4) for current-driven instabilities (Daughton, 1999, 2003; Yoon et al., 2002) were carried out to take into account the effects of the boundary conditions of a current sheet. This followed similar work on the same subject (Huba et al., 1980). It was found that for short wavelengths ($kd_e \sim 1$; $d_e \equiv c/\omega_{pe}$), the unstable modes concentrate at the low-β edge and are predominantly electrostatic lower hybrid drift waves (LHDW). For relatively longer wavelengths ($k\sqrt{d_e d_i} \sim 1$), unstable modes with significant electromagnetic components, which may be explained by an electromagnetic LHDW, develop in the center region. For even longer wavelengths ($kd_i \sim 1$), a drift kink instability (Daughton, 1999) exists but has a slower growth rate at realistic ion–electron mass ratios.

Particle simulations under various limited conditions have been carried out in three dimensions to study the stability of a Harris current sheet (Horiuchi and Sato, 1999; Lapenta and Brackbill, 2002; Daughton, 2003; Scholer et al., 2003; Daughton et al., 2004; Ricci et al., 2005; Silin et al., 2005; Moritaka et al., 2007). It was found that at first the electrostatic LHDW-like instabilities at $kd_e \sim 1$ are active only at the low-β edge. These edge instabilities grow to large amplitudes to heat electrons anisotropically, thin the current sheet, and induce ion flow shear. These modifications to the background state lead to secondary electromagnetic instabilities localized at the center of the current sheet. These instabilities are identified as drift kink instabilities (Horiuchi and Sato, 1999; Moritaka et al., 2007), Kelvin–Helmholtz instabilities (Lapenta and Brackbill, 2002), or collisionless tearing modes (Ricci et al., 2004; Daughton et al., 2004). Combinations of these instabilities are considered to cause substantial increases in the reconnection rate.

High-frequency electrostatic and electromagnetic fluctuations have been detected in reconnecting current sheets in both space (Shinohara et al., 1998; Bale et al., 2002) and the laboratory (Carter et al., 2002a,b; Ji et al., 2004). In agreement with the numerical predictions, it was found that electrostatic fluctuations peak at the low-β edge of the current sheet, while electromagnetic fluctuations peak at the center of current sheet, as shown in figure 5.14.

The measured frequency spectra show that most fluctuations are in the lower hybrid frequency range, but it was found that the electrostatic fluctuations did not correlate with the observed enhanced resistivity or the fast reconnection rate (Carter et al., 2002a). With the use of the hodogram probe (Ji et al., 2004), the observed electromagnetic waves were found in the lower hybrid frequency range, and appeared in an impulsive manner in all three magnetic components when the current sheet formed. They persist as long as the reconnection proceed. The dispersion relation of the waves was measured from the phase shift between two spatial points. The fluctuations have large amplitudes and appear consistently near the current sheet center with peak $\delta B/B_0 \sim 5\%$, where B_0 is the upstream reconnecting magnetic field. A correlation has been found between the wave amplitudes and fast reconnection rates in the low-density regime. The question remains as to how these electromagnetic waves compare to the waves seen in numerical simulations.

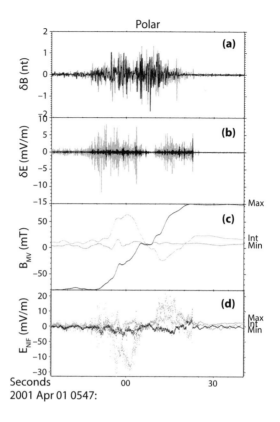

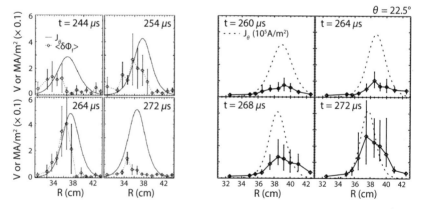

Figure 5.14. Comparison of spatial profiles of electrostatic and electromagnetic fluctuations measured in the magnetopause by the Polar satellite and in the MRX laboratory plasma. In the top panel, magnetic (δB) and electrostatic fluctuations (δE) are shown in (a) and (b) respectively, together with profiles of reconnecting magnetic field components (B_z shown in black; B_y, B_x) and reconnecting electric fields (E_z shown

To find the causes of the observed enhanced dissipation at the center of the current sheets, another step is needed to clarify the interrelationships between laminar Hall dynamics and magnetic fluctuations at the sheet. There is no clear consensus with regard to how the observed waves are excited and how they affect the reconnection rate or dissipation. In a reconnection experiment in the electron MHD (EMHD) regime (Stenzel and Gekelman, 1981), where the electrons were magnetized and the ions were not magnetized, their gyro-orbit exceeded the size of the plasma. Ion acoustic waves were observed in their hot electron plasma ($T_e \gg T_i$) and the observed anomalous resistivity was attributed to them (Gekelman et al., 1982).

A strong guide field can qualitatively alter the kinetic stability properties of a reconnecting current sheet. Due to strong ion Landau damping, electrons need to drift by their thermal speed relative to ions for a Buneman instability to take place. Drake et al. (2003) performed three-dimensional particle simulations of magnetic reconnection with a guide field and found that such instabilities can lead to the development of electron holes, where electron density is substantially depleted in a highly nonlinear state. Such electron holes can be a source of anomalous resistivity. Similar waves were observed in other three-dimensional particle simulations with a strong guide field (Pritchett and Coroniti, 2004), but electron holes were not specifically identified. In space, electron holes have been observed by the Cluster satellite (Cattell et al., 2005), and they propagate rapidly along the current direction in a reconnecting magnetotail current sheet, especially near the separatrices. Lower hybrid drift waves, although not predicted in the simulations, were observed. In the Versatile Toroidal Facility (VTF), electrostatic structures, like electron holes, have been measured near the X-line where a strong guide field was present during driven reconnection (Fox et al., 2008).

5.8.2   Effects of plasma waves excited in the reconnection layer and impulsive reconnection

Although there is clear evidence for the existence of anomalous (much enhanced over the collision-based classical value) resistivity during reconnection, so far no consensus has been found regarding which waves should affect the reconnection rate most. Extensive study has been carried out to determine how the electron diffusion layer is affected by the presence of wave turbulence and how the profile of the electron diffusion layer affects overall reconnection dynamics, including energy dissipation. The lower hybrid

Figure 5.14 (Continued). in black; E_y, E_x shown in gray). In (c), (d) the z-components of B and E are shown in darker gray. The time axis is translated to the spatial distance of the satellite moving against plasma [from Bale et al. (2002)]. In the bottom two panels, measurements in the MRX reconnection layer are shown for electrostatic fluctuation profiles on the left (Carter et al., 2002b) and magnetic fluctuations on the right (Ji et al., 2004) and compared with the profiles of reconnection current. They show clearly that electrostatic fluctuation peaks off the center of the reconnection layer, while magnetic fluctuation tends to become maximum at the center of the layer in both space and laboratory data.

frequency range of waves has been observed and identified both in MRX and in the magnetopause. It was observed that the electron flows in the electron diffusion region often fluctuate on a variety of timescales, causing impulsive and turbulent reconnection. The electron current channel becomes unstable due to a sharp radial gradient of the current density, making the local flux transfer rate fluctuate and generating impulsive reconnection. The reconnection rate measured by the flux transfer rate at the diffusion layer has been compared with the global rate of flux inflow by Ren et al. (2008) and an experimental campaign has been carried out on this topic in more detail on MRX (Ji et al., 2004; Yoo et al., 2019, 2020). It appears that the presence of waves from the lower hybrid frequency range does contribute to the enhanced resistivity, as well as to the enhanced diffusion of electrons without a decisive correlation.

Ren et al. (2008) examined how fluctuations correlate with reconnection rates, based on the observations from MRX. Multiple high-resolution magnetic pickup probes were used to measure magnetic fluctuations at several locations in the reconnection R–Z plane near the electron diffusion region. In a subsequent experiment by Dorfman et al. (2013), impulsive high-frequency magnetic fluctuations were observed concomitantly with a sudden increase in the reconnection rate. The correlation among the magnetic fluctuations, the electron flows, and the reconnection rate at the current sheet center suggests the following picture: As the neutral current sheet narrows, the electron current sheet becomes unstable and suddenly disrupts, generating broader current profiles in both the R- (inflow) and Z- (outflow) directions. The magnetic fluctuations also propagate together with the reconnected magnetic field lines, moving outward in the outflow direction. In other words, the outgoing electrons carry magnetic field lines with them, accelerating the reconnection rate and generating an impulsive reconnection event. In this instant, the local reconnection rate can significantly exceed the flux injection rate by external forcing. After the magnetic flux is ejected out of the reconnection region to the exhaust region with a higher speed than the incoming flux rate, reconnection slows down and a slow flux buildup resumes. These flux buildup phenomena were studied in MRX during driven experiments. The sequence of flux buildup and sudden disruption of the magnetic profile is remarkably similar to the sawtooth reconnection phenomena observed in fusion plasmas that will be mentioned in later chapters. The impulsiveness of reconnection can be related to a drift kink instability expected to occur in the local current sheet, while the cause of the sudden crash of the current profile in a tokamak plasma is considered to be caused by an MHD instability.

5.8.3   Scaling of laboratory data to astrophysics with respect to collisionality

As described earlier in this chapter, there is a distinct transition in the feature of reconnection from the collisional regime to the collisionless regime. While we do not have a fully satisfactory theory that explains the cause of the enhanced resistivity, we could learn from the scaling we obtain from the experimental data. The MRX data suggest that a transition from MHD-like reconnection to collisionless Hall reconnection occurs when the thickness of the reconnection layer becomes comparable to the electrons' mean free path (Yamada et al., 2006). In the two-fluid regime, the sheet thickness is

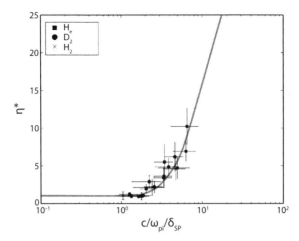

Figure 5.15. MRX scaling. Effective resistivity $\eta^* = \eta_{\text{eff}}/\eta_{\text{SP}}$ ($\eta_{\text{eff}} \equiv E/j$), normalized by the Spitzer value η_{SP} versus the ratio of the ion skin depth to the Sweet–Parker width. Effective resistivity measured in MRX is compared with numerical calculation of the contribution of Hall MHD effects to the reconnection electric field (Yamada, 2007). The simulations were based on a 2D two-fluid code: helium (solid squares), deuterium (solid circles), and hydrogen (asterisks). The solid line is a two-fluid code by Breslau and Jardin (2003).

generally determined by the ion skin depth $d_i = (c/\omega_{pi})$. In the one-fluid collisional MHD regime, on the other hand, the sheet thickness is determined by the Sweet–Parker width $L/S^{1/2}$. The ratio of the ion skin depth to the Sweet–Parker layer thickness δ_{SP} is proportional to the square root of the ratio of the electron mean free path to the system size (length), as shown in Yamada et al. (2006):

$$\frac{c}{\omega_{pi}}\frac{1}{\delta_{\text{SP}}} = 4.5\left(\frac{\lambda_{\text{mfp}}}{L}\right)^{1/2}\left(\frac{m_i}{m_{iH}}\right)^{1/4}, \qquad (5.21)$$

where m_i and m_{iH} are the mass numbers for plasma ions and protons, and we have assumed $T_e \sim T_i$ and $\eta_\perp \sim 2\eta_\parallel$, $v_A \sim v_{\text{thi}}$ ($\beta \sim 1$), and $m_{iH}/m_e \sim 1{,}800$.

In MRX, the classical rate of reconnection with Spitzer resistivity is obtained when the resistivity is large enough to satisfy $\delta_{\text{SP}} > d_i$. When the ion skin depth becomes larger than δ_{SP}, the reconnection rate is larger than the classical reconnection rate determined by Spitzer resistivity and the reconnection layer thickness is expressed as $0.4d_i$. Figure 5.15 presents an MRX scaling for effective resistivity $\eta^* = \eta_{\text{eff}}/\eta_{\text{SP}}$ ($\eta_{\text{eff}} \equiv E/j$), normalized by the Spitzer value η_{SP} in the center of the reconnection region. The MRX data set is compared with a scaling obtained in Hall MHD numerical simulation results using a two-fluid MHD code (Breslau and Jardin, 2003). The horizontal axis represents the ratio of the ion skin depth divided by the classical Sweet–Parker width $\delta_{\text{SP}} = L/S^{1/2}$, where L was set to be 20 cm, the system scale. Figure 5.15 demonstrates an important criterion for two-fluid effects to come into play, namely that the reconnection resistivity (or reconnection speed) takes off from the classical Spitzer value (or the Sweet–Parker reconnection rate) when the ion skin depth d_i becomes larger than twice the Sweet–Parker width δ_{SP}.

The apparent agreement between the MRX scaling and the two-fluid Hall MHD (with resistivity included) code has an important implication. It indicates that anomalous resistivity is primarily accounted for by the laminar Hall effect when the Spitzer resistivity is not large enough to balance the large reconnecting electric field in fast magnetic reconnection. Even with the presence of other energy dissipation mechanisms, the reconnection electric field can be primarily represented by the laminar Hall effect, namely the $j_{Hall} \times B$ term, and this is consistent with the MRX data shown in figure 5.15.

We note that the magnitude of this laminar Hall effect peaks somewhere outside the X-line. Additional effects, such as anomalous resistivity caused by waves and turbulence, are needed to support reconnection around the X-line and separatrices. It can thus be concluded that both mechanisms, one based on the laminar Hall effect and the other including effects related to waves and the electron pressure tensor, are responsible for fast reconnection. Looking into the future, experimental facilities can be utilized more effectively by widening operations into more astrophysics-relevant regimes or by building new devices to address specific physics issues. The study of magnetic reconnection in the wider collisionality parameter regime (from $\lambda_{mfp} \ll L$ to $\lambda_{mfp} \gg L$) is desirable. In addition, toroidal fusion experiments, including tokamaks and RFPs, display strong global reconnection phenomena in highly conductive plasmas in the collisionless regime, and they will be utilized effectively to make more comprehensive magnetic reconnection scaling with and without a guide field. From this perspective, an interesting paper (Ma et al., 2018) was recently published, in which an effective resistivity relevant to collisionless magnetic reconnection in plasma was presented. It is based on the argument that pitch angle scattering of electrons in the small electron diffusion region around the X-line can lead to an effective resistivity in collisionless plasma. It was concluded by them that their result agrees very well with the resistivity (obtained from available data) of a large number of observed cases susceptible to magnetic reconnection: from the intergalactic and interstellar, to solar terrestrial, and to laboratory fusion plasmas. The obtained scaling law agrees well with the MRX scaling shown above in figure 5.15.

5.8.4   New findings from the recent interdisciplinary research and future opportunities

The arrival of new MMS data (Burch et al., 2016b) has greatly increased the prospects of understanding the detailed electron dynamics and the associated dissipation. The combined capability of the unprecedented resolutions in both particle and field measurements as well as the close distances between the four spacecraft in the constellation has enabled measurements of detailed electron dynamics and electromagnetic fields on electron scales, such that each term in the generalized Ohm's law can be quantified. Positive identifications of the electron diffusion region were made by close comparisons with two-dimensional symmetric and asymmetric reconnection simulations, which clearly explain the origin of the observed crescent-shaped electron velocity distributions that carry the reconnecting current. Despite these successes, however, the crucial question on the balancing dissipation of the reconnecting electric field is still open, as none of the laminar terms shown in eq. (5.17) is sufficiently large. This is consistent with the fact that the observed thickness of electron diffusion regions (EDRs) from the MMS data is still considerably larger than the two-dimensional predictions. Moving beyond the standard two-dimensional model of collisionless reconnection, various waves have already been reported by the MMS team from low-frequency drift waves (Ergun et al., 2017), lower hybrid drift waves (Graham et al., 2017), high-frequency whistler waves (Cao et al., 2017; Wilder et al., 2017), to standing quasi-electrostatic whistler waves localized in EDRs (Burch et al., 2016a). Many of these waves are also found in MRX, where wave characteristics were studied as shown in figure 5.14. Whistler waves were detected in EDRs with significant power up to the local $0.5 \times \omega_{ce}$ (where ω_{ce} is the electron cyclotron angular frequency) in frequencies in both MMS (Cao et al., 2017) and MRX.

In the most recent work carried out by Yoo et al. (2020), generation and propagation of LHDW near the EDR during guide field reconnection at the magnetopause is studied with data from MMS. Inside the current sheet, the electron beta β_e determines properties of the excited LHDW. Inside the EDR, where the electron beta is high ($\beta_e \sim 5$), the long-wavelength electromagnetic LHDW is observed propagating obliquely to the local magnetic field. In contrast, the short-wavelength electrostatic LHDW, propagating nearly perpendicular to the magnetic field, is observed slightly away from the EDR, where β_e is small ($\beta_e \sim 0.6$). The observed LHDW features were explained by a local theory model, including effects from the electron temperature anisotropy, finite electron heat flux, electrostatics, and parallel current. The short-wavelength LHDW is capable of generating significant drag force between electrons and ions, which is consistent with the concept mentioned earlier (section 5.8).

Future collaborative research from three approaches (laboratory, numerical, and observational) will be intensified to systematically investigate each of the candidate waves and their effects on the reconnection rate and the dissipation of magnetic field to electrons. The latter question is a key consequence of the collisionless reconnection process and will be addressed in chapter 10.

Chapter Six

Laboratory plasma experiments dedicated to the study of magnetic reconnection

As mentioned in chapter 1, research on magnetic reconnection has advanced significantly in the last three decades by connecting theory with results from numerical simulation codes, space satellites, and laboratory experiments dedicated to the study of magnetic reconnection. The recent dedicated experiments have been carried out in plasma systems that satisfy the global conditions for MHD (magnetohydrodynamic) treatment of the plasma with $V_A/c \ll 1$, $S \gg 1$, and $\rho_i/a \ll 1$, providing realistic examples of magnetic reconnection. In this chapter we highlight laboratory experiments, starting from early electron current sheet experiments, followed by the merging of spheromaks and tokamaks, and then modern dedicated laboratory experiments, which can be interconnected with observations from major satellites and numerical simulation results.

6.1 EARLY LABORATORY EXPERIMENTS ON RECONNECTION

Before 1980, most laboratory experiments dedicated to magnetic reconnection research were carried out in short-pulse current-carrying "pinch" plasmas or fast high-density pulsed plasma discharges of a few microseconds duration (Bratenahl and Yeates, 1970; Ohyabu et al., 1974; Syrovatskii et al., 1973; Frank, 1974; Baum and Bratenahl, 1980; Syrovatskii, 1981). One of the major goals was to generate a current sheet where reconnection takes place as predicted by Dungey (1953). Although diagnostics were not yet advanced, and high spatial and temporal resolution was not available, interesting observations were made. The reconnection rate was recognized to be much larger than the value predicted by MHD in the experiment by Bratenahl and Yeates (1970) and it was attributed to possible micro-instabilities driven by the drift of electrons against ions in the current sheet, although a quantitative analysis was not made. These experiments were carried out in the collision-dominated MHD regime, and their low-Lundquist-number ($S = 1$–10) experiments made it difficult to quantitatively compare the results with MHD theories based on large S.

Despite these difficulties, the current density profiles of the neutral sheet were measured by magnetic probes, and density profiles were measured on the Alfvén transit time ($< 1\,\mu$s) by Frank et al. (Frank, 1974; Syrovatskii, 1981). Figure 6.1(a) presents

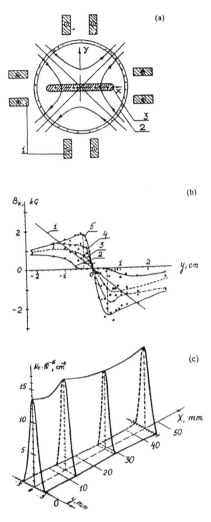

Figure 6.1. Fast reconnection experiments in a linear plasma sheet pinch device. (a) Coil and current sheet geometry, (b) measured magnetic field profiles at different times, and (c) plasma density profiles measured by interferometry. [From Syrovatskii (1981).]

their experimental setup, where formation of a flat current sheet was induced in a Z-pinch discharge along the axis (z-axis) of a straight cylinder. Figure 6.1(b) shows the time evolution of reconnecting magnetic field profiles along their y-axis (not a conventional coordinate system), perpendicular to the sheet. After the magnetosonic waves converged, a current sheet stretched in the x-axis began to form in the vicinity of the neutral line along their y-axis. The final thickness of the sheet appeared to be determined by the balance of the reconnecting magnetic field pressure and the plasma kinetic pressure (Syrovatskii, 1981). This experiment was one of the first laboratory trials

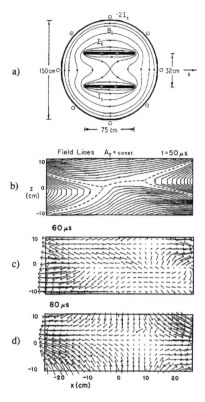

Figure 6.2. LPD (linear plasma device) experiment. (a) Cross-sectional view of the experimental setup without plasma, (b) magnetic flux contours, (c), (d) ion velocity vectors at two different times. The magnetotail coordinate system (y is the out-of-plane direction) is employed in this experiment (A is the vector potential of the magnetic field: $\nabla \times A = B$). [From Stenzel and Gekelman (1981).]

to materialize Dungey's concept of a collapse of the current sheet near the X-point (discussed in chapter 1). They measured profiles of plasma density by interferometry versus the y-direction at four different x-positions. The peak density was as high as $1.4 \times 10^{16}\,\mathrm{cm}^{-3}$, which supports a pressure balance between the reconnecting magnetic field ($\sim 2\,\mathrm{kG}$) and the plasma. The plasma was highly collisional.

In a linear plasma experiment carried out at UCLA between 1980 and 1990 (Stenzel and Gekelman, 1981; Gekelman et al., 1982), magnetic reconnection was a major topic and was studied using parallel conductor plate currents with a strong guide field ($B_G \gg B_{\mathrm{reconn}}$). A reconnection region was created by driving currents in the two parallel sheet conductors shown in figure 6.2(a), and a detailed local study of magnetic reconnection was performed using internal probes based on multiple reproducible plasma discharges.

The experiments were carried out in a cylindrical vacuum chamber (1.5 m diameter, 2 m length) in which a low pressure ($p \sim 10^{-4}$ torr, argon) discharge was produced with a 1 m diameter oxide-coated cathode. The plasma parameters were $n_e \sim 10^{12}\,\mathrm{cm}^{-3}$, $T_e \sim 10\,\mathrm{eV}$, electron–ion collision mean free path ~ 1 m, axial magnetic field 12–100 G, and $\beta_\perp \sim 1$ for a plasma current of about 1.5 kA. The neutral sheet current was generated by a cathode discharge with help from inductive drive by the outside plate.

The electrons were magnetized ($\rho_e \ll L$) while ions were unmagnetized even outside the reconnection region ($\rho_i \gg L$: argon gas); this is a regime often called an electron MHD (EMHD) regime. The plasma was largely resistive and the Lundquist number was small ($1 < S < 10$). Although it was reported that a Petschek slow MHD shock was observed (Stenzel and Gekelman, 1979), it could not be theoretically supported since the experiment was performed in a non-MHD regime.

This experiment was noteworthy in studying EMHD dynamics and wave-related mechanisms in the reconnection region and identifying local microscopic physics associated with neutral sheet formation, in particular particle motions and wave excitation. Profiles of the electron pressure $n_e T_e$, magnetic force density $\boldsymbol{j} \times \boldsymbol{B}$, and ion velocity vectors were measured in the diffusion region. A neutral current sheet was seen to develop in less than two Alfvén transit times ($\tau_A \sim 20\,\mu s$). The neutral sheet became narrower as it was measured further from the cathode. Figure 6.2(b) shows field lines through contours of vector potential A_y at $y = 137$ cm from the cathode and at $t = 50\,\mu s$; here, y is the axial distance from the cathode and $t = 0$ is the start time of the discharge. After a few Alfvén times, a classical plasma flow pattern was observed, with ions jetting from the neutral sheet with velocities close to the Alfvén speed. The two-dimensional features of particle acceleration were measured (Gekelman et al., 1982). Figure 6.2(c), (d) show typical two-dimensional ion flows drifting radially from the diffusion region to outside at $t = 60$ and $80\,\mu s$. The local force on the plasma, $\boldsymbol{j} \times \boldsymbol{B} - \nabla p$, was compared with the measured particle acceleration using differential particle detectors. The ion acceleration was seen to be strongly modified by scattering off wave turbulence and the observed fluctuations were identified as oblique whistler waves. But it was not clear whether the whistlers were solely responsible for the observed large ion scattering rate. It was concluded later that the observed anomalous scattering rate and high resistivity were in large part due to ion acoustic turbulence, although higher-frequency whistler waves were present. However, the role of whistler waves for the observed anomalous resistivity was not conclusively determined.

The physical effects of the strong guide field used in the experiment were not discussed explicitly in their analysis of their data, while they were expected to play a significant role in the force balance. After modifying the shape of the end anode, they found that evolution of the neutral current sheet depended on the strength of the axial magnetic field. As the axial field was raised from 20 G to 100 G, the classical double-Y-shaped neutral sheet topology changed to an O-shaped magnetic island. This result was later reproduced in the MHD regime on MRX (Magnetic Reconnection Experiment) (Yamada et al., 1997b,a). The stability of the current sheet was also investigated. When the current density in the center of the sheet exceeded a critical value, spontaneous local current disruptions were observed with the center of the sheet moving out to the sides. This experiment was extended to a three-dimensional study (Gekelman and Pfister, 1988), in which tearing of the current sheet was observed.

In summary, the LPD experiment was valuable in measuring the local structure and some of the EMHD features of the reconnection region for the first time and in finding the relationship between the reconnection rate and wave turbulence. However, one of the most important questions on reconnection—how the diffusive neutral sheet is formed in a plasma relevant to space and astrophysics—was not answered because the MHD conditions for the global plasma were not satisfied due to the boundary condition in which the reconnection layer is too close to the wall ($\rho_i \gg L$). Also, the effects of the end electrodes in their linear device were not addressed. We note that the short ion mean free path (compared to L) may explain the fluid-like behavior of the ions as shown in figure 6.2. The role of line tying in their linear plasma is not clear. In the following sections, reconnection experiments without end effects, in MHD regimes where the Lundquist number exceeds 100, and with both electrons and ions being magnetized ($\rho_e \ll \rho_i \ll L$) are discussed.

6.2   EXPERIMENTS OF TOROIDAL PLASMA MERGING

Merging of toroidal plasmas would create a magnetic toroidally symmetric reconnection layer not terminated by electrodes, and toroidal plasma merging is fully utilized to study uniform reconnection layers. Local and global MHD issues for magnetic reconnection have been extensively investigated in the toroidal geometry in spheromak merging experiments. The studies showed that a double spheromak geometry is a configuration well suited for basic studies of magnetic reconnection. An advantage of this type of experiment is that we can simultaneously study local and global features of magnetic reconnection.

6.2.1   Plasma merging experiments at the TS-3/4 facility

The TS-3 (Todai Spheromak-3) group carried out laboratory experiments (Yamada et al., 1990, 1991; Ono et al., 1993) to study magnetic reconnection by making two spheromak-type plasma toroids merge together through contacting and reconnecting along a toroidally symmetric line. A spheromak is a spherical- or toroidal-shaped plasma in which near force-free currents ($j \times B \simeq 0$) set up an equilibrium configuration, depending on whether there is a current (flux) hole at the major axis (Yamada et al., 1981; Taylor, 1986). The toroidal magnetic field is generated by its own poloidal current, thus making a spheromak a good candidate for an ideal compact toroid reactor core without external coil systems which interlink with plasma. Two toroidal spheromaks, carrying equal toroidal current with the same or opposite toroidal field, were forced to merge by externally controlled coil currents. This is called co-helicity merging and counter-helicity merging, respectively. As explained by Yamada et al. (1990), counter-helicity merging generates antiparallel magnetic field lines merging at the reconnection sites, while co-helicity merging happens when magnetic field lines merge at an angle, as shown in figure 6.3.

In antiparallel merging, magnetic reconnection was expected to occur very efficiently, and experimental results demonstrated that counter-helicity merging indeed

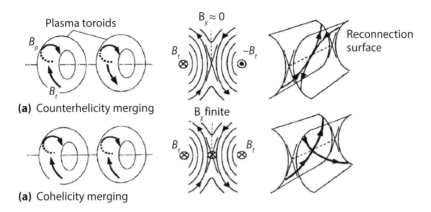

(a) Counterhelicity merging

(a) Cohelicity merging

Figure 6.3. Schematic drawings of the merging geometry of two toroidal plasmas. Merging toroid configurations of the reconnection region, as well as the 3D geometry of reconnecting field lines, are shown for (a) counter-helicity merging and (b) co-helicity merging.

made much faster reconnection globally than co-helicity merging, where the field lines meet with an angle. During this merging, a field reversed configuration (FRC) is generated without a net toroidal flux. The FRC is another good candidate for an efficient reactor core, since it can confine a high-β plasma ($W_B \sim W_P$). Thus, this merging scheme became one of the most popular schemes for generating FRC plasmas.

Figure 6.4 (left) shows an experimental setup for the TS-3 merging experiment. Two plasma toroids are made by coaxial electrode discharges and made to merge coaxially, as shown in the figure. To document the time evolution of the internal magnetic structure of the reconnection on a single shot, a two-dimensional magnetic probe array is placed on an $r-Z$ plane or toroidal cutoff plane. The plasma parameters are $B \sim 0.5$–1 kG, $T_e \sim 10$ eV, and $n_e \sim 2$–5×10^{14} cm^{-3}. The time evolution of the poloidal flux contours shows that counter-helicity merging of plasma toroids of opposite helicity occurs significantly faster (by more than three times) than merging of the same helicity (figure 6.4, right). It was also reported that for counter-helicity merging, the opposite toroidal fields canceled each other after the merging and the total toroidal flux was quickly annihilated. The poloidal flux contours are shown for the same sequence of shots (Yamada et al., 1991; Ono et al., 1993) in figure 6.4 (right) for co- and counter-helicity merging respectively, demonstrating an important difference in the evolution of the two-dimensional features of magnetic reconnection. Strong dependence of the reconnection speed on the global forcing was also observed, i.e., the merging velocity of the two plasmas. It was observed that the global reconnection rate γ, defined by a flux transfer rate $(1/\Psi)\delta\Psi/\delta t$, increased nearly proportionally to the initial colliding velocity v_i. This result could not be explained by the classical two-dimensional MHD theories of Sweet and Parker and/or Petschek, which are based on the local dynamics. This experiment clearly suggests the importance of an external driving force in determining

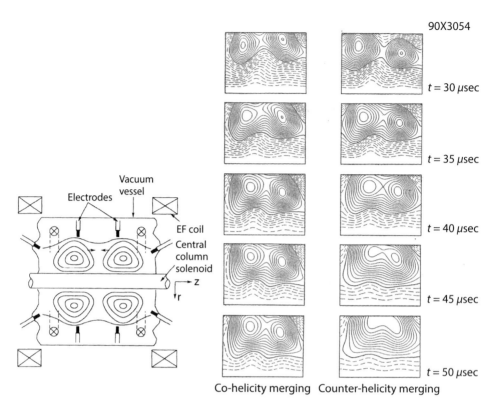

Co-helicity merging   Counter-helicity merging

Figure 6.4. (Left) TS-3 experimental setup. (Right) Evolution of poloidal flux contours for co- and countermerging. Counter-helicity merging proceeds much faster than co-helicity merging. The plasma parameters are kept identical for the two cases shown. [From Yamada et al. (1990); Ono et al. (1993).]

the reconnection rate and supports an important aspect of a driven-reconnection model with external and global forcing.

6.2.2   Plasma heating and acceleration during plasma merging

A violent "sling-shot" plasma acceleration is expected in the toroidal direction as the field lines contract after the merging of two toroidal plasmas of opposite helicity: figure 6.5(a). Let us look at clear evidence of this phenomenon in the TS-3 experiment (Ono et al., 1993).

Figure 6.5(a)–(c) show the time evolution of the profile of the toroidal field B_t versus Z (axial direction) for counter-helicity merging. Note that time proceeds from the top to the bottom of the figure. This result was obtained by a B_t probe array axially inserted at radius $R = 14$ cm (which matches the magnetic axis). Initially, the merging plasmas formed the B_t-profile shown in the figure, positive on the left and negative on the right-hand side. As reconnection progressed, the value of B_t decreased

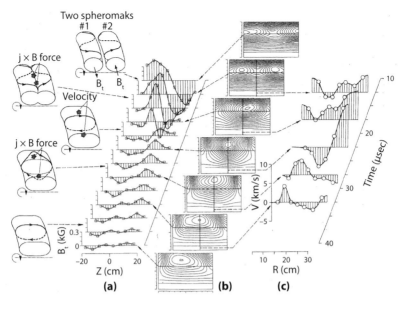

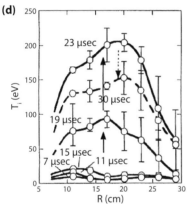

Figure 6.5. Time evolution of plasma parameter profiles (note that time proceeds from the top to the bottom of the figure). (a) Axial profiles of the toroidal magnetic field B_t at $r = 18$ cm, (b) poloidal flux contours on the R–Z plane, (c) radial profiles of the ion global velocity V in the toroidal direction on the midplane, and (d) radial profiles of ion temperature T_i on the midplane, during the reconnection of two merging spheromaks with equal but opposite direction B_t. [From Ono et al. (1996).]

as expected but then the B_t-profile flipped (changed its polarity) between $t = 20$ and $30\,\mu s$. This overshoot is regarded as evidence of the toroidal sling shot effect (Yamada et al., 1990) as shown on the far left of figure 6.5(a). Figure 6.5(b) describes the (three-dimensional) evolution of magnetic field lines through poloidal flux contour plots during and after the reconnection. Energy transfer from magnetic to plasma thermal energy is expected in this dynamic toroidal field annihilation process. Strong plasma acceleration and ion heating were documented during counter-helicity merging (Ono et al., 1996), as shown in figure 6.5(c), (d) respectively. Numerical MHD simulations show similar three-dimensional effects in solar flare processes (Matsumoto et al., 1993) and in magnetospheric physics (Hawkins et al., 1994).

Local ion heating due to reconnection has been also measured (Hsu et al., 2000) in MRX using an Ion Dynamic Spectroscopy Probe (Fiksel et al., 1998) placed inside the neutral sheet. The ion heating rate was found to be much larger than the values predicted by classical dissipation. The SSX experiment (see section 6.2.3) was also utilized to study ion heating during merging (Kornack et al., 1998). While their results are consistent with the results from TS-3 and MRX, a burst of plasma flow at the Alfvén speed was observed in the reconnection plane.

6.2.3   Plasma merging experiments at the SSX facility

The Swarthmore Spheromak Experiment (SSX) facility (Brown, 1999; figure 6.6) also studies magnetic reconnection through the merging of spheromaks. Reconnection physics, particularly its global characteristics, has been studied in a number of geometries with diameters varying from 0.17–0.50 m (Brown, 1999; Cothran et al., 2003; Brown et al., 2006). Different types of flux-conserving conductors consisting of two identical copper containers have been used.

Many optical diagnostics were utilized in this device. The line-averaged electron density is monitored with a HeNe laser interferometer (Brown et al., 2002). Alfvénic outflow has been measured both with electrostatic ion energy analyzers (Kornack et al., 1998) and spectroscopically (Brown et al., 2006; Cothran et al., 2006). The line-averaged ion flow and temperature T_i at the midplane are measured with a 1.33 m ion Doppler spectrometer.

The plasma parameters are similar to those of TS-3: $n_e = (1-10) \times 10^{14}\,cm^{-3}$, $T_i = 40-80\,eV$, $T_e = 20-30\,eV$ (inferred from soft-X-ray radiation), with typical magnetic fields of 0.1 T. The ion gyroradius is much smaller than the radius of the outer flux-conserving boundary of the plasma (defined by a cylindrical copper wall). The Lundquist number S range is 100–500, making the global structure of SSX spheromaks fully in the MHD regime ($S \gg 1$, $\rho_i/L \ll 1$). The merging of a pair of counter-helicity spheromaks generates turbulent three-dimensional magnetic reconnection dynamics at the midplane. In recent decades, more experiments have been carried out to study the global characteristics of various current-carrying plasma configurations with axially elongated force-free equilibrium (Gray et al., 2010, 2013). Recently, a three-dimensional magnetic structure was obtained with 600 individual internal magnetic probes operated at 1.25 MHz.

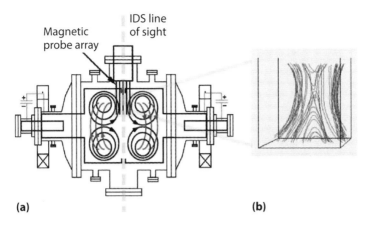

(a)   **(b)**

Figure 6.6. The SSX device to study local and global dynamics of magnetic reconnection. [From Brown et al. (2006).]

6.2.4   TAE experiments and plasma heating during the merging of two FRC plasmas

The main focus of the Tri-Alpha Energy (TAE) project is the non-neutronic fusion reaction of p-11B to create three helium atoms by injecting high-energy ion beams to stabilize an FRC plasma and to obtain a well-confined high-energy-density fusion plasma. While their reactor design is very challenging at the moment, they are making steady progress in obtaining high-temperature plasmas and keeping them for a long time.

In the past ten years, significant effort has been made by TAE to utilize plasma heating during the reconnection process of toroidal plasmas effectively. In their initial phase, two toroidal FRC plasmas were merged together, resulting in notable heating of the merged plasma, as shown in figure 6.7. Magnetic reconnection was utilized to form their target compact toroid plasma. In the initial phase the magnetic energy and the kinetic energy of the fast-moving colliding plasmas are converted to the plasma's thermal energy. In one of the more recent experiments, FRCs with a high ion temperature of 1 keV were obtained in their C-2 device by combining plasma-gun edge biasing and neutral beam injection. A separate plasma gun creates an inward radial electric field that counters the deleterious FRC spin-up phenomena (Binderbauer et al., 2015). The $n = 2$ rotational instability is stabilized by the presence of external gun plasma (could be due to line-tying effects) without applying stabilizing external quadrupole magnetic fields. The FRCs are nearly axisymmetric, which enables fast ion confinement. The plasma gun also produces $E \times B$ plasma flow shear in the FRC edge layer, which may explain the observed improved stability as well as reduced particle transport. It is significant that they have relatively quiet plasma conditions in which they can assess the confinement characteristics of the high-energy-beam-driven FRC plasmas. This result significantly improves the prospect of a compact fusion reactor. The FRC confinement

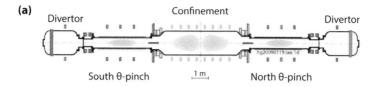

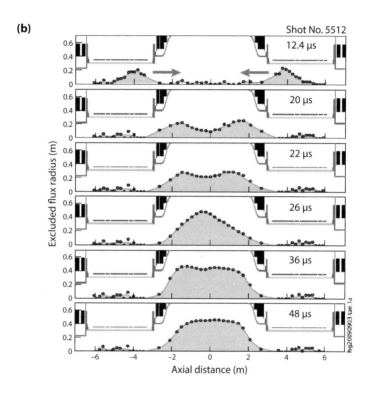

Figure 6.7. Plasma merging in the TAE device for initial FRC plasma formation through magnetic reconnection. (a) Schematics of the C-2 device at the Tri Alpha Energy company. (b) Evolution of the outer radius (measured by the excluded flux radius) of an FRC plasma in the C-2 device obtained from a series of external diamagnetic loops at the formation chamber and main confinement chamber. Time is measured from the instant of FRC plasma formation in the formation chamber, and the distance z is given relative to the center of the confinement chamber. [From Binderbauer et al. (2015).]

times have been improved recently by factors of 4 to 5, and the plasma lifetimes have been significantly extended from 1 ms to up to 30 ms.

In order to generate an optimum initial high-flux target FRC plasma, they plan to construct new midscale spheromak injectors at TAE and then characterize the performance, including studies of spheromak merging and reconnection as well as the stability of a formed FRC. A multipulsed spheromak injection into an FRC plasma on the geometrical axis for effective refueling and refluxing is planned.

6.3   CONTROLLED DRIVEN RECONNECTION EXPERIMENTS

In the past two decades, a series of dedicated laboratory experiments have been performed to investigate the fundamental processes of reconnection by making a prototypical reconnection layer in a controlled manner. The goal is *not* to simulate specific reconnection events in space or in fusion devices, but to provide key data to understand the fundamental process of reconnection. In these experiments, a reconnection layer can be generated in a controlled setting by driving oppositely directed field lines into the neutral sheet, generating a reconnection region with various plasma parameters.

6.3.1   MRX facility

The MRX (Magnetic Reconnection eXperiment) device was built at the Princeton Plasma Physics Laboratory (PPPL) in 1995 to investigate the fundamental physics of magnetic reconnection (Yamada et al., 1997a). Another goal was to gain understanding of self-organization phenomena of fusion plasmas as well as space and astrophysical plasmas. The analysis focuses on the coupling between local microscale features of the reconnection layer and global properties such as external driving force and the evolution of plasma equilibrium. The local features of the reconnection layer have been extensively studied. In most experimental campaigns, the initial boundary geometry is axisymmetric and two-dimensional features of reconnection have been extensively studied. But if necessary, the boundary condition can be made nonaxisymmetric to study the three-dimensional characteristics of merging and also asymmetric reconnection, in which the plasma densities of the two inflowing plasmas are significantly different, as observed in the magnetopause. The global plasma properties can be described by MHD ($S > 10^3$) with the ion gyroradius (1–5 cm) being much smaller than the plasma size ($R \sim 30$–50 cm), while it is necessary to use two-fluid kinetic theory to describe the dynamics of the reconnection region.

Experiments have been carried out in a double annular plasma setup in which two toroidal plasmas with annular cross-section are formed independently around two flux cores and magnetic reconnection is driven in the poloidal field shown in figure 6.8. Each flux core (gray circle in figure 6.8) contains a TF (toroidal field) coil and a PF (poloidal field) coil to inductively generate plasma discharges (Yamada et al., 1981). First, a quadrupole poloidal magnetic field is established by the PF coil currents (which flow in the toroidal direction), and plasma discharges are created around each flux core by pulsing currents into the TF coils (Yamada et al., 1997a) of the flux cores. After the annular plasmas are created, the plasmas are made of two private sections around each flux core and a public section that surrounds both private plasmas.

Two toroidal plasmas carrying identical toroidal current, with the same or opposite toroidal field, is are made to merge to induce reconnection by controlling external coil currents. After the annular plasmas are created, the PF coil current can be increased

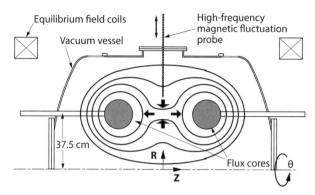

Figure 6.8. Experimental setup for MRX and illustration of "pull" driven reconnection in the double annular plasma configuration. [From Yamada et al. (1997b).]

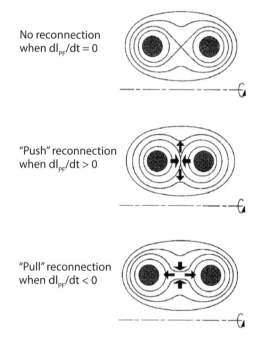

Figure 6.9. Illustration of "push" and "pull" reconnection in the MRX double annular configuration. [From Yamada et al. (1997b).]

or decreased. For rising PF current, the poloidal flux in each plasma increases and is "pushed" toward the X-point (push mode). For decreasing PF current, the poloidal flux in the common plasma is "pulled" back toward the X-point (pull mode) (see figure 6.9). In this way, a current layer or a typical reconnection layer is generated in a well-controlled manner. For standard conditions ($n_e \sim 0.1$–$1 \times 10^{14}\,\mathrm{cm}^{-3}$, $T_e = 5$–$15\,\mathrm{eV}$, $B = 0.2$–$1\,\mathrm{kG}$, $S > 500$), MRX creates strongly magnetized MHD plasmas. The

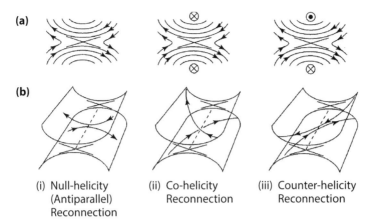

Figure 6.10. 2D and 3D views of magnetic reconnection. When the third vector component is zero, the reconnecting field lines are exactly antiparallel, i.e., null-helicity, case (i). With the presence of the third field component, field lines reconnect obliquely in co-helicity for case (ii) while the reconnecting field lines are again antiparallel in counter-helicity for case (iii). [From Yamada et al. (1997b).]

mean free path for electron–ion Coulomb collisions can be varied in the range 0.1–20 cm.

To measure the internal magnetic structure on a single discharge, a two-dimensional magnetic probe array is placed in the R–Z plane or toroidal cutoff plane, where Z is the axis of the vacuum vessel. The profile of the neutral sheet was carefully measured and different shapes of neutral sheet current layers were identified, depending on the third (toroidal) vector component of the reconnecting magnetic fields. As poloidal flux is driven toward the diffusion region, a neutral sheet is formed, as seen in figure 2.10. Without the third component (called null-helicity reconnection), a thin double-Y-shaped diffusion region is clearly seen in figures 1.5 and 2.10. With a significant third component (co-helicity reconnection), an O-shaped sheet current appears (Yamada et al., 1997b). Recent results from MRX are discussed throughout this book.

The common two-dimensional description of magnetic field-line reconnection is shown in figure 6.10. In actual reconnection phenomena, the magnetic field lines have three vector components as illustrated. For case (i), conventional two-dimensional reconnection is applicable. In the presence of a third component, figure 6.10(ii) shows the field lines reconnecting at an angle when unidirectional toroidal fields exist (the co-helicity case). In (i) and (iii) they reconnect with antiparallel geometry. Note that the reconnecting field lines are antiparallel for both null-helicity and counter-helicity merging. Although cases (i) and (iii) are the same in a two-dimensional description, they can be quite different in the global MHD picture.

6.3.2 VTF facility

The Versatile Toroidal Facility (VTF) magnetic reconnection experiment (Egedal et al., 2000, 2003, 2005, 2007; Egedal and Fasoli, 2001), which is a controlled externally

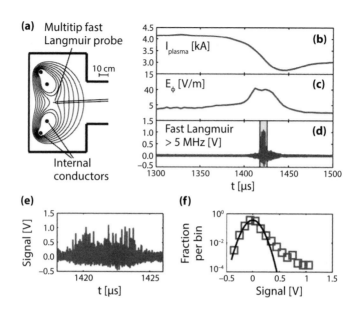

Figure 6.11. VTF experiment. (a) Schematic of the experiment at a poloidal cross-section, (b) total plasma current, (c) toroidal-averaged toroidal electric field, (d) fluctuation trace for the time period $1,300$–$1,500\,\mu\text{s}$, (e) zoom-in on the fluctuation trace from $1,418$ to $1,426\,\mu\text{s}$, (f) log-histogram of voltages measured by a fluctuation probe over this time period, where the solid curve indicates a best Gaussian fit to the central portion of the data. [From Fox et al. (2008).]

driven reconnection experiment, was built at the Plasma Science and Fusion Center of the Massachusetts Institute of Technology.

The VTF experiment explores fast magnetic reconnection in a collisionless plasma environment, where the mean free path between electron and ion collisions is much larger than the dimensions of the plasma for a configuration with a strong variable guide magnetic field. The VTF geometry and a poloidal cross-section are shown in figure 6.11(a). The understanding gained from research on reconnection in the VTF was applied to interpret recent in situ measurements of electron phase-space distribution during reconnection in the deep magnetotail (Egedal et al., 2008, 2009, 2012, 2016). This is of particular relevance to the reconnection event observed by the Wind and MMS (Magnetospheric Multiscale Satellite) satellites, as discussed in chapter 8. They also reported an observation of large-amplitude, nonlinear electrostatic structures, identified as electron phase-space holes, during magnetic reconnection experiments on VTF. The holes are positive electric potential spikes, observed on high-bandwidth (2 GHz) Langmuir probes: see figure 6.11. They observed a localized (three-dimensional) onset of magnetic reconnection in a well-diagnosed laboratory experiment. After the onset, the reconnection spreads toroidally to the rest of the device, connecting their results to recent observations of solar flares. The reconnection is observed only when the magnetic geometry permits a global mode structure outside the reconnection region. We discuss the implications of the onset for current continuity, provide a simple semi-empirical model for the gross features of the onset, and describe the interchange-like character of the global mode.

Table 6.1 on page 114 summarizes major devices dedicated to the study of the physics of magnetic reconnection. The three-dimensional aspects of magnetic reconnection have been studied in two linear devices (Stenzel et al., 2003; Gekelman et al., 2007), the Rotating Wall Experiment (Bergerson et al., 2006) and the Reconnection Scaling Experiment (Furno et al., 2007), which were partially used for the study of magnetic reconnection, in particular line-tying effects.

6.3.3 TREX facility

A relatively new large device was built recently to access the regime of electron pressure anisotropy. The Terrestrial Reconnection EXperiment (TREX) is a newly built dedicated reconnection experiment and is now being operated using the 3 m diameter spherical vacuum vessel of the Madison Plasma Dynamo Experiment (MPDX). The TREX is also leveraged by an earlier MPDX experiment in which dynamo physics was being studied (http://plasma.physics.wisc.edu/trex).

It is considered that in a collisionless plasma, electron pressure anisotropy develops in the electron diffusion layer, which strongly influences the properties of the reconnection process in ways not accounted for in traditional Hall reconnection. This indication has already been found by Ren et al. (2005). Spacecraft observations and kinetic simulations show that large-scale electron jets/current layers are driven by electron pressure anisotropy that builds in the reconnection region. Compared to Hall effects, the pressure anisotropy occurs at a much smaller scale and is more sensitive to micro-fluctuations. An extension of the research is needed to evaluate the role of the pressure anisotropy in particle heating. A good comparative study between laboratory results and space observations, such as by MMS, is expected from this device. Their goal is to expand the operation regime in the phase-space diagram for reconnection experiments. Just like in FLARE (see section 6.3.4) to be built at PPPL, a pulsed operation TREX will reach the regime of turbulent reconnection involving multiple X-lines.

6.3.4 FLARE facility

The FLARE device (Facility for Laboratory Reconnection Experiments; http://flare.pppl.gov) is a new laboratory experiment constructed at Princeton for studies of magnetic reconnection in multiple X-line regimes directly relevant to space, solar, astrophysical, and fusion plasmas, as guided by a reconnection phase diagram (Ji and Daughton, 2011). The first plasma operation was successfully conducted to validate its engineering design and to demonstrate its experimental access to the parameter space beyond its predecessor MRX (http://mrx.pppl.gov). The main goal of this device is to extend the reconnection research frontier to much larger system sizes, measured in units of the ion kinetic scale. Most of the work on reconnection in the past, both numerical and experimental, investigated relatively small systems (≤ 100 ion kinetic scales d_i).

Table 6.1. Dedicated experiments for reconnection research.

Facility names	Main features	Main references
3D-CS at GPI, Russia	Linear geometry, strong guide field	Frank (1974), Frank et al. (2005, 2006)
EMHD reconnection, LAPD at UCLA	Linear geometry, strong guide field	Stenzel et al. (1982), Gekelman et al. (2007)
TS-3/4/6 at U. Tokyo	Plasma merging, compact toroids, spherical torus	Yamada et al. (1990), Ono et al. (1993, 1996)
MRX at PPPL	Driven reconnection, plasma merging	Yamada et al. (1997a, 2010, 2016a), Ji et al. (1998, 2004, 2008)
SSX at Swarthmore	Plasma merging, compact toroids	Brown (1999), Brown et al. (2002, 2006)
VTF at MIT	Strong guide field, particle dynamics	Egedal et al. (2000, 2012), Fox et al. (2010)
RSX at LANL	Plasma merging by injection, generation of 3D current channel	Furno et al. (2007)
RWX at Wisconsin	Linear geometry, reconnection in line-tied plasma	Bergerson et al. (2006)
Laser driven US, UK, China	High-β plasma merging, strong electron heating	Nilson et al. (2006), Fox et al. (2020)
TREX at Wisconsin	Collisionless reconnection	http://plasma.physics.wisc.edu/trex (2018–)
FLARE at PPPL	Large size, multiple reconnection, plasmoid reconnection	https://flare.pppl.gov (2019–)

High-performance computing capabilities have enabled researchers to extend the size of simulation domains, especially in two dimensions, uncovering new secondary (plasmoid) instabilities of thin current sheets that lead to new nonlinear regimes of fast reconnection. However, most of the natural space and astrophysical systems motivating reconnection research have even larger sizes. This huge separation of scales motivates the scaling problem of reconnection research: How can we extrapolate the knowledge gained from studies of relatively small- and intermediate-size systems, both laboratory and numerical, to the universe?

6.3.5 Reconnection in high-β plasmas

Generally speaking, magnetic reconnection takes place when the magnetic energy of a plasma system significantly exceeds the plasma's kinetic or thermal energy ($W_B \gg W_p$), since excess magnetic energy tends to be converted to kinetic energy through reconnection. Reconnection phenomena in solar flares, the magnetosphere, and low-β magnetically confined fusion plasmas are such examples. However, magnetic reconnection can occur in the wide regime of plasma beta values $\beta = 10^{-6}$–10^6. Even in high-β systems ($W_p \gg W_B$), magnetic reconnection often plays an important role. Nilson et al. (2006) presented measurements of magnetic reconnection and strong electron heating at the electron current layer in plasmas created by injecting two closely focused heater beams on a planar foil target. The two plasmas typically collide and stagnate, and for laser spot separations of about seven focal spot sizes, two very distinct, highly collimated jets were observed. The azimuthal magnetic fields that are generated by a Biermann mechanism from a $\nabla T_e \times \nabla n_e$ mechanism around each laser-heated column (the Biermann effect) were also observed using proton deflectometry. The experimentally observed plasma flows and magnetic field convection, high electron temperatures, and jet formation are consistent with magnetic reconnection processes that happen in the current layer between the two columns. The acceleration and heating mechanisms are unique in this experiment, with Thomson scattering measurements in the reconnection layer showing high electron temperatures of 1.7 keV.

The experiment used the Vulcan laser at the Rutherford Appleton Laboratory, UK. The experiment is shown in figure 6.12. Two laser beams, with wavelengths of 1.054 μm, irradiated either an aluminum or a gold target foil. A square pulse of 1 ns duration was used with an average energy of 200 J per heating beam. The targets were 3–5 mm foils of 20–100 μm thickness. Each beam was focused using $f = 10$ optics to a focal spot diameter of 30–50 μm, giving an incident laser intensity of 1×10^{15} W/cm^2. The two laser beams were aligned with varying on-target separations.

The observed electron temperature in the reconnection layer is surprisingly high since there is no direct laser heating in this region. If the midplane interaction consisted of a standard collision, it would be ion heating that dominated the interaction. The electrons gain energy subsequently through electron–ion equilibration. However, the timescale over which this occurs is many nanoseconds. We measure high electron temperatures that cannot be reconciled to electron–ion equilibration alone or compressional heating (this is not an efficient plasma compression geometry). This therefore indicates another energy source. Such a source would need to supply energy to the electrons at a sufficiently high rate that it was not simply radiated away. The only source available with sufficient free energy that could be provided at such a rate is the conversion of magnetic energy into plasma thermal energy through a reconnection mechanism.

Recently, using a similar technique, a very-high-aspect-ratio current sheet was created during forced merging of high-β plasmas generated by laser beam irradiation

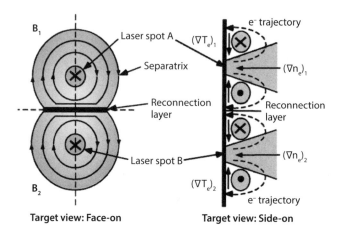

Figure 6.12. The target geometry and field configurations. The two plasmas collide and reconnect, and for laser spot separations greater than about seven focal spot sizes, two very distinct, highly collimated jets were observed. [From Nilson et al. (2006).]

(Fox et al., 2020). Their special experimental geometry allowed a unique reconstruction of the magnetic field, which showed a reconnection current sheet that thinned down to a half-width close to the electron gyroscale ($\beta \sim 100$) with an aspect ratio of 100. Despite the large aspect ratio, the current sheet in this large-β plasma was stable for a long period. A large fraction of the magnetic flux was observed to be reconnected, suggesting that fast reconnection is caused by the electron pressure tensor effects. This experiment will be discussed again in chapter 13 in the context of dynamo action generated by a pressure tensor.

6.4   MAIN FACILITIES DEDICATED TO RECONNECTION STUDY

There have been more than a dozen experimental devices built to study magnetic reconnection. Table 6.1 summarizes major facilities dedicated to the study of the physics of magnetic reconnection and most of them have been described in this chapter.

Chapter Seven

Recent observations of magnetic reconnection in solar and astrophysical plasmas

7.1 FEATURES OF MAGNETIC RECONNECTION IN SOLAR FLARE ERUPTIONS

In this section we present the typical features of magnetic reconnection observed in solar flares, together with the historical development of physical explanations and more recent interpretations. Typical eruptions of solar flares or CMEs (coronal mass ejections) eject 10^{11}–10^{13} kg of mass and 10^{29}–10^{33} erg of energy as the kinetic energy of bulk plasma and radiated photon energy (Chen, 2017). While the energy source was presumed by many researchers to be a magnetic field at the solar surface, the physical mechanisms of solar eruption and CMEs have been a major problem in solar research. The concept of coronal energy buildup and storage can be traced to Carrington (1859) who recorded that the observed sunspots showed no signs of change before, during, or after the white-light flare and that the brightening appeared to occur above the sunspot group. He concluded that the energy of the flare was stored in the corona. The form of stored energy was not specified at that time. With the later discovery that sunspots contained strong magnetic fields, it was conjectured that the energy of eruption was in the magnetic field. Giovanelli (1946) proposed an original idea that the merging or annihilation of oppositely directed magnetic field lines—magnetic reconnection—could release energy stored around the sunspot magnetic field to power solar flares, as mentioned in the introductory chapter of this book. The concepts of merging of sunspots or current loops of opposite magnetic polarity were proposed to explain flare energy release. Giovanelli considered that magnetic energy generated by electric current was caused by the differential rotation of the footpoints. However, modern observations find that their movement and their speed of rotation, or the rate of current injection, are too slow to explain accumulation and a sudden release of magnetic energy.

It is difficult to directly measure magnetic field-line evolution in the solar corona. There is a basis that magnetic field-line reconnection is happening all the time, through time evolution of photographs taken through optical, ultraviolet, and soft-X-ray emissions. Owing to the nature of plasma, which tends to move together with magnetic field lines (flux freezing), magnetic field configuration can be deduced from these pictures. Thus there is ample evidence for a change of magnetic field topology in the solar

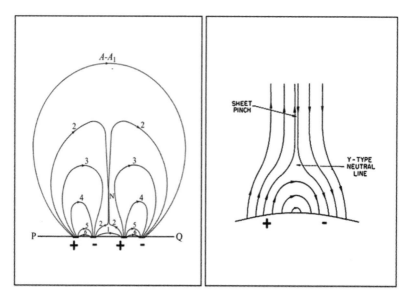

Figure 7.1. Schematic description of the pre-flare magnetic field in early flare models. (Left) Quadrupolar model of Sweet (1958). (Right) Bipolar model of Carmichael (1964), Sturrock (1966), and Sturrock and Smith (1968). The drawings are from Sweet (1958) and Sturrock and Smith (1968). The plus and minus signs are added to mark the flux polarities.

atmosphere. To describe magnetic reconnection in solar flares, let us start with the Sweet model discussed in chapter 3 (figure 3.2). The drawing on the left in figure 7.1 depicts the well-known pre-flare magnetic field configuration considered in Sweet's model. It shows the configuration that Sweet described would be formed by the magnetic fields of two identical dipolar, active flare arcs merging close together. In this model, by pushing the two arches against each other, a current sheet is formed at their interface between the two vertical legs of oppositely directed fields. Figure 7.1(left) portrays the two opposite magnetic fields being compressed and sandwiching the current sheet interface. In this quadrupole pre-flare field configuration, the field lines are closed and connected to the solar surface. Because all the reconnecting field lines are tied to the solar surface, the heating and particle energization by reconnection should make "four" flare ribbons through heat and particle transport, forming a pair of ribbons at the feet of the reconnected field lines on each of the quadrupole foot lines. Recent observations, however, have found that major flares typically have only two ribbons instead of four, indicating that Sweet's idea is not supported by most observations. Moore et al. (2011) concluded that the early flare model sketched on the left-hand side of figure 7.1 was not exactly correct.

So another natural idea was to have only a single bipolar active region, as proposed by Carmichael (1964) and Sturrock (1966). This would build an inverted-Y open field configuration having a vertical current sheet at the interface between the two opposite-polarity opened legs of the magnetic arch, as shown in the right-hand schematic of figure 7.1. The buildup is supposed to continue until there is enough free energy in the stretched opened field for a major flare. Then reconnection takes place in the current sheet, invoking a topology change. The reconnection of the open field lines makes both (1) closed field lines that are released downward, building a growing flare arcade that heats a flare ribbon at each foot and (2) open-field U loops that are released upward, propelling a large chunk of plasma (plasmoid) into the outer corona, generating a solar wind. In the model proposed by Sturrock, the ejected plasmoid was more like what would later become known as a CME.

7.2 DEVELOPMENT OF THE STANDARD SOLAR FLARE MODEL AND MAGNETIC RECONNECTION

Satellite observations indicate that the core magnetic field contains free magnetic energy. When the system becomes globally unstable or is deformed by an external force, the magnetic energy can be transferred to the plasma through transition to another equilibrium state of lower magnetic energy. This situation can induce a magnetic reorganization and cause the core field or even the entire arcade to erupt. Major flares typically occur in strong-field regions where the magnetic pressure is much greater than the pressure of the surrounding plasma.

7.2.1 Pre-eruption configurations

In the chromosphere and low altitude coronae, the pressure of the plasma is negligible in the overall force balance of the pre-eruption field. Figure 7.2 shows a pre-eruption arcade standing alone in the absence of any other appreciable overarching field rooted around it. In this case, the core field is partly restrained from expanding upward by the downward pull of its own field lines. While the eruptions are indeed explosive releases of magnetic energy, no current sheet of a size comparable to that of the overall pre-eruption field configuration was visible. Thus, it was considered by many that the reconnection current sheet was too small compared to the overall size of the erupting field. One question is how the buildup of dipole field is made until there is enough free energy in it for a major flare. The schematics in figure 7.2 depict in three dimensions a popular scenario among solar physicists (Moore et al., 2011) for the eruptions that produce major flares and major CMEs. In most solar eruptions that produce major flares, whether or not the eruption produces a CME, the field that erupts is a bipolar arcade that is basically of the form sketched in the right-hand panel of figure 7.1. When one observes the two-dimensional structure of the pre-eruption arcade from the top, the field lines over the arcade's neutral line (on the surface) appear to be strongly sheared and twisted so that they roughly trace the neutral line and typically have an overall sigmoid form, like that shown in figure 7.2.

The twist in the arcade's field changes with distance from nearly parallel to perpendicular to the central neutral line or the inversion line. The pre-eruption core field often holds a filament of chromospheric-temperature plasma. The filament is suspended along nearly horizontal field lines that thread the filament and are strands of what

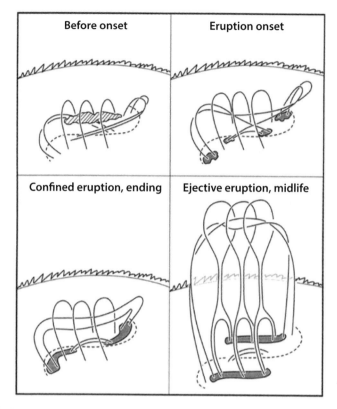

Figure 7.2. Sketches for the evolution of a 3D configuration of the driving magnetic field in major flare/CME eruptions. Only a few representative field lines (solid curves) are drawn. The dashed curve is the magnetic polarity inversion line on the photosurface (when it is observed from the top). The cross-hatched elongated feature in the first panel is a "filament" of high-density cool plasma suspended in the sheared core field. The filament plasma is often seen in the erupting core-field flux rope, but is not shown in the other panels for clarity of the field configuration. The shaded areas are flare ribbons at the feet of reconnection-heated field lines. The ragged arc in the background is the chromospheric limb of the sun. [From Moore et al. (2011).]

basically amounts to a flux rope (not shown) that runs the length of the sigmoid. The core field is directed nearly parallel to the neutral line, while the outer field-line configuration appears nearly orthogonal to the neutral line. A careful study of field lines at typical pre-eruption field sites through images of chromospheric and coronal images shows that the arcade evidently has a magnetic equilibrium just like we observe in low-β toroidal fusion plasmas, that of the RFP (reversed field pinch) as described in chapter 2, figure 2.8. These configurations generally satisfy near force-free equilibrium conditions, $j \times B \sim 0$ (section 3.3). In the middle of the arcade, the field direction near the plasma core is nearly parallel to the neutral line (central line) and the field direction

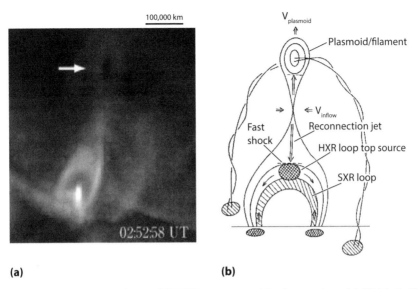

Figure 7.3. Comparison of the (CSHKP) concept with observation. (a) Yohkoh SXT image of a flare. (b) Schematics of the CSHKP configuration for the event [from Shibata et al. (1995)].

changes only gradually toward perpendicular to the neutral line with some distance from the neutral line: a typical profile of a toroidal pinch plasma such as an RFP and spheromak (Yamada et al., 2010). At the early phase of the arcade, there is no evidence of a specific localized current sheet visible inside the arcade. Most of the current corresponding to the arcade's near force-free equilibrium appears to be smoothly distributed throughout the core field (Heyvaerts et al., 1977; Moore et al., 2011).

7.2.2   Standard model of coronal mass ejection

As mentioned in chapter 2, the current prevailing model is attributed to the concept of Carmichael, expanded by Sturrock, Hirayama, and Kopp and Pneuman. Much subsequent work has been made to refine this concept, which is often referred to as the Carmichael–Sturrock–Hirayama–Kopp–Pneuman (CSHKP) model or the "standard model" of solar flares (Carmichael, 1964; Sturrock, 1966; Hirayama, 1974; Kopp and Pneuman, 1976). In this model, the initial structure is typically specified as an arcade. This model is based on an upward surge of the low-β plasma configuration described in figure 7.2. Similar structures have been considered for prominence formation with the sheared arcades producing flux ropes via reconnection. Within this framework, prominences are often thought of as current sheets or flux ropes within arcades, supported by the repulsive Lorent force from the image currents in the photosphere against gravity and the downward magnetic tension of the arcade field, and if the tension is reduced or removed, prominences erupt upward because of this repulsive force. Figure 7.3 shows a qualitative comparison of the CSHKP concept with observation: (a) is a Yohkoh image

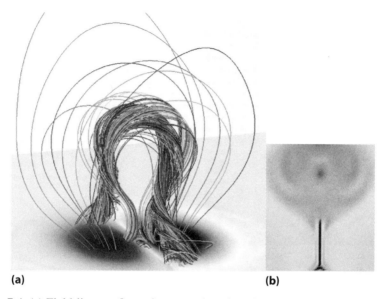

(a)   **(b)**

Figure 7.4. (a) Field-line configuration at a relaxation phase at $t = 450$ (simulation unit time). The configuration experiences a major disruption. (b) Out-of-plane current density (the darker the stronger) in a vertical cut in the central plane during the reconnection phase at $t = 450$. [From Amari et al. (1999).]

of a flare in soft X-rays (Masuda et al., 1994), compared with (b) which is a pictorial interpretation based on the CSHKP concept.

A variety of simulations have been carried out, based on the CSHKP concept, to study the eruption of arcades caused by specified footpoint motions. They showed that such motions can result in a response to the emergence of erupting flux through the photosphere after a slow buildup of magnetic energy. For an example, a three-dimensional resistive MHD (magnetohydrodynamic) simulation was carried out to describe the topological evolution from an initial current-free arcade to a current-carrying flux rope due to a prescribed footpoint motion at the solar surface (Amari et al., 1999). In this model, a current sheet is formed below the flux rope, and reconnection occurs according to specified resistivity, leading to the reduction of the overlying field and allowing the evolving flux rope to rise. An S-shaped configuration, as observed in soft-X-ray sigmoid structures, cannot stay in equilibrium and a considerable amount of magnetic energy is released during its disruption.

As shown in figure 7.4, the magnetic topology of the configuration reveals several interesting features through a set of field lines that form an island through which runs the twisted flux rope, and field lines defining a vertical finite-size reconnecting region, and a set of arcades close to the boundary that reform as the reconnection goes on (and that may represent post-flare loops in this model). Reconnection occurs (within their numerical resolution) in the vertical current sheet. Unlike in two-dimensional axisymmetric studies, in this three-dimensional model, the twisted flux rope is an essential ingredient at the origin of the disruption. In their three-dimensional study, a majority

(90%) of the energy accumulated during the shearing plus emerging flux phase was observed to be released. Although the detailed magnetic field structures may be different from model to model, the basic process described by CSHKP models is basically the same "quasi-static" slow build up of magnetic energy in the corona and a sudden "cut" of the tethering field by magnetic reconnection to release a large component of plasma, a toroidal-shaped "plasmoid." While it is difficult to identify and measure the precise features of the magnetic reconnection layer in the solar atmosphere, two-dimensional neutral-sheet-like patterns have sometimes been recognized through soft-X-ray satellite images of solar flares, but their exact magnetic profiles are unknown. It appears that a reconnection process is underway throughout this area based on the sequence of high-energy-electron flux hitting the footpoints at the photosphere. In order to describe the observed fast reconnection rate (Yokoyama et al., 2001) by the Sweet–Parker model, and to explain the apparent fast flux transfer, strongly enhanced plasma resistivity had to be employed.

7.3 BREAKOUT MODEL WITH A MULTIPOLAR MAGNETIC CONFIGURATION

Another concept often popular among solar flare physicists is a breakout model. In the modeled processes of breakout, a forced reconnection is driven by a complex multipolar magnetic configuration, including an energized low-lying coronal structure with a sheared arcade of coronal loops and a high-lying arcade of opposite polarity, with a magnetic null point sandwiched in between. The null and the surrounding separatrix-fan structure are locations in favor of current concentration and magnetic reconnection. Figure 7.5 shows a multipolar geometry constructed to illustrate the key components of a magnetic-breakout setup. According to both two- and three-dimensional numerical simulations of magnetic breakout, reconnection can occur when the lower-lying sheared arcade starts to rise, thereby compressing the current layer around the null to be thin enough (Antiochos et al., 1999, 2002).

As a result of the reconnection, flux is transferred from the restraining arcade to neighboring, nonrestraining side lobes and effectively reduces the restraining force acting on the sheared arcade, leading to a successful eruption. In the right-hand panel of figure 7.5, a representative soft-X-ray picture of such an event in a CME is shown and it can be described by a breakout model.

7.3.1 Observed chromosphere jets and magnetic reconnection

The Hinode satellite (Shibata et al., 2007) reported the ubiquitous presence of chromospheric jets at inverted-Y-shaped exhausts outside sunspots. As presented in chapter 2, they are typically 2,000 to 5,000 km long and 150 to 300 km wide, and their velocity is 10 to 20 km/s. It was suggested that magnetic reconnection similar to that seen in the corona occurs at a smaller spatial scale throughout the chromosphere and that the heating of the solar chromosphere and corona may be related to small-scale reconnection. Coronal X-ray jets are a subclass of the solar eruptions that can occur when a

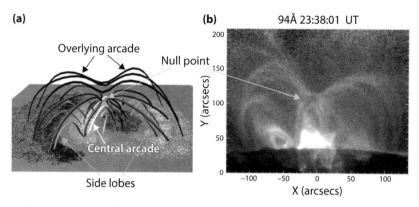

Figure 7.5. (a) 3D image for a typical breakout configuration and (b) a pre-eruption AIA (Atmospheric Imaging Assembly) image of our event. Panel (a) shows key breakout components, including the central arcade, side lobes, and overlying arcade. The straight arrow points to the magnetic null. Field lines undergoing reconnection across the null are cusp-like. [Adapted from Chen et al. (2016b) and a color figure and more detailed description are found in https://aasnova.org/2016/05/18/reconnection-on-the-sun/.]

small bipolar magnetic arcade, a miniature active region, emerges in the feet of a high-reaching unipolar field, such as the ambient field in a coronal hole or in one leg of a large-scale ($\sim 100,000$ km) coronal loop. In two different ways, both involving reconnection with the ambient field, magnetic energy can be explosively released to eject plasma up into the corona along the ambient field. If the eruption makes the ejected plasma hot enough to be seen in coronal X-ray movies, such as from the Hinode X-Ray Telescope (XRT), the eruption is observed as an X-ray jet (Shibata et al., 1992). If the ejected plasma is heated only to subcoronal temperatures, the ejection cannot be seen in coronal X-ray images but can be seen in extreme UV (EUV) images and/or in visible-wavelength chromospheric (e.g., H_α) images and is then called an EUV or chromospheric jet, or macrospicule (Shibata et al., 2007; see also figures 2.2 and 2.3). Many of the X-ray jets that occur in the sun's polar coronal holes are blowout jets, but two-thirds are of another type, the type that was first recognized and, until recently, was generally thought to be the only type (Shibata et al., 1992). In contrast to blowout X-ray jets, in these most common X-ray jets—standard X-ray jets—the interior of the emerging arcade remains quasi-static and stable as the jet is produced (Shibata, 2016).

7.3.2   A recent simulation with a new interpretation

Magnetically driven eruptions on the sun's surface, stellar-scale CMEs, and the small-scale plasma jets shown in figure 7.5 with X-rays, have frequently been observed to involve the ejection of the highly stressed magnetic flux of a so-called "filament." Theoretically, these two phenomena have been thought to arise from different mechanisms: CMEs from an ideal (nondissipative) process and coronal jets from a resistive process involving magnetic reconnection. However, Wyper et al. (2017) have recently proposed

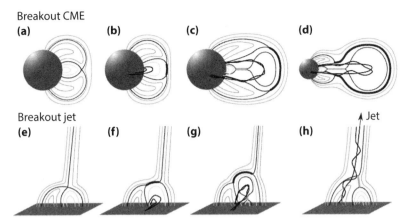

Figure 7.6. Schematic of the breakout process: (a)–(d) in CMEs and (e)–(h) in breakout jets. Time increases from left to right. The gray spheres in (a)–(d) represent the entire solar surface and the gray sheets in (e)–(h) a local patch of the surface. Dark gray lines show separatrices (or quasi-separatrices) dividing different regions of the magnetic field. The black twisting lines show the core of the filament (or flux rope), the thick black lines in the bottom figures (f) and (g) denote the current sheet, and the thick black lines in (b), (c), (d) represent the flare current sheet. [Adapted from Wyper et al. (2017).]

from new observations that all coronal jets are driven by plasma ejection with the formation of a current sheet, just like large mass ejections (Masuda et al., 1994). Based on the breakout model, it was suggested that the two phenomena may have physically similar origins and hence that a single breakout mechanism may explain them; that is, either CMEs arise from reconnection, or jets arise from a breakup of plasma due to an ideal instability. They reported simulation results of a coronal jet driven by filament ejection, whereby a region of highly sheared magnetic field near the solar surface becomes unstable and erupts. The results show that magnetic reconnection causes the energy release via "magnetic breakout," or a close relationship between plasma filament ejection and reconnection. They concluded that if CMEs and jets are of the same physical origin, despite being in different spatial scales, then magnetic reconnection (rather than an ideal process) must also underlie CMEs, and that magnetic breakout is a universal model for solar eruptions.

Wyper et al. (2017) demonstrated through their three-dimensional simulation work how the main stages of their breakout jet compare with their large-scale breakout models for CMEs. The critical physical difference between the two configurations is the role of expansion. They thought that in CMEs it was difficult, with either observations or simulations, to separate the effects of the ideal and resistive processes and to determine definitively the mechanism of eruption. In contrast, in coronal jets the background field is strong and the ideal expansion was thought to be suppressed. The closed-field region in their simulation simply lacks sufficient free energy to push aside the surrounding field and open ideally. We note that in figure 7.6 the closed-field region expands only marginally throughout the event. In their simulation, the sharp increase in kinetic energy and the explosive jet are due solely to a rapid magnetic reconnection caused by ideal motion of unstable plasma inside the half-sphere-shaped plasma. However, as we will learn in the next section, a global MHD "tilt" mode would also induce such reconnection inside the separatrix, a conclusion that can be tested by the recently launched Parker Solar Probe.

7.3.3 A new Princeton model

A well-accepted reconnection scenario for standard X-ray jets was first suggested by Shibata et al. (1992) from the anemone structure and development of X-ray jets observed in coronal X-ray snapshots and movies from Yohkoh. Recently, Wyper et al. (2017) carried out a numerical simulation based on a similar concept. A new model has been recently developed by the author of this book. This new model is schematically shown by the sequence of drawings in figure 7.7, the first of which shows the pre-eruption field configuration for a half-sphere-shaped force-free plasma (for a spheromak, see Yamada et al., 1981) confined in the background of a coronal hole field. A polarity inversion line appears around the magnetic axis (toroidal null line) of the emerging arcade. Then, as the inner sphere is elongated upward due to the slow emergence of flux from the solar surface, this spheromak-like configuration becomes unstable to a tilt mode (Latham et al., 2021). As the inner configuration tilts, a current sheet develops between the emerging closed field lines and the ambient open field, and magnetic reconnection takes place near the top null point. The reconnection site is characterized by a tilted X. In the final schematic, the reconnected field lines expand upward, and a plasma jet is generated with bursts.

A current sheet develops between the emerging closed field and the opposing ambient open field where reconnection takes place. During this phase, a burst of reconnection occurs and produces the jet. In figure 7.7(b), a bipolar arcade is emerging into an ambient unipolar field of opposite polarity. Due to a tilt motion, the direction of the internal field lines becomes opposite to the outside ambient field impacted against it, a current sheet is formed at their interface, and reconnection occurs. When the current sheet grows to be extensive enough, the global self-organization (reconnection) begins as depicted. The reconnection releases plasma upward, sending the reconnected field lines that make the spire of the jet upward, and releases downward closed reconnected field lines that build an arcade of hot "flare" loops over the neutral line at the top of the original arcade.

7.3.3.1 *Recent numerical simulation work*

A spheromak is one of the well-explored fusion configurations which, in the low plasma pressure limit, is basically a simply connected force-free magnetic vortex. In a simply connected closed plasma system, the turbulent plasma tends to relax into a spheromak configuration (Taylor, 1986) with an equilibrium given by $\nabla \times \boldsymbol{B} = \mu \boldsymbol{B}$, where $\mu = (\boldsymbol{j} \cdot \boldsymbol{B})/B^2$ is a spatial constant along and across field lines (Yamada et al., 1981). Experimentally, a spheromak can be formed by inductive coils or a magnetized coaxial

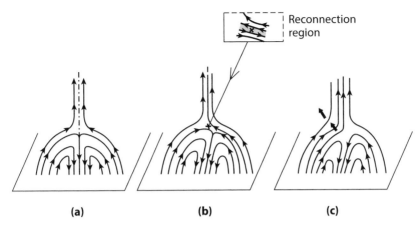

(a)   **(b)**   **(c)**

Figure 7.7. Princeton model: A sequence of schematics illustrating the field configuration, tilting, reconnection, and plasma ejection for X-ray jets. (a) The field configuration is a half-sphere-shaped force-free plasma confined in the background of a coronal hole field. A polarity inversion line appears around the circular null line of the magnetic axis of the emerging arcade. (b) As the inner sphere configuration is extended and tilts, an unstable current sheet develops between the emerging closed field and the ambient open field (where reconnection takes place), where oppositely directed field lines merge. The reconnection site is shown in an expanded insert. (c) As the reconnected field lines expand upward, a plasma jet is generated with bursts. [Yamada et al. (2020)]

gun. This spheromak is unstable against a tilt mode, as it tends to flip to align its magnetic moment with the background magnetic field. This mode is stabilized when the spheromak is partially embedded onto the solar surface just like the magnetic topology, similar to that of the solar anemone shown in figure 7.7.

Recently, using a three-dimensional numerical simulation code (HYM code, Belova et al., 2000), Latham et al. (2021) have investigated the time evolution of the characteristics of the tilt mode for a line-tied spheromak, relevant to coronal jet formation. The vector plots in figure 7.8 show plasma flow velocities, and the contours show velocity magnitude normalized to Alfvén velocity. The lines display the magnetic field depicted by (B_R, B_Z) components in the plane. In this simulation, plasma jets are formed with up to the Alfvén velocity as an inner magnetic configuration tilts against the surrounding background field. This work is a good example of a past approach in laboratory experiments contributing to the understanding of astrophysical phenomena.

7.4   MAGNETIC RECONNECTION OCCURS IMPULSIVELY

There is a commonality among all these eruptive reconnection phenomena. Namely, these global reconnection (magnetic self-organization) phenomena almost always occur unsteadily or impulsively. In laboratory fusion plasmas, the magnetosphere, and solar flares, reconnection is seen to occur suddenly with very fast semi-Alfvénic speed after

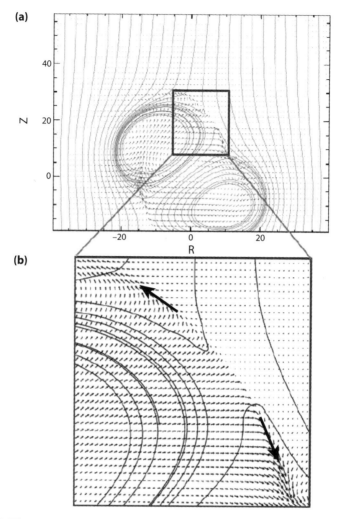

Figure 7.8. The vector plots show flow velocity and the contours show velocity magnitude normalized to Alfvén velocity. Black lines display magnetic field depicted by (B_R, B_Z) components in the plane. (a) The top reconnection region is marked with a rectangular box, and the X-point is located near the center of the box. (b) The zoomed-in plot shows the reconnection region in higher resolution. The outflow velocity reaches as high as V_A. [From Latham et al. (2021).]

a long flux buildup phase. Fast reconnection leads to an impulsive global topology change or global magnetic self-organization phenomena. Impulsive global reconnection takes place after a gradual change of equilibrium that builds up sufficient free energy to induce motion of the plasma or topological changes. It is conjectured that this phenomenon can occur in active solar arcade flares. A slow change of equilibrium drives a plasma to an unstable regime and then drives a global magnetic self-organization, leading to eruptions. In solar flares, reconnection sites are often identified with hard-X-ray emissions near the top of solar flare arcades during CMEs or near the top of a half-sphere-shaped magnetic configuration (such as in a coronal hole). The reconnection speed was almost always measured to be much faster than the Sweet–Parker rate.

We can hypothesize that global magnetic self-organization phenomena in both tokamak sawtooth crashes and solar flares share a common process. After a long period of flux buildup, a plasma configuration becomes unstable and a sudden change of magnetic configuration occurs, resulting in a newly connected global plasma. This leads to a large electric field along the magnetic field lines and acceleration of electrons to superthermal energy. Indeed, in reconnection events in both solar flares and tokamak sawteeth, we observe a significant amount of high-energy (runaway) electrons and X-rays. A careful comparative study of tokamak sawteeth and RFP relaxation events should illuminate this important magnetic self-organization phenomenon. We will discuss magnetic self-organization in tokamak and RFP plasmas in chapter 9.

7.4.1 Limitation of MHD models for reconnection in solar flares

It should be noted here that the dynamics of the reconnecting current sheets can be kinetic in nature even on the sun's surface. In the recent MHD simulations for solar flares, including those with adaptive mesh refinement, the reconnection mechanisms and speed are due to prescribed resistivity or numerical diffusion. A fundamental gap between such simulations and real situations may be that the known physical length scale of collisionless reconnection can be as small as the ion skin depth or ion gyroradius of plasma (10–100 m), which are smaller by many orders of magnitude than the scales of observed eruptions (length scale of 10^4–10^5 km) or even apparent current sheets (10^2–10^4 km). A timescale of tens of minutes in eruptive processes in the corona compares to a kinetic timescale of $1/\omega_{ci} \sim 10^{-7}$–$10^{-3}$ s. Dynamics on the even smaller electron inertial length can also be important. However, since it is almost impossible to verify the local kinetic physics of the reconnection region in the solar flare, using an effective resistivity is the only way at the moment to describe fast magnetic reconnection in solar coronae. Also, we note that while the detailed physical mechanisms are described in two dimensions in the above-mentioned MHD models, the actual reconnection dynamics are expected to be different in a three-dimensional description.

7.4.2 Initial results from a new solar satellite, the Parker Solar Probe

In August 2018, NASA's Parker Solar Probe (PSP) launched into space and has become the closest ever spacecraft to the sun. With new advanced instruments to measure the environment around the spacecraft, the PSP has completed the first few of 24 planned passes through never-before-explored parts of the sun's atmosphere and extended coronae. Many new data sets are being analyzed now. It was found that the majority of the ejected plasma jets encountered around the closest approach to the sun were Alfvénic structures associated with bursty radial jets. Although it has been suggested that these

Alfvénic structures may be driven by reconnection in the lower corona, the majority of these current sheets appeared not necessarily tied to local reconnection sites. The initial observations from the PSP from the first perihelion pass revealed large numbers of positive spikes in radial velocity that may be revealing the underlying structure of the development of the solar wind. The spikes were bursty and had an Alfvénic structure with a dominant directionality of the perturbed magnetic field, as reported at the 2019 and 2020 American Geophysical Union meetings.

On the other hand, the spacecraft can be magnetically connected to a coronal hole during the closest approach to the sun, and one possible source of these spikes is magnetic reconnection between the open field lines in the coronal hole and an adjacent region of closed flux. Collaborating with particle-in-cell simulations, more detailed analysis will be carried out.

Most recently, Phan et al. (2020) have analyzed the data from the PSP, which shows the presence of heliospheric current sheets (HCSs). They are represented by a wide spectrum of switchbacks of the radial components of solar magnetic fields emitted from the sun, which should often imply a toroidal current on site: $\nabla B_R = \mu J_\theta$. The data show that the occurrence of reconnection in the inner HCS near the sun (29.5–107 R_S, where R_S is the radius of the sun) appears to be much more frequent than at 1 AU near the earth, and reconnection seems to be very active at close distances from the sun. Five out of six well-defined full HCS crossings displayed accelerated flows, as well as field-line-topology signatures that are consistent with active reconnection in the HCS (Szabo et al., 2020). It remains to be seen how the inner heliosphere reconnection properties change as PSP gets even closer to the sun.

7.5   A MODEL OF MAGNETIC RECONNECTION IN THE CRAB NEBULA

In an intriguing observation reported in *Science News* (Tavani et al., 2011) in 2011, the Fermi Gamma-Ray Space Telescope recorded repeated outbursts of high-energy gamma rays emanating from the center of the Crab Nebula over a period of several days (see figure 7.9). The observed energy range of up to 10^{12} eV challenges the conventional theory for particle acceleration through shocks and the gamma-ray emissions from astrophysical plasmas that result.

Magnetic reconnection is often considered to play a significant role in the acceleration of charged particles to ultra-relativistic energies on astrophysical scales. If it happens, the reconnection electric field should be induced by the flows of merging plasmas and would be of order $v \times B$, where v is a typical global MHD plasma flow velocity and B is a typical magnetic field strength. The maximum energy of particles accelerated by this field should be expressed as vBL, where L is a characteristic length scale of the reconnection layer. Figure 7.10 presents the maximum electron energy observed in laboratory plasma experiments and in space astrophysical plasmas with respect to vBL. The scaling works for reconnection experiments in which (i) $E = vBL = (1\text{–}5 \times 10^4 \text{ m/s})(0.03 \text{ T})(0.2 \text{ m}) = 60\text{–}300$ volts for the solar corona, (ii) $E = vBL = (10^5 \text{ m/s})(0.01 \text{ T})(10^6 \text{ m}) = 10^9$ volts is consistent with observed 1 GeV

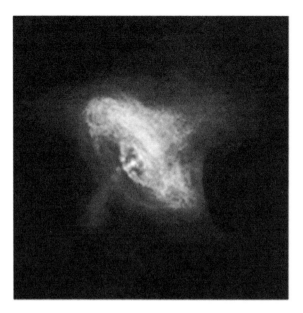

Figure 7.9. X-ray photo of the central core (toroidal shape) of the Crab Nebula. The inner central core is made of electron–positron plasmas. [From NASA/CXC/SAO. Copyright: http://chandra.harvard.edu/photo/image_use.html.]

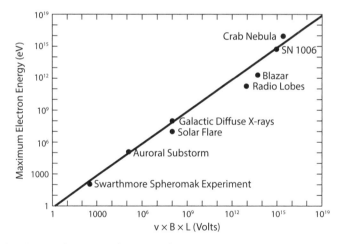

Figure 7.10. The maximum particle energies observed or estimated for lab and astrophysical systems plotted versus vBL, where v, B, and L are typical values of velocity, magnetic field, and linear size of the plasma system, respectively. The original figure was presented at the US–Japan Workshop on Magnetic Reconnection in Tokyo (2000) by M. Makishima. Laboratory experimental data are from MRX, SSX, and TS-3.

gamma rays, and (iii) $E \approx 10^{16}$ volts $\approx vBL = (10^8 \text{ m/s})(10^{-7} \text{ T})(10^{15} \text{ m})$ for measured emission from high-energy electrons in the electron–positron (pair) plasma near the core of the Crab Nebula.

It was proposed that charged particles can be accelerated near the center of the Crab Nebula (shown in figure 7.9) by magnetic reconnection when magnetic fields are violently rearranged. Nonthermal electron–positron (pair) plasmas are known to be abundant in many astrophysical environments from pulsars to quasars, as well as in our own galaxy and in supernovae remnants such as seen in the Crab Nebula. Electron–positron pair production has been the subject of many studies in astrophysics, as well as in theoretical, computational, and experimental physics (Hibschman and Arons, 2001). There is a plausible possibility that magnetic reconnection in the periphery of the toroidal core plasma in the Crab Nebula could generate a sufficient electric field to produce high-energy particles (Uzdensky et al., 2011). We can consider the possibility that magnetic reconnection is responsible for acceleration of plasma particles at a reconnection layer on the outer edge of the Crab Nebula. Magnetic field lines, like Parker spirals stretched from the sun, can form a reconnection layer as a result of merging of oppositely directed field lines. While reconnection layers can be formed between different solar winds, they should extend in the radial direction and should accelerate particles to the 10^{16} eV range, as schematically shown in figure 7.11.

Let us describe the situation in more detail. A pulsar wind nebula is sitting in the core of the Crab Nebula remnant (Rees and Gunn, 1974; Kennel and Coroniti, 1984a,b; Uzdensky, 2011; Uzdensky et al., 2011). The plasma there is not a remnant as it is of the original supernova explosion but is continuously being supplied and energized by the pulsar—the neutron star that is left over after the original massive star has exploded. By rapidly rotating (~ 30 ms in the Crab Nebula), the strongly magnetized pulsar is considered to continuously produce highly relativistic (electron–positron) pair plasma that forms an outgoing, ultra-relativistic, magnetized pair-plasma wind from the pulsar (similar in many ways to the solar wind, with its Parker-spiral magnetic field and a ballerina-skirt equatorial current sheet; see figure 7.11). This winds flows out to about 0.1–0.3 parsec, where it goes through a termination shock (similar to the heliospheric termination shock). This shock marks the inner boundary of the pulsar wind nebula (Uzdensky et al., 2011). Magnetic reconnection is proposed there as shown in figure 7.11 (Yamada, 2012). The pair flow beyond the shock slows down to subsonic speeds, heats up, and accelerates nonthermal particles. The result is the toroidal pulsar wind nebula (the core of the photo shown in figure 7.9), a cloud of ultra-relativistically hot pair plasma (which came from the pulsar), turbulent and magnetized, and about a parsec across. It shines brightly across the whole electromagnetic spectrum, mostly via synchrotron radiation. This nebula, namely its magnetic field, expels regular, colder electron–ion plasma from its volume, and does not let the electron–ion plasma from the surrounding region enter (Uzdensky, 2011).

By carefully comparing results from simulations, space satellites, and laboratory experiments, we may be able to achieve real progress in determining the role of magnetic reconnection physics in distant astrophysical objects. For this goal, we look for a new generation of magnetic reconnection experiments on a large-scale device that will enable investigation of magnetic reconnection in collision-free plasmas and

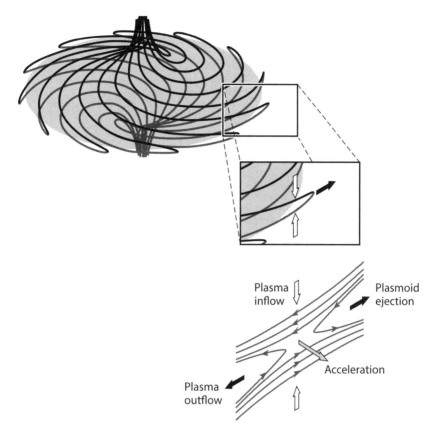

Figure 7.11. Schematic views of a magnetic reconnection layer considered in an accretion disk of the Crab Nebula. Multiple neutral sheets extend radially toward the outside, generating high-energy particles by the reconnection electric field.

simultaneously help us to understand the physics of energy conversion from magnetic to particles.

7.6 NOTES ON FAST COLLISIONLESS RECONNECTION IN SPACE ASTROPHYSICAL PLASMAS

Notable progress has been made in the last several decades in understanding the dynamics of magnetic reconnection in solar flares as well as in astrophysical plasmas, based on MHD theory. However, in solar flare plasmas, the ion skin depth is in the range 1–100 m, with the mean free path of electrons ranging between 100 and 1,000 km. This satisfies the conditions for a collision-free plasma. In an idealized collision-free reconnection layer such as seen in the magnetosphere, the driven reconnection layer becomes comparable to the ion skin depth (c/ω_{pi}) and ions become demagnetized while electrons remain magnetized. The relative flows of electrons against ions in the reconnection plane can generate a strong $j_e \times B$ force due to the Hall effect. This Hall field, which contributes to the enhanced electric field perpendicular to the reconnection plane, can be considered responsible for speeding up the rate of reconnection, providing a partial answer to the very important question of why reconnection occurs so fast. We will address this problem in the next chapter.

Let us take another example: magnetic energy release in the solar corona. The large number of electrons accelerated or heated in solar flares, far more than could be supplied by the neighborhood of a single X-point, is evidence for local–global plasma coupling (see chapter 9). We have seen that many of these problems could be addressed by a mechanism that broadens the reconnection region, which is small in both MHD and two-fluid models. Thus we are looking for reconnection dynamics beyond the idealized, classical, single quasi-stationary X-line geometry, and exploring more realistic, highly dynamic reconnection regimes characteristic of large systems, such as those found in most space and astrophysical environments. These complex regimes feature multiple X-lines, plasmoid and flux rope formation due to secondary instabilities, and the self-consistent emergence of turbulence and accompanying coherent structures under a variety of plasma conditions. This theme has emerged in the last several years as the new paradigm of how magnetic reconnection really happens in natural plasmas. Understanding the generation and influence of secondary reconnection instabilities is one of the important goals in modern reconnection research, as described in chapter 14.

Chapter Eight

Recent observations of magnetic reconnection in space astrophysical plasmas

8.1   MAGNETIC RECONNECTION LAYER IN THE MAGNETOSPHERE

Since the magnetosphere is a very important test ground to investigate the characteristics of two-fluid reconnection, let us study here more in detail this feature of the earth's magnetospheric reconnection. Magnetospheres are magnetic structures discovered during the space age by satellite-borne instruments that made possible physical measurements in distant regions previously not accessible. The earth's magnetosphere was the first one discovered: it laps around the earth with a radius of 60,000–120,000 km, which is 10–20 times the earth radius. It then became clear that magnetospheres are ubiquitous in space. In our solar system, the sun's coronal atmosphere is typically 10^6 K or 100 eV, and dynamic, so it expands into space. The expanding solar coronal atmosphere is called solar wind and consists mostly of hydrogen ($\sim 95\%$ H$^+$) and helium ($\sim 5\%$ He^{2+}) ions and an equal number of electrons in a plasma state. Since this plasma is a good electrical conductor and magnetic fields decay slowly in conductive plasma, it was immediately verified that the solar wind carries solar magnetic fields with it into space. Space is therefore filled with magnetized plasma. All of the planets immersed in the solar winds are interacting with them all the time. The electromagnetic interaction induces large-scale currents and forms magnetic cavities around magnetized planets. These cavities are called magnetospheres. Except for Mars and Venus, which do not have intrinsic magnetic fields, the planets in our solar system all have magnetospheres. This chapter will focus on magnetic reconnection in the earth's magnetosphere, which has all of the elements to characterize a planetary magnetosphere.

Magnetic reconnection in the magnetosphere, particularly at the dayside magnetopause, is especially important for a number of reasons. One is that it occurs naturally at one of the closest space places to the earth. It can be measured in situ using satellites with the greatest detail of any naturally occurring reconnection phenomenon. In collaboration with dedicated laboratory studies of reconnection and its numerical simulations, significant progress has been made in understanding collisionless magnetic reconnection, which is observed at the magnetopause and in the magnetotail. Another reason magnetospheric reconnection is important is because of its direct impact on the earth through its role in space weather. Without reconnection, the transfer of material and energy from interplanetary space to the magnetosphere would be minimal, and

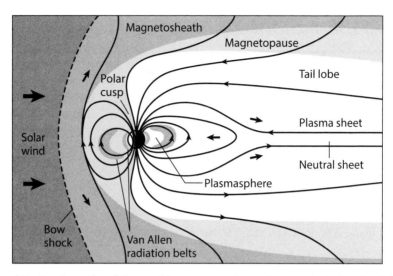

Figure 8.1. A schematic of the earth's magnetosphere showing both the dayside and the nightside. Reconnection layers are formed at the magnetopause and the tail lobes, which are occupied by the magnetic field lines that connect to the two polar regions of the earth. The energy and plasma in the outer terrestrial region are intermittently released into the inner magnetosphere during magnetic substorms. [Figure from https://ase.tufts.edu/cosmos/print_images.asp?id=29.]

the earth's magnetic field would largely shield the earth from the charged particles in space. Actually, reconnection happens here and changes the connectivity of magnetic field lines, allowing the transfer of material and energy from interplanetary space to the magnetosphere. This transfer of energy and material is known to cause many problems for people on the earth, including power outages; satellite failures impacting global positioning system (GPS), communications through cell phones, and navigation; increased drag on satellites; harm to astronauts; and negative impacts on airline communication for planes. The characteristics of reconnection at the dayside, therefore, play a crucial part in determining how strongly interplanetary space couples to the magnetosphere. Therefore, predicting space weather to the level where it can be mitigated requires a thorough understanding of reconnection at the dayside magnetopause (Cassak and Fuselier, 2016).

The magnetosphere is generated when solar wind meets the dipole field of the earth. Figure 8.1 is a schematic of the magnetosphere, showing it on both the dayside and nightside. The magnetopause is the boundary that separates the geomagnetic field and the solar-wind plasma, as described in chapter 3. On the dayside magnetopause, pressure balance is maintained between the incoming solar winds and the earth's magnetic field. Ampère's law applied across the boundary shows that currents have to flow (out of the plane of the page) in the boundary sheet shown in the figure. On the nightside of the magnetosphere, there is a magnetotail in which the magnetic field lines stretch behind the earth in a direction away from the sun. A current sheet is also formed between the tail lobes and is occupied by the magnetic field lines that connect to the two polar regions of the earth. It is considered that solar-wind plasma and energy are injected into the magnetosphere and then released from it through magnetic reconnection processes.

8.2 OBSERVATIONAL STUDIES OF MAGNETIC RECONNECTION IN THE MAGNETOSPHERE WITH THE AID OF NUMERICAL SIMULATIONS

Observations of reconnection by satellites in the magnetosphere had already begun in the 1970s. The twin-spacecraft International Sun–Earth Explorer (ISEE) mission carried plasma instrumentation with the necessary time and space resolutions. Initial in situ measurements were made (Russell and Elphic, 1979; Paschmann et al., 1979; Sonnerup, 1981) which showed evidence of magnetic reconnection at the magnetopause and also indicated that it could be localized in space and intermittent in time in the form of so-called flux transfer events or FTEs. Similarly, early evidence for the occurrence of reconnection in the magnetotail was provided by in situ measurements in the plasma sheet by the Vela and IMP satellites (Hones Jr, 1977).

For direct comparison with observations, it is convenient to use a boundary normal (LMN) coordinate system, as defined in the upper left of figure 8.2 (Fuselier and Lewis, 2011). The maximum change in the magnetic field occurs across the current sheet in the N-direction, which is normal to the current sheet. The M-direction is along

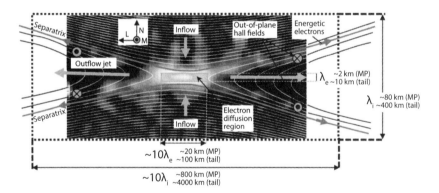

Figure 8.2. Structure of the two-scale diffusion region of the reconnection layer in the magnetotail. The features are from a 2D guide field simulation that shows the parallel electric field, which is high in the electron diffusion region (Fuselier and Lewis, 2011). There is some amount of guide field in this situation. Estimated thicknesses and widths of electron and ion diffusion regions are shown in skin depths and kilometers for typical conditions at the magnetopause (MP) and in the magnetotail (tail). Plasma and magnetic field inflows symmetrically from the top and the bottom and are accelerated out of the two sides. Other features include out-of-plane Hall fields and energetic electrons flowing along the separatrices. This figure should be self-similar to Figure 5.6.

the reconnection line, often called the X-line. Plasma flows into the reconnection region at equal rates from the top and bottom (parallel and antiparallel to the N-direction) and "jets" of plasma on reconnected field lines flow out of the left and right sides (parallel and antiparallel to the L-direction).

It is very difficult to directly measure the spatial profiles of the reconnection region in the magnetosphere because the number of measuring locations by satellites is limited. However, thanks to the recent advanced data analysis of cluster satellites in coordination with numerical simulations, more precise features of the reconnection layer have been measured, as we will see in later sections. Surprisingly, the measured characteristics are consistent with the two-dimensional profile calculated by recent PIC (particle-in-cell) simulations.

8.2.1   Observation of Hall effects in the magnetosphere

As already described in detail in chapter 5, the two-fluid dynamics of reconnection, which are illustrated in figure 5.6, predict the presence of strong Hall effects due to the decoupling of electron flow from ion flow. In a collisionless neutral sheet such as is seen in the magnetosphere, this situation is equivalent to magnetized electrons pulling magnetic field lines in the direction of the electron current, thus generating an out-of-plane quadrupole field. In the magnetopause, the two-fluid physics of magnetic field reconnection was analyzed in terms of the ion diffusion region of scale size $c/\omega_{pi} \sim$ 100 km in the subsolar magnetopause (Mozer et al., 2002). The detailed data are shown in figure 5.11 in which a sizable Hall field $B_Y \sim 0.55 B_{X0}$ was measured (where B_{X0} is the reconnection magnetic field). More recently, a reconnection electric field was carefully studied to deduce a reconnection speed when a satellite flew through the ion diffusion region (Mozer and Retinò, 2007).

About the same time as the aforementioned Mozer et al. (2002) report, evidence of Hall effects was reported through the detection of a quadrupole B_y field after analyzing the data from Geotail skimming along the dayside magnetopause in January 1997 (Deng and Matsumoto, 2001). Another report of the out-of-plane quadrupole field was made from the data from the Wind satellite, which traveled in the reconnection sheet of the magnetotail (Øieroset et al., 2001).

Since we have already described the observation of two-fluid reconnection at the magnetopause in chapter 5, let us look at the case for a magnetic reconnection layer in the magnetotail in this chapter. Øieroset et al. (2001) reported on a direct encounter with an ion diffusion region in the magnetotail by the Wind spacecraft. Figure 8.3(a), (b) illustrate the diffusion region and the Wind trajectory schematically, which was deduced by comparing magnetic and plasma parameters with two-dimensional numerical simulation. The right-hand side presents measured ion density, ion velocity, and magnetic field components obtained as the spacecraft traversed the diffusion region. The Cartesian coordinate system is commonly used in space data analysis as shown in figure 8.3: X is toward the earth, Y is the direction along the reconnection line (normal to the page), and Z is the direction normal to the (reconnecting) magnetotail current sheet. In figure 8.3 (right), the trajectory across the entire diffusion region is recognized

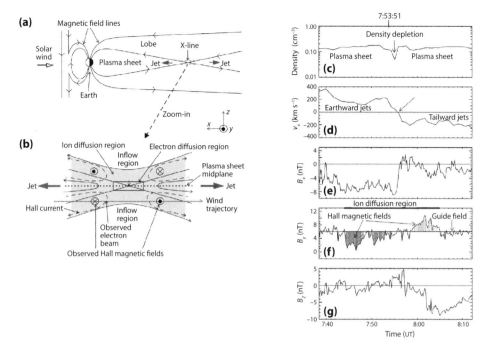

Figure 8.3. In situ detection of collisionless reconnection in the earth's magnetotail. (a) Reconnection geometry in the magnetotail where bidirectional reconnection jets were observed. (b) Blow-up of the reconnection region showing the spacecraft trajectory through the ion diffusion region, including the observed Hall magnetic fields. (Right column: top to bottom) Observations of an ion diffusion region in the tail: plasma density, earthward–tailward flow velocity, three components of the magnetic field. As the spacecraft passed through the ion diffusion region, out-of-plane Hall fields were observed. This magnetotail reconnection event had a substantial guide field. [Adapted from Øieroset et al. (2001).]

in the reversal of V_X as the spacecraft first encounters an earthward flowing jet and then a tailward flowing jet. The Hall magnetic field structure is highlighted in the B_Y panel, first in the reduction of the average B_Y field (noted as out-of-plane components) and then in its increase. The Hall fields are not symmetric about zero because of the presence of a guide field. Because this measurement involved only a single spacecraft, the only way to estimate the size of the diffusion region is to determine the duration of the encounter and to assume that the reconnection X-line was stationary with the spacecraft moving through it at its known velocity of ~ 1 km/s. Using this assumption, Øieroset et al. (2001) estimated that the ion diffusion region width was $1,300$ km or about 2 ion skin depths. Since the actual width depends on the motion of the reconnection line relative to the spacecraft during the encounter, there is some ambiguity in this measurement.

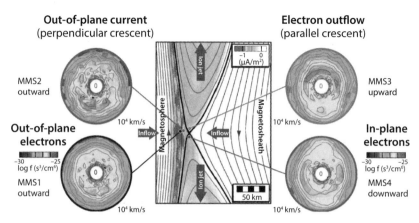

Figure 8.4. See Color Plate 4. Electron distributions measured by MMS. With fast particle measurements, the MMS mission has measured how electron dynamics evolve in the reconnection layer. The data in the circles show the distribution of electrons (in color), with velocities from 0 to 10^7 m/s, carrying current out of the page on the left-hand side of the X-line and then flowing upward and downward to the exhaust regions along the reconnected magnetic field on the right-hand side. The most intense fluxes are red and the least intense are blue. The plot in the center shows magnetic field lines and out-of-plane currents derived from a 2D numerical plasma simulation using the parameters observed by MMS. [From Burch et al. (2016b).]

8.3   ELECTRON-SCALE MEASUREMENTS OF THE RECONNECTION LAYER IN THE MAGNETOPAUSE

To further advance understanding of magnetic reconnection up to the electron scale in space, NASA developed and launched the Magnetospheric Multiscale Satellite (MMS) mission in March 2015. Flying in a tightly controlled formation of four satellites, the MMS spacecrafts collected data from the magnetopause, where the IMF (interplanetary magnetic field) and the earth dipole field reconnect. They successfully made detailed measurements of the plasma properties and the electric and magnetic fields in the reconnection region. Because the reconnection dissipation region at the magnetopause is considered to be so thin (a few kilometers) and moves rapidly back and forth across the spacecraft (10 to 100 km/s), high-resolution measurements were needed to capture the microphysics of reconnection. They made critical measurements to provide three-dimensional electron distributions with a timescale of 30 ms, or a space resolution of less than 3 km, as shown in figure 8.4.

In order to make major progress in the study of the two-fluid physics of collisionless reconnection in space, they extended the measurements to the electron scale and made accurate three-dimensional measurements of electric and magnetic fields together. Electron-scale kinetic physics in the region around the reconnection site (or the X-line), where field-line breaking and reconnection occur, has not previously been investigated experimentally in space, owing to insufficiently detailed measurements.

Our knowledge of this region at the electron scale mainly came from computer simulations (Hesse et al., 2014; Chen et al., 2016a) and laboratory experiments (Yamada et al., 2014). The higher resolution of MMS measurements in both time and space relative to previous missions provided major advantages for investigating the cause of reconnection by resolving the structures and dynamics within the X-line region.

Here, let us look in detail at their magnetopause measurements, made during the first science phase of the mission, following the content of their first report (Burch et al., 2016b). For this phase, the region of interest was identified at geocentric radial distances of $9\text{–}12R_E$ (70,000 km from the earth), during which all instruments were operated at their fastest cadence. The four spacecraft were maintained in a tetrahedral formation, with separations variable between 160 and 10 km in the initial phase. By 14 December 2015, the spacecraft had crossed the magnetopause more than 2,000 times. On the basis of the detection of plasma jetting and heating within the magnetopause current sheets, they concluded that at least half of the crossings encountered magnetic reconnection regions. Most crossings occurred in the reconnection exhaust downstream of the X-line, but a few of them passed very close to the X-line. The data for one of these events (16 October 2015, 13:07 UT) are presented here as an example of the electron-scale measurements of the reconnection diffusion/dissipation region around an X-line.

The data set obtained by MMS generated the following important progress in the understanding the electron dynamics: (i) By cross-correlation of the sequence of signals from 48 diagnostic components from the four satellites, they could make a plausible estimate of their flight path with the aid of two-dimensional numerical simulations. This worked surprising well and was an important achievement! (ii) Three-axis electric and magnetic field measurements with accurate cross-calibrations allowed measurement of spatial gradients and time variations. (iii) All-sky plasma electron and ion velocity–space distributions were measured with a time resolution of 30 ms for electrons and 150 ms for ions. The mission was conducted in two phases, the first (2015–16) targeting the dayside outer boundary of the earth's magnetosphere (the magnetopause) and the second (2017–18) targeting the geomagnetic tail, for which the apogee is raised to $25R_E \sim 150,000$ km.

Figure 8.4 (see Color Plate 4) presents typical electron distribution functions measured by MMS in the reconnection layer during a typical magnetopause crossing. With high-resolution fast particle measurements, MMS measured how the electron dynamics shown by the electron velocity distribution profile evolve in the reconnection layer. The data in the circles show electrons with velocities from 0 to 10^7 m/s carrying current out of the page on the left-hand side of the X-line and then flowing upward and downward along the reconnected magnetic field on the right-hand side. The most intense fluxes are red and the least intense are blue (figure 8.4 in Color Plate 4). The plot in the center shows expected magnetic field lines and out-of-plane currents derived from a two-dimensional numerical simulation using the parameters observed by MMS. The observed features agree remarkably well with the electron flow vector data obtained in MRX (Yamada et al., 2014). We will discuss this comparison in chapter 12 in more detail.

Figure 8.5 (Color Plate 4) shows data from the MMS2 satellite during two encounters with the magnetopause over a period of about two minutes. The assumed

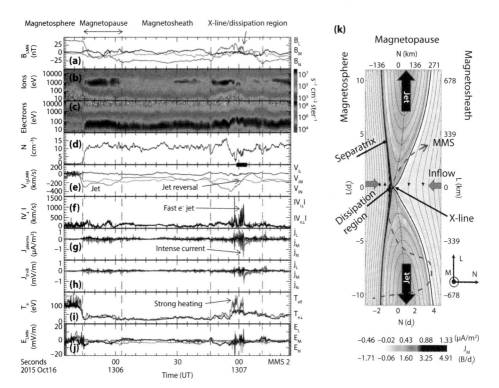

Figure 8.5. See Color Plate 4. Summary data for two magnetopause crossings of MMS2 on 16 October 2015. The crossings are shown by the vertical blue dashed lines. Panel data include (a) magnetic field vector components, (b) energy–time spectrogram of ion energy flux, (c) energy–time spectrogram of electron energy flux, (d) total plasma density, (e) ion flow velocity vectors, (f) magnitudes of electron and ion convection velocities, (g) current computed from velocity moments of ions and electrons, (h) current computed from $\nabla \times \boldsymbol{B}$, (i) parallel and perpendicular (to $\boldsymbol{B}$) electron temperatures, and (j) electric field vectors. In the very-low-density region to the left of the first vertical blue dashed line, spacecraft charging effects on plasma moment calculations may affect the data. Diagram (k) to the right is the result of a numerical plasma simulation using parameters from the magnetopause crossing centered on 13:05:52 UT. Spatial coordinates in the diagram are shown both in kilometers and in ion diffusion lengths L (d_i). The color scale indicates J_M, the current density (current in the out-of-page plane). [From Burch et al. (2016b).]

magnetopause crossing times are denoted by two pairs of vertical blue dashed lines. The diagram on the right-hand side shows the "estimated" two-dimensional structure of a magnetopause in which asymmetric reconnection is occurring, taken from a two-dimensional numerical plasma simulation and shown for the observed magnetosheath and magnetospheric conditions of the entire MMS magnetopause crossing at 13:05:30 UT. The diagram shows the northward magnetic field on the magnetosphere (earth side) side of the boundary and the southward magnetic field on the magnetosheath side. The shear angle between the magnetosphere and magnetosheath magnetic fields was very large (~ 170 degrees), implying a crossing with a low guide field or almost antiparallel.

The converging plasma flows carry the two nearly oppositely directed magnetic field domains toward each other. An X-line directed normal to the plane of the diagram denotes the small region in the reconnection plane where the field lines are expected to reconnect, and this X-line can extend from hundreds to thousands of kilometers in the east–west direction (Phan et al., 2000), which is why a large number of exhaust regions are typically crossed by spacecraft near the magnetopause. Another reason why reconnection events are routinely observed is the presence of the exhaust jets (red arrows) flowing northward and southward from the X-line and the nearby dissipation region (or diffusion region). Although the results of reconnection are readily (Burch et al., 2016b) observed with measurements at the fluid and ion scales, field-line reconnection is considered to occur within the electron dissipation region. The color scale in the plasma simulation result in figure 8.5 shows the plasma current normal to the plane of the picture (J_M), which is nearly all due to fast-moving electrons generated by the reconnection process. Strong J_M values (shown in green) are highly localized at the dissipation region and X-line.

The approximate path of the MMS tetrahedron, based on the plasma and field measurements, is shown by a blue dashed curve. According to the space physics convention (see figure 8.2), boundary-normal coordinates (L, M, N) are used, with L being the reconnection field-line direction, N the normal to the boundary and away from the earth, and M normal to the L–N plane (east- or westward). These directions were determined from a minimum variance analysis of the magnetic field data during the flight time between 13:05:40 and 13:06:09 UT. Because the velocity of the magnetopause is approximately 100 times the spacecraft velocities, the MMS path shown is produced entirely by the motion of the magnetopause along L and N. For the magnetopause crossing centered at 13:07 UT, it is concluded that the spacecraft traversed both exhaust jet regions and passed through the dissipation region between them. To confirm this, flow reversal of the ion jet flow was observed near 13:07 UT when the reconnecting magnetic field (B_L) component was close to zero, suggesting that the spacecraft was in close proximity to the X-line. The red highlight bar at the top of figure 8.5(e) shows this reversal. Another important indicator for a dissipation region is the enhancement of E_M (the out-of-plane reconnection electric field), which is shown by the green trace in figure 8.5(j). The size of the E_M bursts at more than $10\,\text{mV/m}$ is substantially (10 times) larger than the correction due to X-line motion. There are also strong E_N components bracketing 13:07 UT, which are electric fields pointing outward and normal to the magnetopause, as predicted by simulations.

Figure 8.6 shows the 4 seconds of data marked with the red bar in figure 8.5(e) of MMS2 data near the X-line. Figure 8.6(a) shows that a deep magnetic field minimum occurred just after 13:07:02.4 UT and (b) shows a strong plasma current (j_M) starting at 13:07:02.1 UT (on the magnetosphere side of the X-line) and extending through the minimum magnetic field. Panel (c) shows vector electric fields. Inside the j_M current layer, the E_N component, which points outward from the magnetopause

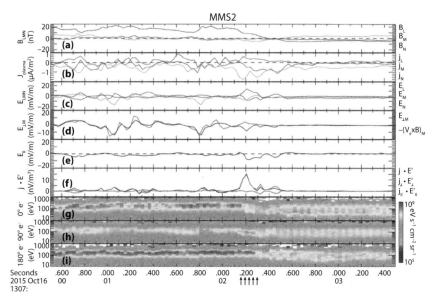

Figure 8.6. See Color Plate 5. MMS2 plasma and field data on 16 October 2015. (a) Magnetic field vector components, (b) currents from plasma measurements, (c) electric field vector, (d) comparison of the M-component of E and $V_e \times B_M$, (e) $E_\parallel$, (f) $j \cdot E'$. Plot (f) shows clearly that the reconnection dissipation is caused by the strong j_M current multiplied by the E_M electric field, which are perpendicular to B. The quantity $j \cdot E' > 0$ provides a form of signature for a reconnection dissipation region as shown in MRX laboratory experiments (Yamada et al., 2014, 2018). Panels (g) to (i) are energy–time spectrograms of electrons moving parallel, perpendicular, and antiparallel to the local magnetic field direction, respectively. [From Burch et al. (2016b).]

as described above, is the strongest. It is also noteworthy that E_M (the reconnection electric field) is negative, as is the j_M current. Panel (d) shows a comparison between E_M and $V_e \times B_M$. There is excellent agreement except near the dissipation region. *This verifies that in the electron frame just outside the X-region, electrons are frozen to magnetic field lines, namely flux freezing is working for the electron fluid, as discussed in chapter 5: $E_M \approx (V_e \times B)_M$.*

Figure 8.6(e) shows the electric field component parallel to B, which is strongest in the region of the J_M plasma current. Panel (f) shows $j \cdot E'$, where $E' = E + V_e \times B$, along with its parallel and perpendicular components. The plot in panel (f) shows clearly that the reconnection dissipation is caused by the strong j_M current and E_M electric field, which are perpendicular to B in the dissipation region as B is dominated by B_L in that region. As reconnection is expected to convert magnetic energy to heat and the kinetic energy of electrons, the observation that $j \cdot E' > 0$ provides a form of signature for a reconnection dissipation region as shown in the MRX laboratory experiment (Yamada et al., 2014, 2018). Shown in panels (g) to (i) are energy–time spectrograms of electrons moving parallel, perpendicular, and antiparallel to the local magnetic field direction, respectively. In the region of dissipation (13:07:02.15 to 13:07:02.29 UT), the parallel fluxes shift to lower energies, the perpendicular fluxes rise in intensity and shift to lower energies, and the antiparallel fluxes remain at high energies. All of the fluxes drop to lower magnetosheath levels after exiting the dissipation region.

8.3.1   Summary of messages from MMS data in the magnetopause

Using measurements of plasma currents and reconnection electric fields, MMS data have shown that the energy dissipation expressed by $j_e \cdot E'$ spikes up in the vicinity of the X-line, as predicted for the dissipative nature of reconnection. Their results are remarkably in agreement with earlier, laboratory results obtained on MRX (Yamada et al., 2014, 2015). Another important finding from MMS is that the flux freezing principle, $E_M \approx (V_e \times B)_M$, holds well for the electrons outside the narrow electron dissipation region.

Electron distribution functions obtained by MMS were found to contain characteristic crescent-shaped features in velocity space, indicating a drift of electrons in the out-of-plane direction. This was predicted by the results from MRX (see figure 5.7(d)) and is evidence for the demagnetization and acceleration of electrons by an intense electric field near the reconnection X-line. MMS has directly determined the current density based on measured ion and electron velocities, which allowed the resolution of currents and associated dissipation on electron scales. These scales are smaller than the spacecraft separation distances and hence smaller than currents that can be determined by $\nabla \times B$. The X-line regime exhibits a region of electron demagnetization and acceleration (by both E_N and E_M), which results in intense J_M current that is carried by the crescent-shaped electron distributions. Kinetic simulations, as well as laboratory results (Yamada et al., 2014), had predicted some elements of the crescent distributions near the X-line.

The MMS measurements have led to discoveries about the evolution of electron acceleration in the dissipation region, as well as the escape of energized electrons away from the X-line into the downstream exhaust region. The latter was detected by the two MMS spacecraft located on opposite sides of the X-line, as shown in figure 8.4. The observed structures of the normal electric field and electron dynamics near the X-line by the four spacecraft are highly variable spatially and/or temporally, even on electron scales. Among the implications of this MMS observation is the confirmation that the X-line region is important not only for the initiation of reconnection (breaking of the electron frozen-in condition), but also for electron acceleration and energization, leading to much stronger electron heating and acceleration than seen in the downstream exhaust. The details of the electron distribution functions, which show the rapid transition (within 30 ms) of the perpendicular crescent distributions to parallel crescents, provide experimental evidence for the opening up of reconnected magnetic field lines while also demonstrating that it is the electron dynamics that drive reconnection of magnetic field lines. Through the actual three-dimensional analysis using advanced space technology and laboratory experiments, MMS and MRX together have demonstrated that the essence of three-dimensional magnetic reconnection occurring in

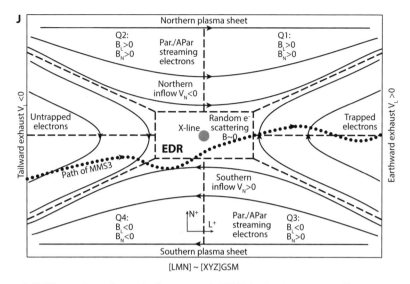

Figure 8.7. Illustration of a typical symmetric EDR in the *LMN* coordinate system and the expected properties in the various quadrants (Q1 to Q4), together with the inferred relative path of the MMS satellites as the X-line retreated tailward. [From Torbert et al. (2018).]

space plasmas can be well described by two-dimensional analysis made by numerical simulations and laboratory experiments. Further cross-disciplinary study should lead to a more accurate picture of magnetic reconnection.

8.4   ELECTRON-SCALE DYNAMICS OF THE SYMMETRIC RECONNECTION LAYER IN THE MAGNETOTAIL

The MMS mission focuses on investigating two reconnection regions that exist around the earth: the dayside magnetopause and the nightside magnetotail, which are in quite different plasma parameter regimes. As described in the previous section, during the first phase of MMS (2015–16), its four spacecraft investigated the reconnection region in the dayside magnetopause, where the inflow conditions are highly asymmetric, with different plasma and magnetic pressures in the two inflow regions. In its second phase (2017–18), MMS explored the kinetic processes of reconnection (Torbert et al., 2018) in the earth's magnetotail, where the inflow conditions are nearly symmetric, and the available magnetic energy per particle is more than an order of magnitude higher than on the dayside. While the plasma density is much smaller, we note that the amount of magnetic energy per particle in the magnetotail is comparable to that of the solar corona, where magnetic reconnection also occurs.

On 11 July 2017, MMS encountered an EDR (electron diffusion region) where it detected tailward-directed ion and electron jets, followed by earthward-directed jets, spanning a reversal of essentially the north–south component of the magnetotail

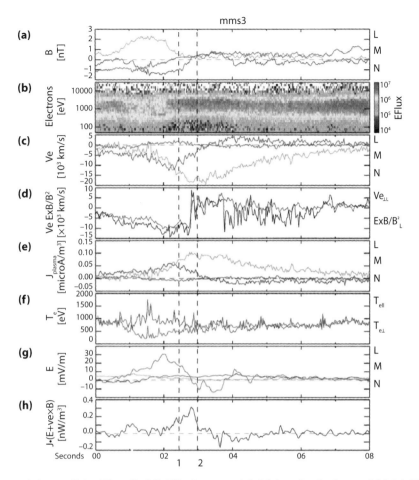

Figure 8.8 (see Color Plate 5). MMS3 plasma and field data for the interval 22:34:00 to 22:34:08 UT on 11 July 2017 are shown. (a) Magnetic field components in the LMN coordinate system. (b) Electron omnidirectional spectrogram, with minimum energy set at 50 eV (to avoid the lower-energy spacecraft photoelectrons). (c) Electron bulk velocity. (d) L-components of V_e and $E \times B/B^2$. (e) Current from plasma measurements. (f) Parallel and perpendicular electron temperatures $T_{e\parallel}$ and $T_{e\perp}$. (g) Electric field. (h) $J \cdot E'$. (Here, electric and magnetic omnidirectional frequency spectrograms are omitted from the original paper.) In this MMS data, current density is expressed by J, while lowercase j is used for most of this book unless specifically defined. [Adapted from Torbert et al. (2018).]

magnetic field B_N (figures 8.7 and 8.8) in an intense current sheet (large out-of-plane electron velocity V_{eM}). The LMN coordinate system was again used to orient the data to the usual two-dimensional view of the magnetic field near a reconnection X-line (figure 8.7(J)), with L in the outflow direction, M along the X-line, and N normal to the current sheet (north-south direction). The out-of-plane guide field ratio B_M/B_L for this event is estimated to be small (≤ 0.1). The spacecraft were in the magnetotail at a radial distance from the earth of 22 earth radii. Four-spacecraft timings of the flow and field reversals indicate that the structure moved away from the earth with velocity $V_L \sim 170$ km/s. These are signatures of a tailward retreat of the reconnection X-line past the spacecraft, as indicated by the MMS path in figure 8.7.

Figure 8.8 (Color Plate 5) shows MMS3 field data for the interval 22:34:00 to 22:34:08 UT on 11 July 2017. Except for a brief excursion to the edge of the inflow region, seen in a small perturbation in magnetic field components (beginning at 22:34:00 UT) due to a flapping of the current sheet, the spacecraft stayed close to the neutral sheet ($B_L = 0$ plane), indicated by small values of B_L (~ 0 to $2nT$) during the flow and field reversal. These observations are consistent with crossing both ion and electron diffusion regions, an identification that is supported by the profiles of the ion and electron flows: V_{eM} peaked at $\sim 15{,}000$ km/s, within an order of magnitude of the electron Alfvén speed, which is approximately 20,000 to 25,000 km/s. Starting from the X-line (at the V_{eL} and B reversal location) and going left and right in figure 8.8, the electron perpendicular outflow speed V_{eL} increased and greatly exceeded the ion speed. While the ion outflow speed V_{iL} increased with increasing distance from the X-line, V_{eL} reached a peak ($\sim 7{,}000$ km/s), before slowing and approaching the ion flow speed at $\sim$ 22:33:50 before, and $\sim$ 22:34:20 after, the X-line. Thus, at the ends of the ion diffusion region, the ion and electron outflow velocities are expected to match. The end of the EDR, on the other hand, marked by the departure of V_{eL} from $\mathbf{E} \times \mathbf{B}/B^2$, was confined to a much smaller distance from the X-line, where the electron density reached a symmetric minimum of 0.03 cm^{-3}. This means that the diffusion is confined to the EDR of 70–100 km in length (the electron inertial length $d_e \sim 30$ km).

The above results are again consistent with the experimental results obtained in MRX (Yamada et al., 2014) for symmetric reconnection, as well as recent simulations. Although $J_e \cdot E'_{\perp}$ is mostly positive throughout the period shown in figure 8.8, there are some regions with negative values, indicating that the electrons are transferring energy to the electromagnetic field, as also seen in simulations (Zenitani et al., 2011). A value of $E_M \sim 1$ to 2 mV/m (as seen in figure 8.8(g)) is notably smaller than E_N and fluctuation components. The electrons were eventually turned toward the L- (or exhaust) direction by B_N as they exited the EDR, forming the electron jet seen in figure 8.8(c) on either side of the X-line. The electron temperature profile in (f) shows strong anisotropy from 1.0 to 2.8 s, due to magnetic-field-aligned electrons in the inflow region. During the EDR crossing, there was a small rise (of a few hundred electron volts) in parallel and perpendicular temperatures (the parallel and perpendicular pressures divided by n_e), unlike the case of asymmetric reconnection (Burch et al., 2016b), implying that a substantial fraction of the energy conversion went into the strong electron flows in the M- (out-of-reconnection-plane) and L- (outflowing) directions.

In summary, the MMS observations of the magnetotail reconnection EDR show that it differs from the dayside because it involves symmetric inflow. The aspect ratio of the diffusion region (0.1 to 0.2) determined by MMS is consistent with two-dimensional simulations of collisionless reconnection carried out earlier. It is rather surprising to observe that MMS observations of electron dynamics in the diffusion region match predictions that are nearly laminar, as if we can assume that the effects of three-dimensional turbulence and associated fluctuations on the electron dynamics are small. It was found that electrons can be accelerated by the reconnection electric field to higher values than the case of the magnetopause, possibly as a consequence of longer confinement in the symmetric magnetic structure or due to lower-density conditions. Taken together with MMS observations at the magnetopause, it is remarkable to note that the two-dimensional concept of describing magnetic reconnection works so well.

Chapter Nine

Magnetic self-organization phenomena in plasmas and global magnetic reconnection

9.1 MAGNETIC SELF-ORGANIZATION IN PLASMAS

In the preceding chapters of this book, the physical mechanisms of magnetic reconnection inside or in the vicinity of the reconnection layer have been the primary topic. Plasma dynamics in these narrow diffusion regions are extremely important in determining the rate at which magnetic fields reconnect and magnetic energy is released. However, the dynamics of the magnetic reconnection layer do not alone decide the features of global magnetic reconnection. Neither does global reconnection necessarily start from these spatially localized reconnection layers. Rather, magnetic reconnection takes place when there is a condition for a magnetic field configuration to release the excessive energy stored in it, requiring a change of topology. Magnetic reconnection thus invokes magnetic self-organization on global scales, often generating reconnection layers in some parts.

When an external force is applied to a plasma system, the magnetic configuration (often slowly) changes to a new equilibrium while plasma parameters gradually adjust. When this new state becomes unstable, the plasma reorganizes itself suddenly into a new MHD (magnetohydrodynamic) equilibrium state of lower magnetic energy. This process often drives magnetic reconnection by changing magnetic topology and often forming current sheets or reconnection layers. The excess magnetic energy is converted to plasma kinetic energy or thermal energy, and the plasma magnetically relaxes or self-organizes to a lower magnetic-energy state. In this chapter, let us take a side step to discuss global aspects of magnetic reconnection, focusing primarily on results from laboratory fusion plasmas in which the global conditions are well defined and the global and local plasma parameters are quantitatively monitored. This global view of magnetic reconnection phenomena, magnetic self-organization, can be applied to almost all other natural cases, such as magnetospheric reconnection phenomena, solar flares, and some magnetic relaxation phenomena in distant astrophysical plasmas.

As shown in figure 9.1, let us consider a case in which an external energy source is applied to a globally stable plasma. The plasma often becomes unstable and goes through self-organization processes such as magnetic reconnection, dynamo, and magnetic instabilities, and then the plasma system settles into a new equilibrium state. Typically, the self-organization processes are driven by the free energy contained in

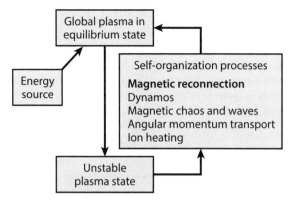

Figure 9.1. Evolution of magnetic self-organization. When an external perturbation or energy source is applied to a globally stable plasma, the plasma often becomes unstable and, by going through self-organization processes such as magnetic reconnection, dynamo, and wave turbulence, the unstable state settles into a new equilibrium state.

gradients of large-scale quantities of the plasma system. These processes involve various nonlinear phenomena that alter, or feed back on, the large-scale structure. The processes can transport, generate, and restructure magnetic field, plasma flow, thermal energy, and other quantities settling into a new equilibrium state. These process often occur impulsively or violently. Dynamo is a plasma process opposite to magnetic reconnection. When there is abundant kinetic energy in a plasma with respect to magnetic energy, magnetic fields are considered to be generated through a converse self-organization process, a dynamo mechanism in plasma. The other magnetic instabilities and ion heating processes involve magnetic self-organization of plasma.

It is generally believed that when the magnetic energy of a global MHD equilibrium state is larger than the kinetic or thermal energy (low-β plasma), it can be lowered by a reorganization of plasma topology, in which process magnetic reconnection takes place. Reconnection will stop when it no longer lowers the total magnetic energy. It is recognized that global reconnection (magnetic self-organization) phenomena almost always occur fast or unsteadily. Fast reconnection generally leads to an impulsive global topology change or global magnetic self-organization phenomenon. Impulsive reconnection typically occurs after the gradual evolution of the global equilibrium builds up sufficient free energy in order to induce the motion of plasma or a topological change.

Solar eruptions are driven by the sudden release of magnetic energy stored in the solar corona. In many cases, it is believed that the magnetic energy that drives the CME (coronal mass ejection) is slowly accumulated in arched structures called line-tied magnetic flux ropes. We analyze our data based on the storage-and-release model for solar eruptions. In this model, eruptions are triggered by a global MHD instability in the corona rather than by dynamic fast-flux injection at the solar surface. For an arched flux rope, the relative invariance of the solar surface condition translates to a slow driving mechanism at the two "line-tied" footpoints. The observed eruption of the flux rope shows clear evidence for plasma motion caused by an ideal MHD instability, which often generates currents sheets in and around the flux rope.

The underlying global instability for magnetic reconnection is determined by magnetic structures and boundary conditions. Magnetic fields in toroidal fusion plasmas in the laboratory consist of those produced by both external and internal currents. The magnetic energy of internal origin is free energy and is released when the plasma is unstable. In tokamak plasmas, the internal magnetic field generated by its plasma current is typically much smaller than the external one, while in RFP (reversed field pinch) and spheromak plasmas they are comparable. Magnetic reconnection due to these instabilities can cause only relatively small change in the magnetic field profile (or a localized change in the magnetic shear) of tokamaks, while it can reorganize the whole magnetic structure of RFP and spheromak plasmas. Significant effort has been devoted to studies of sawtooth relaxation of these current-carrying plasmas. In the following subsections, the relaxation phenomena in tokamak, RFP, and spheromak plasmas are examined. The common paradigm is that "magnetic energy is stored in a magnetic equilibrium configuration via slow adjustment of an external parameter or slow injection of free energy, and plasma often reorganizes itself suddenly into a new MHD equilibrium state, forming current sheets and driving magnetic reconnection." The effects of global boundaries on local reconnection are discussed and applications to astrophysical plasmas are briefly discussed. In the final section of this chapter, we will present a special study of magnetic self-organization in a toroidal plasma arc generated in a laboratory, and we will discuss its application in understanding solar flare dynamics.

9.2 MAGNETIC SELF-ORGANIZATION IN LABORATORY PLASMAS

9.2.1 Sawtooth reconnection in tokamaks

Magnetic reconnection can be observed in fusion research devices by measurements of field-line rearrangement, which is caused by breaking and reconnection of magnetic field lines. Here we observe the evidence for reconnection during the process of self-organization of magnetic field configuration, which consists of multiple layers of magnetic flux surfaces made of *equally* pitched magnetic field lines as shown in figure 9.2. When magnetic reconnection occurs in a certain flux surface, the pitch of the field lines changes through breaking and reconnection of field lines. Most fusion laboratory experiments are carried out in toroidal (donut-shaped) plasma systems that satisfy, for the most part, the conditions for an MHD treatment of the plasma. Typical experimental examples of magnetic reconnection are found in "sawtoothing" tokamak fusion plasmas with large Lundquist numbers of as much as 10^8. They are also found in magnetic self-organization in spheromak and RFP plasmas. Many experiments have been carried out to investigate the physics of magnetic reconnection phenomena in these devices, to find better control of the current-carrying plasmas. As discussed in chapter 3, a sawtooth relaxation oscillation in a tokamak is characterized by a periodic peaking and sudden flattening of the electron temperature (T_e) profile. It presents a typical example of global magnetic reconnection in a laboratory plasma (Kadomtsev, 1975; Wesson, 1987).

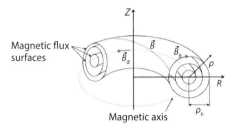

Figure 9.2. Magnetic flux surfaces in a toroidal plasma. The axis of the toroidal plasma core is called the magnetic axis ($\rho = 0$) and (R, Z) define the radial and vertical coordinates in the poloidal plane.

Evolution of magnetic flux and electron temperature profile. As shown in figure 9.2, an axisymmetric tokamak plasma consists of nested flux surfaces on each of which T_e can be assumed constant because of high parallel heat conductivity of electrons. Utilizing black-body radiation from the electrons, which contains information on the local magnetic field and local electron temperature, electron cyclotron emission (ECE) diagnostic systems were developed to measure the T_e-profile as a function of radial position. Since the predominant toroidal field varies as $B_t \propto 1/R$ with plasma major radius R, this diagnostic provides the features of flux surfaces or electron temperature contours using an equilibrium code.

q-profile measurement. The motional Stark effect (MSE) diagnostic developed by Levinton et al. (1993) was employed to measure the magnetic pitch angle profile. This diagnostic utilizes polarized light emission from injected neutral beam ions to measure the poloidal components of internal magnetic fields. Hence the $q(r)$-profile was measured, where $q(r)$ is a safety factor of the flux surface located at the specific minor radius r, which is defined as $q(r) = 2\pi/\iota$, where ι is the rotational pitch angle at r, based on an equilibrium for a circular tokamak (Wesson, 1987). The MSE diagnostic system is based on polarimetry measurements of the Doppler-shifted D_α emission from a neutral deuterium-beam injection (NBI) heating line (Levinton et al., 1993). This technique is noninvasive and nonperturbative. The field-line pitch is localized to the geometric intersection of the field of view, with the neutral beam lines leading to good spatial resolution of $\delta r = 3$–5 cm. If the plasma has good axisymmetric flux surfaces, the measured field-line pitch profile can be translated into a radial profile of the field-line pitch, namely the reverse rotational transform, or $q(r)$, making use of tokamak equilibrium calculations (Yamada et al., 1994).

T_e-profile measurement. The electron temperature profile in a poloidal plane of the plasma has been derived using a rigid-body rotation model for a circular-cross-section tokamak (Edwards et al., 1986; Nagayama et al., 1991). The sawtooth crash phase, which takes 100–500 μs, has been studied extensively with this technique (Yamada et al., 1992), as shown in figure 9.3. By color coding the change of the electron temperature (transfer of heat), a fast electron heat transfer was documented. Just before

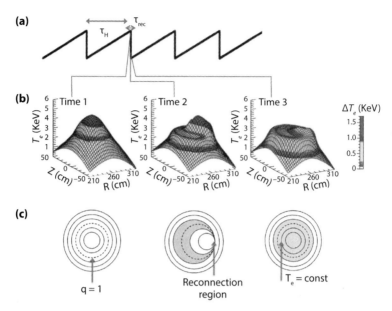

Figure 9.3. See Color Plate 6. Evolution of a measured central electron temperature T_e-profile on a poloidal plane during a short crash phase, and expected flux surfaces during the same period based on MSE diagnostics. (a) Flux buildup time is typically $100\,\mathrm{ms}$ and crash (reconnection) time is 100–$150\,\mu s$. (b) Crash phase evolution of $T_e(R, Z)$ over $150\,\mu s$. (c) Shaded (gray) area shows the region of constant T_e, indicating field lines that are reconnected through the reconnection region. Broken lines show the original radius of the $q = 1$ flux surface. [From Yamada et al. (1994). https://mrx.pppl.gov/mrxmovies/Sawtooth.mov]

the crash, a shrinking circular hot peak shows up and a crescent-shaped flat island grows inside the $q = 1$ region with a kink structure of $m/n = 1/1$. During the crash phase, fast heat transfer from inside to outside the $q = 1$ surface was observed and was attributed to magnetic reconnection, i.e., heat was transferred through reconnected field lines. The T_e-profile inside the $q = 1$ radius becomes flat after the crash, consistent with Kadomtsev's prediction (Kadomtsev, 1975). Does this then mean that q becomes uniform inside the $q = 1$ flux surface as he suggested? We had to wait for an independent magnetic profile measurement to answer this question.

The measured q-profiles on the TFTR (Tokamak Fusion Test Reactor) indicate that central q-values increase by 5–10%, typically from 0.75 to 0.80, during the sawtooth crash phase but do not relax to unity even while the pressure gradient disappears inside the $q = 1$ region. In this case, as well as other tokamak sawtoothing discharges (Soltwisch, 1988), q_0 stays below unity throughout the sawtooth cycle, contrary to Kadomtsev's model. The increase in the q_0-value is more than the statistical error of the measurement. Because only field-line breaking and rearrangement can make $q(r)$ change on such a short timescale, this verifies a magnetic field-line reconnection. We note this result is consistent with the earlier experimental result obtained by Osborne

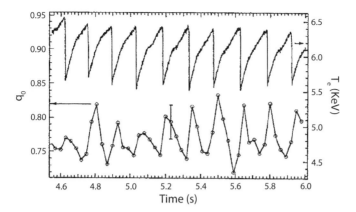

Figure 9.4. Time evolution of peak electron temperature and central q-value associated with sawtooth crash. [From Yamada et al. (1994).]

et al. (1982) using internal magnetic probes in their lower-temperature tokamak configuration of noncircular cross-section.

Physical processes during sawtooth reconnection. Figure 9.4 shows time evolution of central q-value together that of peak electron temperature. The observations raise an important question as to why the magnetic field lines inside the $q = 1$ region do not form a flat $q \sim 1$ inner region after the crash as suggested by Kadomtsev (1975), while the temperature gradient diminishes to zero as predicted by his full reconnection theory. Simultaneous measurements of $T_e(r, \theta)$- and $q(r)$-profile evolutions (Levinton et al., 1993; Yamada et al., 1994) were made in the TFTR. It appears that the $T_e(r, \theta)$-profile does not necessarily coincide with the $q(r)$-profile. Also, the central q-value never reached unity. Based on these results, a heuristic model was proposed for the sawtooth crash. The plasma is viewed as two concentric toroidal plasmas separated by the $q = 1$ flux surface as seen in figure 9.2. A kink mode develops due to a strong peaking of toroidal current and displaces the pressure contours on an ideal MHD timescale with a helical ($m = 1, n = 1$ poloidal and toroidal mode numbers) structure, inducing a forced reconnection at the $q = 1$ surface in both toroidal and poloidal directions.

Simultaneously, a rapid transfer of thermal energy occurs through the reconnection region along newly connected field lines which connect the inside and outside of the $q = 1$ surface (Lichtenberg, 1984). The precipitous drop of the pressure gradient, which occurs within a short period of $100-200\,\mu s \ll \tau_{\text{Sweet–Parker}}$ removes the free energy to drive the kink instability, inhibiting the full reconnection process proposed by Kadomtsev.

Similar changes of central q-values were measured in the sawtoothing plasmas of circular-cross-section tokamaks by the two groups Soltwisch (1988) and Levinton et al. (1993), Yamada et al. (1994), Nagayama et al. (1996). Although the final values of the central q after the crash are different between the experiments, all reported a relatively small change of q ($\Delta q < 0.1$) during sawtooth crash. Magnetic reconnection in

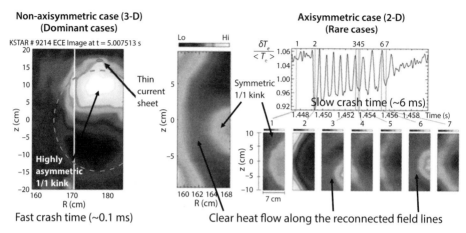

Figure 9.5. See Color Plate 6. Illustrations of a 2D imaging system of T_e-profiles during a sawtooth crash. (Left) Profile of T_e change during a fast crash measured in the KSTAR (Korean tokamak) and (right) TEXTOR tokamaks. The $q = 1$ surface (broken green line) of the KSTAR plasma is shown on the left with the center of the plasma at $R \sim 178$ cm and $Z = 0$. A ballooning-type bulge with a "finger" is clearly seen, together with a distortion of the plasma and harmonic generation of the 1/1 kink mode at the top. On the right, a sawtooth slow crash in TEXTOR is seen with a relatively long reconnection zone. In the slow crash cases, a hot spot is seen to go around in the minor cross-section. [From Park (2019).]

tokamak plasmas is driven by an internal MHD mode (driven reconnection) and is determined by the growth rate of the MHD instabilities. The plasma's stability depends on the plasma parameters ($n_e(R)$, $T_e(R)$, and $T_i(R)$), current profiles (q-profiles), and three-dimensional boundary conditions.

Park et al. (2006b,a) measured two-dimensional electron temperature profiles in the TEXTOR tokamak using sophisticated two-dimensional arrays of electron cyclotron emission spectroscopy, as shown in figure 9.5(a). In most cases, magnetic reconnection occurs very fast, in $< 100\ \mu$s, much shorter than the Sweet–Parker time.

A ballooning-type bulge with a "finger" is seen at the top of the left-hand illustration of figure 9.5. A distortion of the hot region and harmonic generation of the 1/1 kink mode can be recognized at the top. In the right-hand illustration of figure 9.5, a sawtooth crash in TEXTOR is seen with a relatively long reconnection zone. It was observed that when the reconnection region (current layer) is long and flat, a relatively slow reconnection occurred. The observed nonaxisymmetric deformation of toroidal plasma was considered to destroy nested flux surfaces inside the $q = 1$ flux surface, making the T_e-profile uniform inside $q = 1$. It was also found that the reconnection region was distributed equally both on the inner and outer field sides, or both high and low toroidal-field sides of tokamaks, contrary to the ballooning-based models, which predict reconnection occurs predominantly on the lower field side (Park et al., 2006a; Park, 2019).

While it appears more detailed study is needed, we could say that when a sharp edge reconnection layer (site) appears, a fast reconnection results. Most recently, Park (2019) reviewed the relationship between the evolution of multiple high-n, high-m modes near the $q = 1$ flux surface and sawtooth crash. Generally, higher m/n number modes (3, 3), (2, 2), and (1, 1) appear in sequence before a crash.

As a result of the study of sawtooth relaxation in tokamaks, the following summary can be made, while further investigations will reach more conclusive statements.

(1) During the crash phase of sawtooth oscillation, magnetic reconnection takes place rapidly. The change of field-line pitch around the $q = 1$ flux surface Δq is relatively small but with a significant change of the electron temperature profile. It appears that fast electron heat transport occurs through the reconnection region due to fast electron parallel heat conduction along reconnected field lines.

(2) Magnetic reconnection is often driven by an ideal kink-type MHD instability, which is excited after a gradual change of tokamak equilibrium to reach a critical condition. The reconnection time is much shorter than the Sweet–Parker time.

(3) With the recent understanding of two-fluid physics in collision-free plasmas, the observed fast reconnection is not surprising because the Sweet–Parker model is only applicable to collisional two-dimensional symmetric MHD plasmas, while tokamak plasmas are collisionless ($\lambda_{\mathrm{mfp}} \gg R$) and we expect fast two-fluid dynamics to be in play during reconnection.

(4) Generally speaking, the central q-value tends to return to unity after sawtooth crash in tokamaks of noncircular cross-section, while it does not go back to unity in tokamaks of circular cross-section. It can be hypothetically conjectured that a disruptive crash in a noncircular tokamak generates a bigger distortion to the internal flux surfaces, thus inducing a complete reconnection of field lines.

(5) Heat diffusion transport can occur much faster than magnetic reconnection, namely in the timescale of parallel electron heat conduction, and can influence the evolution of global reconnection phenomena or magnetic self-organization. In circular-cross-section tokamaks, Kadomtsev-type full reconnection can be truncated because the free energy in the high pressure gradient that drives a kink mode is reduced due to fast heat conduction through the reconnected region. This is a good example of the case in which local reconnection processes and global plasma dynamics/characteristics interact with each other.

9.2.2   Magnetic reconnection in RFP and spheromak plasmas

A toroidal pinch configuration is one of the most used systems for confining hot fusion plasma by magnetic field. In these systems, a toroidal current is induced to heat and confine plasma through the well-known pinch effects. Tokamaks, reversed field (toroidal) pinch, and spheromak configurations belong to this category. The magnetic field produced by internal current provides an inward pinch force by generating a force balance with outward plasma pressure. Figure 9.6 presents schematics for these three configurations. While all these configurations generate self-pinching poloidal fields, toroidal fields are supplied differently: A tokamak's toroidal field is very strong and

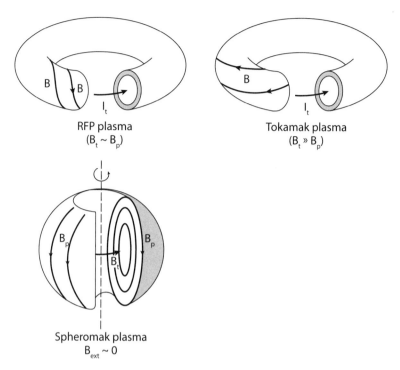

Figure 9.6. Schematic comparison of three toroidal fusion plasmas: tokamak, RFP, and spheromak plasmas.

is primarily created by external coils, while the toroidal field of an RFP is created by the combined effects of internal current and an external field that is much weaker than that of a tokamak. A spheromak does not have an external toroidal field and its internal toroidal field is solely generated by its own internal current. There is a remarkable feature common to all of these toroidal pinch discharges. It has been found that, after an initial highly turbulent state, the plasma settles into a quiescent stable state in which fluctuations are suppressed.

The energy of the internal magnetic field is comparable to that of the external magnetic field in RFP and spheromak plasmas. This internal magnetic energy can be released through magnetic reconnection once the plasma is unstable. Global magnetic structures are reorganized into a state with lower magnetic energy. This is called magnetic relaxation, flux conversion, or dynamo activity (Taylor, 1974, 1986). Magnetic reconnection does not occur arbitrarily in these global relaxation processes; it must satisfy certain global constraints.

9.2.2.1 *Magnetic helicity and Taylor's minimum energy state*

Considering the global aspects of local magnetic reconnection, the first question is what global quantity should be conserved during the formation or relaxation processes. In

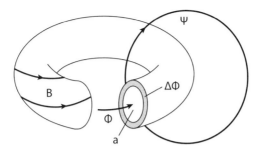

Figure 9.7. Schematic description of magnetic helicity for toroidal plasma systems.

order to quantitatively describe the characteristics of current-carrying laboratory plasmas, magnetic helicity was introduced and studied as an important physics parameter. Magnetic helicity represents a net twist or linkage of magnetic flux, which can
be regarded as a measure of the "knottedness" or the "twistedness" of a magnetic field
(Woltjer, 1958). It is defined by $K = \int A \cdot B \, d^3x$, where A is the vector potential of the
magnetic field B and the integration is over a specified volume V. Magnetic helicity K
can also be expressed as

$$K = \int A \cdot B \, d^3x = \int A \cdot dl \, B \, ds = \int \psi \, d\phi, \qquad (9.1)$$

because magnetic flux can be expressed as $B \cdot ds = d\phi$, $\int A \cdot dl = \psi$, where the inte-
grals extend over a suitable boundary (figure 9.7).

Now we can express the magnetic helicity as the product of two linked fluxes, as
shown in figure 9.7, where $\Delta\Phi$ represents the differential cross-section of the torus.
For a system made of two simply connected loops of flux ψ_0 and ϕ_0, the entire volume
integration leads simply to (Berger and Field, 1984)

$$K = 2\psi_0\phi_0. \qquad (9.2)$$

For a laboratory plasma, we often define the magnetic helicity after subtracting the
helicity associated with vacuum magnetic fields, thus making it a gauge-invariant quantity. But in a open system such as a space plasma, it is often difficult to define gauge-
invariant magnetic helicity because of a difficulty in defining a boundary. The magnetic helicity is an invariant within a flux tube in a perfectly conducting plasma. It is
questioned whether magnetic helicity is still conserved in a highly conducting plasma
undergoing reconnection.

Based on his careful examination of RFP formation processes, J. B. Taylor proposed a conjecture that magnetic helicity K, defined as above, tends to conserve during
plasma relaxation as the plasma rapidly approaches the minimum energy state. The
essence of the Taylor relaxation theory (Taylor, 1974, 1986) is that the plasma has a
tendency to relax toward the minimum (magnetic) energy state while conserving total
magnetic helicity. Experimentally, magnetic helicity was observed to change a little,

but by a much smaller amount compared to magnetic energy, which decreases substantially during RFP relaxation (Ji et al., 1995; Anderson et al., 2004). A simple estimate of helicity change due to magnetic reconnection is given by Ji (1999). It is argued that the total magnetic helicity is a well-conserved quantity, with $\delta K \ll K$ during magnetic reconnection, if the thickness of the diffusion region is much smaller than the global size.

Magnetic reconnection is perhaps the only process (other than self-similar expansions) that can release magnetic free energy while conserving flux and helicity. Thus, magnetic reconnection is strongly implied, although not explicitly specified, in the process of Taylor relaxation.

The predicted relaxed states (Taylor, 1974) are described by the force-free equilibria given by $\nabla \times \boldsymbol{B} = \mu \boldsymbol{B}$, where $\mu = (\boldsymbol{j} \cdot \boldsymbol{B})/B^2$ is a spatial constant along and across field lines. This prediction explains a remarkable common feature of both RFP and spheromak plasmas, that after an initial highly turbulent state, the plasma settles into a more quiescent state in which μ tends to be spatially uniform (Bodin, 1990; Bellan, 2000).

9.2.2.2   Formation of an RFP configuration

An RFP is an axisymmetric toroidal pinch in which plasma is confined by the combined effects of a poloidal magnetic field B_p, created by a toroidal plasma current, and a toroidal field B_t, generated by a poloidal plasma current and external coil currents. By carefully examining RFP discharges, Taylor postulated that the RFP configuration should originate from a process of plasma relaxation in which plasma settles into a state of minimum energy with the constraint of constant helicity. If this state is described simply by $\nabla \times \boldsymbol{B} = \mu \boldsymbol{B}$ with constant $\mu = (\boldsymbol{j} \cdot \boldsymbol{B})/B^2$ throughout the large-aspect-ratio toroidal vacuum vessel, it would naturally lead to the well-known Bessel function solution

$$B_t(r) = B_{\text{to}} J_0(\mu r), \quad B_p(r) = B_{\text{to}} J_1(\mu r), \tag{9.3}$$

where B_{to} denotes the toroidal field at the minor axis and $J_0(\mu r)$ and $J_1(\mu r)$ denote Bessel functions of order 0 and 1 respectively. This formulation is valid as long as the large-aspect-ratio toroidal plasma can be approximated by a cylinder of radius a.

The relaxed force-free states are independent of the initial state and described by the dimensionless parameters F and θ, where $\theta = B_t(a)/\langle B_t \rangle$ is the ratio of the toroidal field at the wall ($r = 0$) to the average toroidal field, and $F \, (= B_p(a)/\langle B_t \rangle)$ is called a pinch parameter, defined as the ratio of the poloidal field at the wall to the average toroidal field. A value $F < 0$ implies a field reversal at the edge. In practice, this force-free state does not exactly describe the RFP configuration since the assumption of $\mu = \text{const.}$ holds only in the interior of the plasma and it goes down to zero at the edge. But the field profiles described by eq. (9.3) agree remarkably well with those for quiescent plasma obtained in RFP devices as shown in figure 9.8. In general, when the toroidal field reverses at the edge ($F < 0$), a more stable plasma is obtained and thus this configuration is maintained as the RFP configuration.

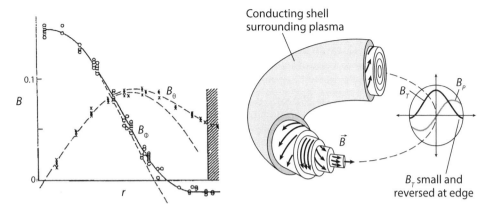

Figure 9.8. Experimental and theoretical magnetic profiles of an RFP configuration. Solid lines are interpolated from experimental data and broken lines are from eq. (9.3). [From Sarff et al. (2005).]

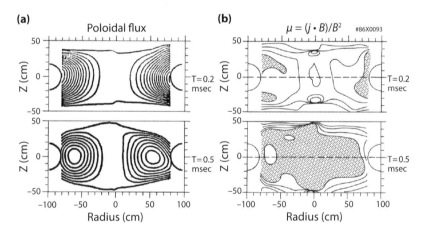

Figure 9.9. Contours of poloidal flux (left) and μ (right) during and after formation of an S-1 spheromak, measured by internal probes. The shaded regions refer to the value of $\mu = 5.5 \, \mathrm{m}^{-1}$ (Taylor value) within 15% error. At the initial formation phase, the current density and μ-value are high near the plasma-forming flux-core area but slowly relax into the "constant-μ" state, supporting the Taylor conjecture. [From Hart et al. (1986).]

9.2.2.3 Formation of a spheromak configuration

A spheromak configuration is also generated by relaxation processes after an internal plasma current is generated either by induction or electrode discharges in a plasma gun. An example is shown in figure 9.9, where a constant-μ Taylor state is experimentally verified (Hart et al., 1986). The turbulent initial state undergoes violent reconnection to form a spheromak configuration, the minimum energy state. After this initial formation process, the relaxation or reconnection activity occurs in a cyclic or continuous fashion. The plasma is driven away from the relaxed state, and the relaxation opposes this tendency. During the period in which the configuration is maintained, the plasma often evolves away from the relaxed state as μ becomes spatially peaked. Then plasma instability sets in and the plasma configuration rapidly returns to a relaxed state as μ becomes flat again. Peaking and flattening of the μ-profile over the relaxation cycles has been experimentally verified in both RFP (Ji et al., 1995) and spheromak plasmas (Yamada, 1999).

The underlying instabilities for reconnection and relaxation are driven by excess internal magnetic energy associated with a peaked current profile within the plasma. In spheromaks, a kink instability can be destabilized. The instability is no longer localized to the central region, but occupies the entire plasma, and causes global reorganization (Bellan, 2000). In RFP plasmas, the underlying instability is a tearing mode instability (Furth et al., 1963) occurring at multiple radii, with each radial location corresponding to a rational surface in which the safety factor is m/n. During a relaxation event, impulsive reconnection takes place at a single radial location or at multiple locations simultaneously to reorganize the plasma back to a relatively stable state (Ortolani and Schnack, 1993).

9.2.2.4 Sawtooth relaxation in RFP plasmas

Once an RFP plasma configuration is established, stable discharge can be maintained as long as the plasma current is sustained by inductive drive. Magnetic helicity is supplied by poloidal flux injection into the plasma. During the discharge the poloidal flux often becomes too excessive to maintain the Taylor state and a relaxation occurs as a sawtooth oscillation. When the current configuration is too peaked and deviates from the Taylor state, it quickly comes back to that of the relaxed state. During this relaxation process, a conversion of flux from poloidal to toroidal has been seen in many RFP discharges and the total magnetic helicity $K = 2\psi_0\phi_0$ is kept roughly constant. As we saw in figure 2.9, magnetic helicity and energy evolution during a sawtooth cycle in RFP plasma were studied (Ji et al., 1995) in the MST (Madison Symmetric Torus) RFP by monitoring the global values of total helicity K and magnetic energy W. Sawtooth oscillations consist of a fast crash phase and a slow recovery phase. The plasma rapidly relaxes toward its minimum energy state during the crash phase. With a lack of measurement of the exact magnetic field profiles, the μ-profiles (and hence energy and helicity) were assumed from other measured quantities by employing equilibrium models of $\mu = \mu_0\left(1 - (r/a)^{\alpha}\right)$. With this model assumption, the changes of K and W during a sawtooth crash were monitored. Figure 2.9 presents an inventory of magnetic helicity, magnetic energy, poloidal flux, and toroidal flux during a sawtooth oscillation. It was found that during the relaxation event, the magnetic helicity decreases by 1.3–5.1%, while the magnetic energy decreases by 4.0–10.5%. Hence, the helicity conservation conjecture is satisfied in that the helicity decay is less than the energy decay by a factor of 3, instead of orders of magnitude. This result indicates that helicity conservation is only a conceptual approximation. The helicity change is

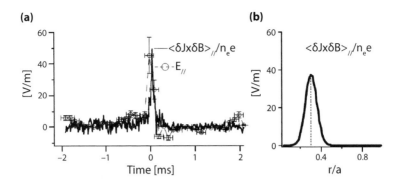

Figure 9.10. (a) Time evolution of the Hall dynamo due to an ($m = 6, n = 1$) mode (solid line) and inductive electric field (dashed line) during a relaxation cycle. (b) Amplitude of the Hall dynamo as a function of radius. [From Ding et al. (2004).]

larger than a simple MHD prediction. Determination of detailed mechanisms for possible anomalous helicity dissipation during relaxation awaits further investigation.

The process of the flattening of the μ-profile due to these instabilities is a redistribution of the current parallel to the mean magnetic field over the plasma radius. Insights can be gained by examining the parallel component of Ohm's law,

$$\langle E \rangle_\| + \langle \widetilde{V} \times \widetilde{B} \rangle_\| = \langle \eta j \rangle_\|, \tag{9.4}$$

where $\langle \cdots \rangle$ denotes an ensemble average over the fluctuations associated with reconnection processes. The fluctuation-induced electromotive force (EMF) $\langle \widetilde{V} \times \widetilde{B} \rangle$ is called the α dynamo effect (Ji and Prager, 2002), derived from the same notion in mean-field theory (Krause and Rädler, 1980) of dynamo action or generation of magnetic field from a turbulent flow. In the MHD frame, a nonzero component of $\langle \widetilde{V} \times \widetilde{B} \rangle$ along the mean field was predicted by a nonlinear computation (Ortolani and Schnack, 1993), experimentally detected by Ji et al. (1994) and Fontana et al. (2000) in the edge of an RFP plasma and by al Karkhy et al. (1993) in a spheromak.

One question of interest is whether two-fluid effects are still important during magnetic reconnection on global scales. Theoretically, it was found that the tearing mode structures and growth rates can be significantly modified by two-fluid effects (Mirnov et al., 2003, 2004). The Hall effect enters eq. (9.4) as a new term on the left-hand side, $-\langle \widetilde{j} \times \widetilde{B} \rangle_\|/en$. It has been experimentally shown (figure 9.10) for a particular mode ($m = 6, n = 1$) in the core of MST plasmas (Ding et al., 2004). The amplitude and time dependence of this term are just what is required to explain the flattening of the μ-profile near the center of the plasma. Other two-fluid effects, such as the electron diamagnetic effect theoretically predicted (Lee et al., 1989) and experimentally explored (Ji et al., 1995), can play a role in global relaxation.

In the central region where μ is peaked before relaxation, $\langle \widetilde{V} \times \widetilde{B} \rangle$ has the opposite sign to that of the parallel current, while it has the same sign in the edge region, and the

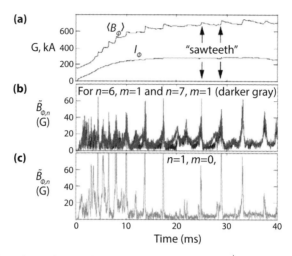

Figure 9.11. Time dependence of (a) toroidal flux and plasma current, (b) core-resonant $m = 1$ magnetic fluctuations, and (c) edge-resonant $m = 0$ magnetic fluctuations. [From Prager et al. (2005).]

μ-profile flattens after a relaxation event. This process can be viewed as a flux conversion of poloidal flux to toroidal flux. Reducing parallel current at the center, where the field lines are mostly toroidal, means reducing poloidal flux, while increasing parallel current at the edge, where field lines are mostly poloidal, means increasing toroidal flux. However, it is not yet completely understood how exactly magnetic reconnection accomplishes this dynamo or flux conversion process in three dimensions. Important observations were made a few decades ago in spheromaks. It was observed by Ono et al. (1988) in the S-1 spheromak that the kinked part of the plasma can twist itself so much that field lines change their orientation significantly, and through subsequent three-dimensional reconnections, field lines restore their axisymmetric state, but with a different ratio of toroidal to poloidal fluxes, resulting in an increase of toroidal flux at the expense of poloidal flux, namely flux conversion. In RFP plasmas, magnetic islands growing out of tearing modes at each rational surface can flatten the current profile in its vicinity, as shown in the quasi-linear calculations of Strauss (1985) and Bhattacharjee and Hameiri (1986), again resulting in the same flux conversion from poloidal to toroidal.

Some important clues have been reported from the MST experiment on the nonlinear aspect of tearing mode interactions. During relaxation, there are several unstable $m = 1$ tearing modes resonant near the center. When the $m = 0$ mode is absent or weak, the resulting relaxation is milder, as shown in figure 9.11. Around $t = 20$ ms, the $m = 0$ mode amplitude is small and the relaxation is much weaker, as indicated in the changes in the toroidal flux. The nonlinear interactions cause rapid momentum transport (Hansen et al., 2000), which is related to charge transport (Ding et al., 2007). Accompanying this rapid momentum transport during relaxation events, anomalous ion heating is observed (Den Hartog et al., 2007). No significant ion heating is observed without the nonlinear $m = 0$ mode. A plausible scenario emerges in which multiple, and interacting, reconnection processes cause an efficient global relaxation to release magnetic energy under the constraints of flux and helicity conservation. Primary instabilities drive localized reconnections, resulting in transient flows and magnetic fields, which can lead to secondary reconnections. These secondary reconnections accelerate the rate of energy release and other nonlinear processes, such as momentum transport and ion heating. A similar physics mechanism was proposed for solar flares to explain their impulsive nature (Kusano et al., 2004). It is interesting to compare the relationship and time sequence between the spatial structures of spontaneous and driven reconnection regions of RFP plasmas and those of solar flare eruptions.

9.2.3 Relationship of current layers with global reconnection phenomena

During magnetic self-organization, magnetic reconnection occurs through current sheets. It is major question how a large-scale system generates local current structures: Are they formed spontaneously or are they forced by change of external boundary conditions?

The aspect ratio of the current layers has generally been taken as the global length over some microscopic length, such as the ion skin depth or Sweet–Parker layer thickness. For such aspect ratios the classical reconnection rate is much too small to account for the observations. When a current layer develops, it is of considerable importance how its actual aspect ratio is estimated.

There has been research on the origin and nature of current layers (Becker et al., 2001; Rosenbluth et al., 1973; Syrovatskii, 1971; Waelbroeck, 1989; Jemella et al., 2004; see also Biskamp, 2000, p.60), but these studies were made on asymmetric situations where the global length of the system is the natural length for the current layer. Situations such as that of the solar flare are far from symmetric. For these cases there is no reason to believe that current layers are as long as the global size. Parker (1979) attempted to show that, in a fully three-dimensional equilibrium, current layers are inevitable and their length is comparable with the local scales of the equilibria, the length over which the ambient field changes by a finite amount.

Parker and Rappazzo (2016) investigated the final equilibrium of a common interlaced field-line topology, starting from an initial smooth, continuous, and bounded interlaced field with no further deformations introduced from outside the region, and letting the magnetic stresses in the field topology alone determine the final equilibrium state; see figure 9.12. The result is that current sheets, i.e., internal sites for rapid reconnection, are intrinsic to interlaced field-line topologies wherever they occur. They conclude that since the reconnection velocity goes inversely as the square root of the current length, these shorter lengths should lead to faster local reconnection.

Much less magnetic energy is released in each of the reconnections associated with these shorter current lengths. There could be so many of them that the total released energy could be very large. To get a rapid release of a lot of energy, the local reconnections have to interact and proceed almost simultaneously. Lu and Hamilton (1991) and Lu (1995) have suggested such a rapid sequence of releases, as a self-organized effect, such as happens in avalanches or sand piles. Their idea is that one local reconnection

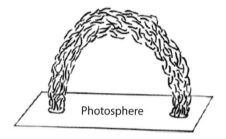

Figure 9.12. Parker's sketch of the interlaced magnetic-field-line topology of a bipolar solar magnetic field rooted in the convective solar photosphere. [From Gonzalez and Parker (2016, fig. 5.1, p. 183).]

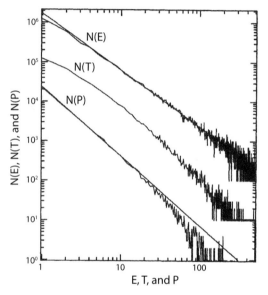

Figure 9.13. Frequency distributions of avalanches as functions of energy E, peak flux P, and duration T. [From Lu and Hamilton (1991).]

happens first, and then the magnetic energy released triggers reconnection in neighboring current layers. Their released energy triggers more reconnection, and so on. This model was inspired by the observation that the distribution function for the number of solar flares as a function of their peak power is a nearly perfect power law over many decades (figure 9.13). Such a power law cannot be produced in any other way.

This bears on two important puzzles: How are current layers formed, and with what length? How is reconnection triggered in them? The details of the physics in a reconnection layer are now fairly well understood, and attention is turning to these more global questions.

9.3   IMPULSIVE SELF-ORGANIZATION IN SPACE AND LABORATORY PLASMAS

Here we discuss the impulsive nature of global magnetic self-organization phenomena in space and laboratory plasmas. In tokamak discharges, reconnection often occurs quite suddenly after a slow evolution of plasma equilibrium and magnetic flux buildup. Generally, the flux buildup phase is significantly longer than the reconnection time, $\tau_H > \tau_{rec}$. This creates a sawtooth-shaped evolution of the central electron temperature. We note that this is a good example of the case in which evolution of the global plasma configuration forces fast local reconnection. In low-q pinch discharges in other laboratory fusion devices such as spheromak and RFP, we observe similar sawtooth events which also consist of a slow flux buildup phase through a slow reconnection and then a fast reconnection/relaxation phase. In the former phase, the current density in the center core gradually increases, while in the latter an impulsive current profile flattening occurs with reconnection. Generally, reconnection occurs in the resonant flux surfaces (which implies that the rotational pitch of a mode coincides with the rotational transform of a flux surface) in the plasma core and, under some conditions, at the edge. In some cases, two unstable tearing modes in the core region are observed to couple each other to nonlinearly drive reconnection at a third location in the outer plasma edge region (Kusano et al., 2004). It is conjectured that similar phenomena occur in active solar arcade flares where spontaneous reconnection at one location can drive reconnection at other locations, leading to eruptions (Kusano et al., 2004). In solar flares, reconnection sites are identified with hard-X-ray emissions near the top of solar flare arcades during CME and coronal eruptions (Masuda et al., 1994; Shibata and Magara, 2011). Reconnection speed has been measured to be much faster than the Sweet–Parker rate. We could hypothesize that global magnetic self-organization phenomena in both tokamak sawtooth crashes and solar flares share a common process. Klassen reported sawtooth phenomena in solar flares (Klassen et al., 2001). When reconnection occurs in a certain region of the globally connected plasma, a topology change results. A sudden change of magnetic flux over a short time is induced in a newly connected part of the global plasma. This leads to a large electric field along the magnetic field lines and acceleration of electrons to superthermal energy. Indeed, in reconnection events in both solar flares and tokamak sawteeth, we observe a significant amount of high-energy (runaway) electrons. A careful comparative study of tokamak sawteeth and RFP relaxation events should illuminate this important energy flow channel. We will see such an example of study in a laboratory plasma.

9.4   MAGNETIC SELF-ORGANIZATION IN LINE-TIED MAGNETIC FLUX ROPES: LABORATORY STUDY OF SOLAR FLARE ERUPTION PHENOMENA

In active solar arcade flares, impulsive global reconnection takes place after a gradual change of equilibrium that builds up sufficient free energy to induce motion of plasma or topological changes. A slow change of equilibrium drives a plasma to an unstable regime and then drives a global magnetic self-organization leading to eruptions. We call it a "storage-and-release mechanism." To expand our study of magnetic reconnection in a laboratory plasma beyond the local reconnection layer, here we consider magnetic self-organization phenomena on the global topology in a laboratory plasma experiment relevant to solar flare dynamics. The particular plasma configuration studied here is that of an arched, line-tied magnetic flux rope. This configuration is of particular interest due to its central role in gradual storing and sudden releasing of magnetic energy in the solar corona (Kuperus and Raadu, 1974; Chen, 1989; Titov and Démoulin, 1999). As mentioned in Chapter 2, a series of laboratory experiments were carried out to demonstrate the dynamics of eruption phenomena (Hansen and Bellan, 2001; Soltwisch et al., 2010). But in these arched-flux-rope experiments, it was difficult to satisfy the storage-and-release condition by relying on the dynamic injection of either plasma or magnetic flux at the footpoints to produce an eruption. In contrast, the MRX experiment we discuss here enforces a clear separation of timescales between the energy storage time, and the relaxation time, the dynamic Alfvén time, such that the observed eruptions are driven by storage-and-release mechanisms.

In this section we look at MRX flux rope experiments, including the key features that make them uniquely relevant to the solar eruption problem. We then see results from a study of the flux rope instability and its relationship to the features of solar eruption as described by Myers et al. (2015). The experiment was equipped with a two-dimensional, well-covered magnetic probe array, which was used effectively here to investigate the detailed evolution of the flux rope during storing and eruptive events.

9.4.1   Basic approach of the MRX experiment for solar eruption

Ideal MHD instabilities such as the torus and kink instabilities are central to the standard storage-and-release model of solar flares and CMEs. MHD simulations have shown that the torus instability plays a crucial part in driving magnetic flux ropes to erupt, even in the presence of magnetic reconnection. Thus the MRX flux rope experiments were designed to identify the stability boundaries for the triggering of candidate ideal instability eruption mechanisms. The torus instability is triggered by an imbalance in the vertical forces acting on the flux rope plasma (Kliem and Török, 2006). The traditional forces considered for the torus instability are (1) the upward "hoop" force, which is the Lorentz force between the toroidal (axial) flux rope current and its own poloidal (azimuthal) magnetic field and (2) the downward "strapping" force, which is the Lorentz force between the same toroidal current and the potential strapping field.

On the other hand, the kink instability occurs when the twistedness of toroidal field lines exceeds a certain value, as also explained in this chapter. The kink instability (Sakurai, 1976; Hood and Priest, 1981; Mikić et al., 1990; Török et al., 2004) is triggered when the magnetic twist at the edge of the flux rope (i.e., the poloidal angle through which an edge magnetic field line rotates as it transits the toroidal length of the flux rope) exceeds a critical threshold. The analytical kink onset condition is often given in terms of the edge safety factor q_a, which is defined as the inverse of the edge magnetic twist.

In addition to such ideal instabilities, the nonideal process of magnetic reconnection is often invoked to explain various observed solar flare and coronal mass ejection features through topology changes. For example, reconnection produces flare emission beneath the rising flux rope and contributes to the evolution of the flux rope height. Reconnection is also the central driving mechanism in some CME initiation models as explained in chapter 7.

9.4.2 Apparatus and major results

To experimentally study the storage-and-release model for solar eruptions, the flux rope laboratory experiment shown in figure 9.14 was carried out in MRX. Primarily, the global features of the flux rope evolution that lead to global magnetic reconnection were studied. The magnetic probe data were used to directly measure magnetic field

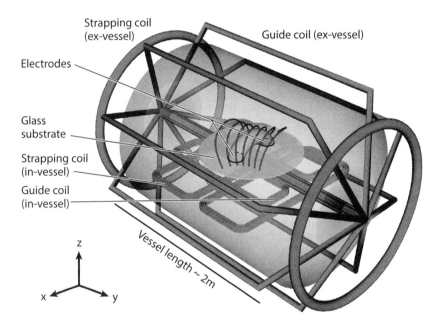

Figure 9.14. Experimental setup. A plasma arc (pink) is maintained between two electrodes that are mounted on a glass substrate. The electrodes, which serve as the flux rope footpoints, are horizontally separated by $2x_f = 36$ cm and they have a minor radius of $a_f = 7.5$ cm. Note that the (x, y, z) coordinate system used in these experiments differs from the local reconnection coordinate system used in previous sections; we are following the convention for the solar surface. The vertical distance from these footpoints to the vessel wall is $z_w \sim 70$ cm. Four magnetic field coil sets (two inside the vessel, two outside) work in concert to produce a variety of vacuum magnetic field configurations. More specifically, the two orange coil sets are used to produce the guide vacuum field, while the two blue coil sets are used to produce the strapping vacuum field. [From Myers et al. (2015).]

configuration and the $J \times B$ acting on the flux rope plasma (Myers et al., 2015, 2016). It was shown that a tension force derived from a self-generated paramagnetic toroidal field exerts a restoring force on the line-tied plasma and suppresses eruptive behavior in a significant portion of the parameter space. This result provides a new mechanism for understanding the dynamics of solar flares and CMEs. From these measurements, we conclude that the toroidal field tension force, which has been neglected in traditional flux rope force balance studies (Kliem and Török, 2006; Démoulin and Aulanier, 2010), was dynamically enhanced by the impulsive reconnection processes identified here. It is this dynamically enhanced tension force that prevents many flux rope eruptions in the failed torus regime.

9.4.3   Detailed experimental setup

The line-tied flux rope experimental setup in MRX is shown in figure 9.14. The arched plasma is formed between the two copper electrodes mounted on a glass substrate. Four vacuum field coil sets, two inside the vessel and two outside, are used to produce a wide range of vacuum magnetic field configurations in the plasma region. This flexibility was necessary to conduct the instability parameter space. The vacuum field configuration consists of two components: (1) the "guide" magnetic field that runs toroidally along the flux rope and (2) the "strapping" magnetic field that runs orthogonal to the flux rope. The guide field is equivalent to the toroidal field in a tokamak, while the strapping field is equivalent to the "vertical" equilibrium field in a tokamak. These two field components combine to produce the obliquely aligned vacuum magnetic field lines shown in figure 9.14. Note that the (x, y, z) coordinate system used in these experiments differs from the local reconnection coordinate system used in previous sections. Here, z is aligned with the vertical axis above the footpoints, as is commonly found in the solar literature.

Once a given vacuum magnetic field configuration has been selected, the various coil sets are ramped to their selected currents and held there for the duration of the discharge. In practice, the vacuum magnetic field configuration is established several milliseconds before the plasma breakdown in order to allow vessel and electrode eddy currents to decay away. The plasma discharge is then initiated by connecting a pre-charged capacitor bank across the electrodes. The discharge driving time is governed by the characteristics of the combined RLC capacitor bank and plasma circuit.

9.4.4   Key findings on the stability criteria for flux rope

First, the major criteria for torus instability and kink instability were obtained for the flux rope instability parameter space (figure 9.15). The parameter space plotted here is defined by the onset criterion for the kink (Gold and Hoyle, 1960; Sakurai, 1976; Hood and Priest, 1981; Mikić et al., 1990; Török et al., 2004) and torus (Kliem and Török, 2006; Fan and Gibson, 2007; Liu, 2008; Démoulin and Aulanier, 2010) instabilities. More specifically, the edge safety factor (Kruskal and Schwarzschild, 1954; Shafranov, 1956) q_a was used to measure the (inverse) twist in the flux rope at the plasma edge $r = a$. Here $q_a = 2\pi/\iota$, with ι being defined as the rotational transform of field lines in

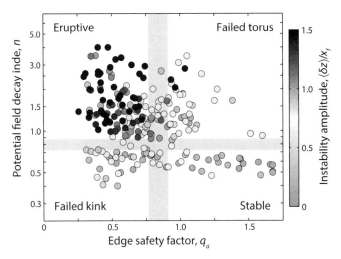

Figure 9.15. The experimentally measured torus versus kink instability parameter space. The horizontal x-axis represents the kink instability criterion through the edge safety factor q_a (the inverse magnetic twist), while the vertical y-axis represents the torus instability criterion through the potential field decay index n. Each data point is the mean of 2–5 flux rope plasma discharges with the same experimental parameters. The instability amplitude is represented by the darkness of each circle (proportional to the amplitude as shown in the right-hand bar). A total of 806 flux rope plasma discharges are represented. The metric used here to quantify the eruptivity of each flux rope is the normalized spatial instability amplitude $\langle \delta z \rangle / x_f$. A value of $\langle \delta z \rangle / x_f < 0.5$ is stable, while $\langle \delta z \rangle / x_f > 1$ is clearly eruptive. The shaded boundaries, which are empirically identified, delineate the four distinct instability parameter regimes described in the text. [From Myers et al. (2015).]

a flux surface (Wesson, 1987). Low q_a corresponds to high twist and therefore helps kink instability. On the other hand, n is the field decay index (Kliem and Török, 2006), which measures how steeply the vacuum magnetic field decays with height above the electrodes. High n corresponds to a steeply decaying profile and therefore leads to torus instability. The darkness of each circle represents the normalized instability amplitude $\langle \delta z \rangle / x_f$. Here, δz is the spatial amplitude of instability-driven motion of the flux rope magnetic axis, and $2x_f$ is the footpoint separation distance. The specifics of how q_a, n, and $\langle \delta z \rangle / x_f$ are extracted from each discharge are detailed elsewhere (Myers thesis, 2015). The data points in figure 9.15 are directly measured from more than 800 laboratory flux rope discharges in MRX.

Four distinct stability regimes are shown in figure 9.15. First, the stable regime at high q_a and low n appears as expected in the absence of both the kink and torus instabilities. Likewise, the eruptive regime appears at low q_a and high n when both instabilities are present. Next, the "failed kink" regime at low q_a and low n reveals that the kink instability can drive motion of the flux rope without causing an eruption. This result is consistent with existing numerical work (Török and Kliem, 2005), and it highlights the primacy of the torus instability in driving eruptions. The fourth and final regime at high q_a and high n, which we call the "failed torus" regime, is also noneruptive. This refutes the notion that the torus criterion is a necessary and sufficient condition for eruption. Instead, flux ropes that exceed the torus criterion in this regime fail to erupt. The salient point is that the stable, eruptive, and failed kink regimes can be explained within the framework of ideal MHD instabilities, but the failed torus regime cannot. Therefore, a nonideal process such as magnetic reconnection must be involved. In the following subsections, we provide experimental evidence of the formation of current sheets in both the eruptive and the failed torus regimes.

9.4.5   Magnetic self-organization during eruptive events

We proceed by discussing magnetic self-organization during eruptive events in these experiments. Before doing so, we must introduce the magnetic probe array (Myers thesis, 2015) used to diagnose the internal structure of the plasma. The probe array consists of seven long, thin magnetic probes that are aligned in a two-dimensional plane and inserted vertically into the flux rope plasma (figure 9.16(a)). Each probe contains up to 17 "triplets" of miniature magnetic pickup coils, spaced at 4 cm intervals along the length of the probe. Each triplet measures the vector magnetic field at one location in space. Thus, in total the probe array contains ~ 300 pickup coils that measure the magnetic field at more than 100 locations distributed throughout the plasma. The probes are separated horizontally by 4 cm so that the triplets form a 4 cm $\times$ 4 cm grid covering an area of 64 cm $\times$ 24 cm. The probe array can be rotated about the z-axis to measure

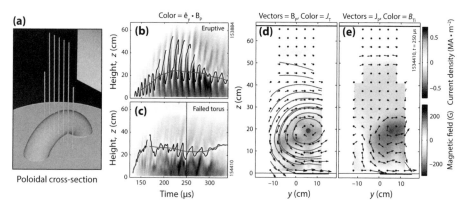

Figure 9.16. See Color Plate 3. Magnetic probe configuration for the MRX flux rope experiments. (a) Seven long, thin magnetic probes (yellow) are arranged in a 2D plane and inserted vertically into the flux rope plasma. (b), (c) Height–time histories of characteristic eruptive and failed torus discharges. (d), (e) Representative magnetic measurements of the internal structure of the flux rope at one point in time. For all of the 2D data presented here, the probes are aligned in the "poloidal cross-section" shown here. All three components of B and J, are plotted here, with J computed from the spatially resolved measurements of B. [From Myers et al. (2015); Yamada et al. (2016b).]

different cross-sections of the flux rope plasma, but all of the data are presented with the probe array in the poloidal cross-section (figure 9.16(a)).

Figure 9.16(b) shows how the magnetic probe data are used to track the height–time history of an erupting flux rope plasma. The magnetic axis position is determined by tracking the reversal of the out-of-plane poloidal magnetic field $B_y \equiv \hat{e}_y \cdot \boldsymbol{B}_P$, where $\hat{e}_y$ is a unit vector. We see that for the representative erupting flux rope in (b), the plasma begins with small amplitude kink oscillations that transition to eruptive oscillations which extend nearly to the wall of the vacuum vessel at $z_w \simeq 70$ cm. These eruptive oscillations commence when the flux rope enters the torus-unstable regime. Figure 9.16(c) shows an analogous height–time trace for a failed torus discharge. Note that only small-scale oscillations are observed.

Figure 9.16(d), (e) show spatially resolved magnetic measurements from a single point in time. In (d), the poloidal (in-plane) magnetic field vectors $\boldsymbol{B}_P$ are shown along with the toroidal current density $J_T = \hat{e}_T \cdot (\nabla \times \boldsymbol{B}_P)$, as defined and used in this section. In (e), the "internal" toroidal magnetic field B_{Ti} is shown. This internal field does not include the vacuum toroidal guide field B_g. In a low-β flux rope, the internal toroidal field should be paramagnetic, or codirected, with the guide field (Myers thesis, 2015; Myers et al., 2015, 2016). Using the toroidal field and the assumption of *local* toroidal symmetry, vectors of the local poloidal current density $J_P = \hat{e}_T \cdot (\nabla \times \boldsymbol{J}_T)$ are computed. Finally, the contours in (d) are contours of a local poloidal flux function. The red contour is the minor radius of the flux rope, which is defined here as the contour that encloses 75% of the total current injected at the electrodes. The measurements of $\boldsymbol{B}$ and $\boldsymbol{J}$ presented here facilitate direct measurement of the forces acting on the flux rope (see section 9.4.7), but first we use them to track the evolution of the flux rope plasma during characteristic eruptive and failed torus events.

Figure 9.17 shows a sequence of J_T and B_{Ti} measurements that capture the Alfvénic rise of a flux rope during a characteristic eruptive event. Since this event is driven by the ideal kink and torus instabilities, current sheets and magnetic reconnection play only a secondary role. Distinct current sheets are visible in figure 9.17. First, a strong, coherent flux rope forms at low altitude (strong current channel) and then rises steadily toward the wall of the machine. The total current in the rope drops as the inductance of the growing loop increases. Notably, a reversed current sheet (with reversed current direction) forms above the flux rope as the rope pushes through the surrounding vacuum magnetic field, although it is hard to recognize in the figure. More features are presented in color figures by Myers et al. (2015) and Yamada et al. (2016a). It is likely that this current sheet mediates the speed with which the flux rope rises. This is just one event in a sequence of eruptive events (see figure 9.16(b)). The remnant of the previous event is visible at high altitude in the first few frames of figure 9.17 and then again in the last few frames. Due to the inductive voltage provided by the capacitor bank, a new flux rope readily forms behind the erupted rope. It is likely that reconnection plays a role in the transfer of flux from the erupted rope to the newly formed rope at low altitude.

9.4.6 Magnetic reconnection and the self-organization processes

The failed torus regime, unlike the eruptive regime, cannot be explained in the context of ideal MHD instabilities. Instead, impulsive magnetic reconnection that reconfigures

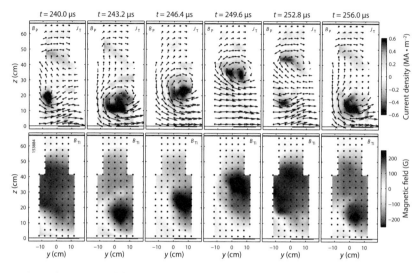

Figure 9.17. Sequence of internal magnetic field measurements during a characteristic eruptive event. The top row plots poloidal magnetic field vectors $\boldsymbol{B}_P$ and the corresponding toroidal current density $J_T = \hat{\boldsymbol{e}}_T \cdot (\nabla \times \boldsymbol{B}_P)$. The bottom row plots the measured out-of-plane paramagnetic toroidal field B_{Ti}. A coherent flux rope forms with a strong forward current channel at low altitude ($t = 243.2\,\mu$s) and then begins to rise. As the flux rope rises through the surrounding poloidal flux, a reversed current sheet forms above the rope, although it is hard to recognize in the figure. More detailed features are presented in color figures by Myers et al. (2015) and Yamada et al. (2016a). As the flux rope reaches the wall, a new flux rope begins to form behind it ($t = 252.8\,\mu$s), and the process repeats itself.

the global topology of the flux rope plays a central role. The magnetic evolution of a characteristic failed torus event is shown in figure 9.17. The sequence of frames shown here, which is analogous to the sequence of frames for the eruptive event in figure 9.16, reveals a very different evolution. Instead of a coherent current channel rising in the vessel as in the eruptive case, the flux rope in the failed torus case undergoes substantial internal reconfiguration. More specifically, the flux rope rises from a low-lying rope with uniform current density to an elevated rope with a "hollowed-out" current profile ($t = 240.0\,\mu$s). Instead of continuing to rise, this hollowed-out flux rope collapses back downward in just two Alfvén times and reforms with a relatively uniform J_T-profile at low altitude ($t = 258.0\,\mu$s). It is this sudden reconfiguration and collapse that characterizes the failed torus regime.

Upon closer examination, the current profiles during the downward collapse ($t = 244.0\,\mu$s, $t = 245.6\,\mu$s) are comprised of multiple sharp current sheets, including a reversed current sheet in the middle two frames. Such current sheets are clear evidence of the transient magnetic reconnection that facilitates the rapid topological reconfiguration of the flux rope. In this context, it is useful to examine the evolution of B_{Ti} in the second row of figure 9.17. We see that B_{Ti} rises substantially in magnitude as the

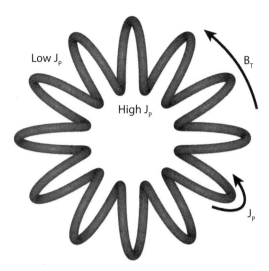

Figure 9.18. Heuristic wound-coil model for the toroidal field tension force. [From Yamada et al. (2016b).]

current profile hollows out. This sharp rise in B_{Ti} indicates that poloidal magnetic flux (associated with J_T) is being converted to toroidal magnetic flux (associated with B_{Ti}) (Myers et al., 2015). This type of flux conversion conserves helicity and is a classical example of magnetic self-organization (Taylor, 1986). The physical mechanism at the heart of helicity-conserving self-organization is magnetic reconnection. As we show in the next subsection, a key consequence of the transient increase in toroidal flux is a large magnetic tension force that causes the flux rope to collapse back downward and fail to erupt.

9.4.7 Dynamically evolving hoop and tension forces

It was concluded by the Myers et al. (2015) experiment that the toroidal field tension force is the dominant restraining force during failed torus events. This term had been neglected in traditional analysis of the torus instability. While it is true that the tension force is small in the failed kink and eruptive regimes, the self-organization and reconnection processes in the failed torus regime elevate its importance. The toroidal field tension force can be heuristically explained by considering a torus-shaped coil with helical windings (see figure 9.18). The toroidal curvature of the coil makes the density of windings (per unit length) higher on the inboard side of the coil than on the outboard side. Since the toroidal magnetic field is also stronger on the inboard side than on the outboard side, which is typically the case, the contraction force on the inboard side ($J_P \times B_T$) will be stronger than the corresponding expansion force on the outboard side. This produces a net contraction force. If we calculate this force for a large-aspect-ratio toroidal plasma discharge, the final result takes the form of a tension term with $F_t \sim B_T B_{Ti}/R$ (Myers et al., 2016). Thus, this force becomes larger for a low-aspect-ratio (a/R, where a is the minor radius) coil. As the profile of the internal paramagnetic toroidal field strengthens and broadens, the total paramagnetic toroidal flux and therefore the toroidal field tension force grow in magnitude.

The relaxation process we see here is typically described by the format shown in figure 9.1 and resembles the relaxation processes observed in RFP and spheromak fusion plasmas. In the central region of a plasma torus where μ is peaked before relaxation, the current profile or μ-profile flattens after a relaxation event. This process can be viewed as a flux conversion of poloidal flux to toroidal flux. Reducing parallel current at the center, where the field lines are mostly toroidal, means reducing poloidal flux and reduced hooping force and thus eruption force. Increasing the parallel current at the edge, where field lines are mostly poloidal, means increasing toroidal flux and results in a contraction force. Thus we can attribute our failed eruption to magnetic self-organization or magnetic relaxation of the plasma torus.

Chapter Ten

Studies of energy conversion and flows in magnetic reconnection

10.1 EXPERIMENTAL STUDY OF MAGNETIC ENERGY CONVERSION IN THE RECONNECTION LAYER IN A LABORATORY PLASMA

The most important feature of magnetic reconnection is that significant acceleration and heating of plasma particles occurs at the expense of magnetic energy. An example of this efficient energy conversion is the observation of large amounts of high-energy electrons associated with the reconnection of magnetic field lines in solar flares (Krucker et al., 2010). In some solar coronae, a third to a half of the magnetic energy is considered to be converted to high-energy electrons. On the other hand, evidence of very-high-energy ions of multiple megaelectron volts have been observed from solar eruptions (Lin, 2006). In the reconnection region of the earth's magnetosphere and solar wind (Phan et al., 2000; Wygant et al., 2005), convective outflows have been documented by in situ satellite measurements, but the exact physical mechanisms for bulk plasma heating and particle acceleration and for energy flow channels remain unresolved. This chapter addresses this unresolved question: How is magnetic energy converted to plasma kinetic energy? Here we study quantitatively the mechanisms of the conversion of magnetic energy to plasma electrons and ions in a laboratory reconnection layer. Based on experimental findings, we will develop an analytical model for the conversion of the magnetic energy to plasma particles in chapter 11.

In the classical Sweet–Parker model based on resistive MHD (magnetohydrodynamics), the energy dissipation rate is small $(\sim (B^2/2\mu_0) V_A L/S^{1/2})$ owing to the slow reconnection rate, where $S \gg 1$ is the Lundquist number (Parker, 1957; Priest and Forbes, 2000; Yamada et al., 2010). It is important to point out that in the Sweet–Parker model, the outgoing magnetic flux energy through the thin diffusion region is much smaller than the incoming magnetic energy, as shown in figure 3.3. In this model, almost all of the incoming magnetic energy is expected to be converted into particle energy within the narrow diffusion region $(S \gg 1)$. The plasma is heated (slowly) by classical resistive dissipation (ηJ^2) in the diffusion region and is accelerated to the Alfvén velocity due to both the pressure gradient and magnetic tension forces. In the exhaust, there is an equal partition between the flow and thermal energy increase,

$\Delta(5nk_B T/2) \sim nmv_{\text{out}}^2/2$, indicating that magnetic reconnection generates Alfvénic flows of heated plasma at the end of the very narrow exhaust (Priest and Forbes, 2000). Recent data from space and laboratory show, however, that these predictions do not hold during collisionless reconnection (Yamada, 2007; Eastwood et al., 2013). The main reason for this is now considered to be that two-fluid physics is dominant in the reconnection layer.

In a collisionless magnetic reconnection layer, electrons and ions move quite differently due to two-fluid dynamics (Sonnerup, 1979; Drake et al., 2009; Yamada, 2007; Yamada et al., 2010); differential motion between the strongly magnetized electrons and the unmagnetized ions generates strong Hall currents in the reconnection plane, as shown in chapter 5; see figures 5.1 and 5.6. Let us take the case of a prototypical reconnection layer. As magnetic reconnection is induced with oppositely directed field lines being driven toward the X-point ($B = 0$ at the center of the layer), ions and electrons also flow into the reconnection layer. The ions become demagnetized at a distance of the ion skin depth ($d_i = c/\omega_{pi}$, where ω_{pi} is the ion plasma frequency) from the X-point where they enter the so-called ion diffusion region, and they change their trajectories and are diverted into the reconnection exhaust, as we learned in chapter 5. The electrons, on the other hand, remain magnetized through the ion diffusion region and continue to flow toward the X-point. They become demagnetized only when they reach the much narrower electron diffusion region. In this two-fluid model, the expanding exhaust region becomes triangular in shape and the outgoing magnetic flux through this region is expected to be sizable, while the incoming magnetic energy is converted much faster to particle energy in this X-shaped reconnection layer. Note that this geometry of triangular exhausts allows fast exhausting fluxes of magnetic field and plasma, just as the Petschek MHD model does, as explained in chapter 3.

In the past, ion heating in plasma during reconnection was observed in a wide range of magnetic configurations in the laboratory such as the RFP (reversed field pinch) (Scime et al., 1992; Fiksel et al., 2009) and spheromak (Ono et al., 1996; Brown et al., 2002). Local heating in the reconnection layer of dedicated reconnection experiments has also been observed for both ions (Hsu et al., 2000; Stark et al., 2005) and electrons (Ji et al., 2004; Yoo et al., 2014a). However, the exact physical mechanisms behind the observed heating have not been well understood.

Observations in space and laboratory plasmas suggest that a significant fraction of the energy released during reconnection is converted into ion thermal energy (Eastwood et al., 2013; Yamada et al., 2014) in the reconnection layer. Recently, a more quantitative analysis of the energy conversion rate has been carried out, together with more accurate identification of energy flow processes (Yamada et al., 2015). The energy partition measured in the magnetotail is remarkably consistent with the recently obtained MRX (Magnetic Reconnection Experiment) data. It was found that a half of the magnetic energy flux is converted to particle energy flux with a high speed ($0.1 V_A$) and then branched off to the ion and electron enthalpy fluxes with a 2 to 1 ratio. This chapter describes the recent experimental investigation in detail, based on accurately measured data from a prototypical laboratory reconnection layer generated in MRX.

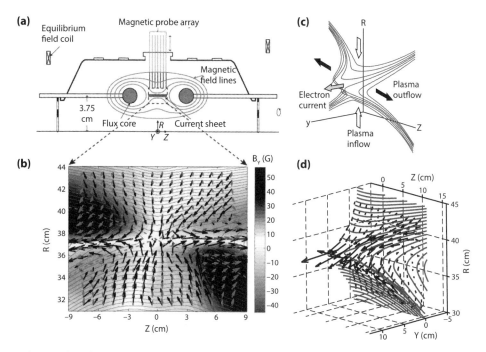

Figure 10.1. See Color Plate 2. (a) MRX apparatus and reconnection drive. (b) Measured flow vectors (the length represents velocity) of electrons (red arrows) and ions (blue arrows) in the full reconnection plane, together with poloidal flux contours (which represent reconnecting field-line components projected in the reconnection plane) and out-of-plane field contours. A 1 cm vector length stands for 2×10^6 cm/s, color contours represent out-of-plane field strength, and broken green lines depict (experimentally identified) separatrix lines. An azimuthal symmetry is assumed. For standard conditions, $n_e = 2\text{–}6 \times 10^{13}$ cm^{-3}, $T_e = 5\text{–}15$ eV, $B = 0.1\text{–}0.3$ kG, $S > 400$ in helium plasmas. (c) Conjectured 3D view of magnetic field lines moving together with plasma flows. (d) Measured magnetic field lines together with electron fluid flow vectors. [From Yamada et al. (2015).]

10.2   EXPERIMENTAL SETUP AND PLASMA PARAMETERS

The MRX facility was used to experimentally study the conversion of magnetic energy to particle energy in a nearly collision-free reconnection layer. Figure 10.1(a) shows a schematic of the MRX apparatus, wherein two oppositely directed field lines merge and reconnect. Experiments are carried out in a setup in which two toroidal plasmas with annular cross-section are formed around two flux cores as shown in figure 10.1; see also section 6.3.1.

Each flux core (darkened section in figure 10.1(a)) contains both TF (toroidal field) and PF (poloidal field) coils. After a poloidal magnetic field is created by the PF coil currents, an inductive helium discharge is created around each flux core by pulsing the

TF currents in the coils (Yamada et al., 1997a,b). After the annular plasmas are created, the PF coil current can be increased or decreased in order to drive different modes of reconnection. For decreasing PF current, the poloidal flux in the common plasma is pulled back toward the X-point (pull mode); this mode was used for the present experiment. For standard conditions of $n_e = 2$–6×10^{13} cm^{-3}, $T_e = 5$–15 eV, $B = 0.1$–0.3 kG, $S > 400$, the electrons are well magnetized ($\rho_e \ll L$, where ρ_e is the electron gyroradius) while the ions are not. The mean free path for electron–ion Coulomb collisions is in the range 5–20 cm (> the layer thickness) and, as a result, the reconnection dynamics are dominated by two-fluid and kinetic effects (Yamada, 2007; Yamada et al., 2010, 2016a). As usual, here we employ a geometry (R, Y, Z), where B_Z is the reconnecting field component and Y is the out-of-plane axis.

Figure 10.1(b) (see figure 5.7 in Color Plate 2) depicts the measured flow vectors of ions (in blue) and electrons (in red) in the whole reconnection plane, together with poloidal flux contours (representing magnetic field lines) and colored contours of the out-of-plane magnetic field component. There are clear differences between the ion and electron flow patterns, which demonstrate the two-fluid dynamics in the MRX diffusion layer. Here we can experimentally verify what was described in the previous section as features of the two-fluid reconnection layer.

Various diagnostics were used to study the comprehensive dynamics of plasma particles and mechanisms for energy conversion in the reconnection layer. Triple Langmuir probes were used to measure the electron temperature and density. The density measurements were calibrated by data from a CO_2 interferometer. The radial profile of the floating potential was obtained from a 17-tip floating potential probe with maximum resolution of 7 mm. Local ion temperature was measured by ion dynamics spectroscopy probes (IDSPs) (Fiksel et al., 1998), which obtained the spectrum of the He II 4686 Å line. The data was subsequently fitted to a sum of 13 Gaussian functions in order to take fine structure effects into account; without considering fine structure, the ion temperature would be overestimated by 15–25%. The time and spatial resolutions of the IDSPs are 5–6 μs and 3–4 cm, respectively. Mach probes were used to measure the ion flow velocity due to their better spatial and temporal resolutions. The data from the Mach probe are calibrated by spectroscopic measurements from the IDSPs. The electron flow vectors in the reconnection plane were derived by the electron current profile from the magnetic profile, measured by fine-scale magnetic probes, using $\mu_0 \boldsymbol{J} = \nabla \times \boldsymbol{B}$ and $\boldsymbol{V}_e = -\boldsymbol{J}/en_e + \boldsymbol{V}_i$.

10.3 ELECTRON FLOW DYNAMICS STUDIED BY MEASURED FLOW VECTORS

From measurements of profiles of magnetic field lines, the electric field, and the local flow vectors of electrons and ions, the dynamics of the plasma particles can be studied in significant detail. As the $E_y \times B_z$ (E_y is the reconnection electric field and B_z the reconnecting magnetic field) drift motion drives electrons toward the X-point together with field lines (figure 10.1(b)), the magnetic field strength weakens. As a result, the electron drift (E/B) velocity in the reconnection plane becomes very large near the X-point and electrons are ejected out to the exit. Figure 10.2(a) presents a more detailed three-dimensional description of the electron flows in one half of the reconnection plane. Ions, which become demagnetized as they enter the ion diffusion region whose width is $\sim d_i$ (5–6 cm), are accelerated across the separatrix lines (shown by blue vectors) while moving in the ion diffusion region and flow outward to the exhaust Z-direction (as seen in figure 10.1(b)). In contrast, the magnetized electrons flow inward toward the X-point along field lines, which are almost parallel with the separatrix lines at the edge of the inflow region. This electron flow pattern generates net circular currents in the reconnection plane, and it thus creates an out-of-plane magnetic field with the quadrupole structure shown in figure 10.1(b) and represented in three dimensions in figure 10.1(c) and figure 10.2(a). This is a signature of the Hall effect and our experimental data show very good agreement with typical PIC (particle-in-cell) simulations (Shay et al., 1998; Pritchett, 2001; Ji et al., 2008). The measured amplitude of this Hall quadrupole magnetic field is of order 40–60 G (Ren et al., 2005; Yamada et al., 2006), compared with 100–120 G reconnecting field strength. The increased reconnection electric field, caused by the strong Hall term ($\boldsymbol{J} \times \boldsymbol{B}$) and a steady current of electrons, leads to the observed fast motion of flux lines ($E = -d\Psi/dt = v_R d\Psi/dR = v_R B z$) in the reconnection plane, or the fast reconnection rate. Here we note that in a steady state reconnection, the reconnection field E_y is induced by a steady inflow and outflow motion of the reconnecting field lines of flux.

As the incoming field lines are stretched toward the y-direction (out of plane), as shown in figure 10.2(a), magnetic field lines break and electrons flow out rapidly to the exhaust direction. In the upstream (inflow) section of the MRX reconnection layer, a slow electron inflow velocity ($V_e \sim V_i \ll V_A$) is seen, while a much faster electron flow velocity is measured ($\sim 5V_A$) in the y-direction near the X-point region, as shown in figure 10.2(a). It should be noted that electrons flow out almost orthogonal to the magnetic field lines near the X-point region. While electrons flow out of the X-point region to the outflow direction (z), the reconnection of magnetic field lines occurs and electrons pull newly reconnected field lines toward the exhaust in the outflow region. The magnetic field lines in the inflow region move quickly, as reconnection occurs near the X-point, while in the exhaust region they slowly cross the separatrices.

10.4 OBSERVATION OF ENERGY DEPOSITION ON ELECTRONS AND ELECTRON HEATING

Distinctly different motions of magnetized electrons and demagnetized ions in the reconnection layer create a notable electric field profile. Near the current sheet, a strong in-plane electric field is expected, with electrons being frozen to the field lines as described section 5.5. In the center of the ion diffusion region, electrons primarily flow in the out-of-plane direction (y) perpendicular to $\boldsymbol{B}$, and we expect a sizable electric field toward the X-point. This situation would generate a strong potential well around the electron diffusion region with respect to inflow direction (R). This prediction was verified experimentally in MRX (Yoo et al., 2013), as well as in the magnetosphere by the MMS (Magnetospheric Multiscale Satellite) (Burch et al., 2016b)

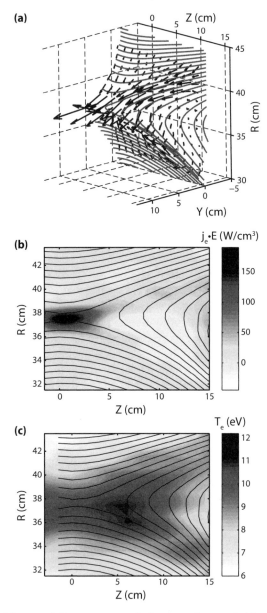

Figure 10.2. (a) Measured flow vectors of electrons in one half of the reconnection plane in a bird's eye view (in 3D geometry). While ions and electrons move together with field lines before entering the ion diffusion region, electrons move much faster as they reach toward the X-point region. (b) The energy deposition to electrons $\boldsymbol{j}_e \cdot \boldsymbol{E}$ is concentrated near the X-point. (c) A strong electron temperature rise is observed in the wide area of the exhaust region. Strong parallel heat conduction is considered to cause the high T_e at the exhaust region. [From Yamada et al. (2015, 2016b).]

satellite observation. At the same time we expect that a notable energy deposition to electrons occurs in the electron diffusion region, thus here we investigate first the value of $J_e \cdot E$ in this region.

Measurements show that the energy deposition rate to electrons $J_e \cdot E$ is concentrated near the X-point, as seen in figure 10.2(b), but in a much wider region ($\sim 10 d_e$) than predicted by numerical simulations (Pritchett, 2010). Furthermore, our data indicate that electron heating takes place in an even wider region of the exhaust, as seen in figure 10.2(c). The measured two-dimensional electron temperature profile shows that the region of high electron temperature expands along the magnetic field lines in the exhaust. We observe that electrons are heated in a wide region, with width $\sim 0.5 d_i$. Strong parallel heat conduction is considered to cause this wide observed region of high T_e. Based on two-dimensional energy transport analysis, we note that Ohmic dissipation based on the perpendicular Spitzer resistivity accounts for less than 20% of the observed deposition power (Yoo et al., 2017). Magnetic and electrostatic fluctuations in the lower hybrid frequency range are observed (Ji et al., 2004; Ren et al., 2008) near the X-point and throughout the downstream region, and are believed to cause the observed anomalous electron heating. However, more quantitative analyses on wave–particle interactions are required to determine the exact cause of the anomalous heating.

While the magnitude of the magnetic field decreases toward the X-point, the total electron kinetic and thermal energy increases substantially with respect to magnetic energy. As the electron beta $\beta_e = 2\mu_0 n_e T_e / B^2$ is initially 0.1 (in the inflow section) before reaching the reconnection region, it is well over unity inside the broad electron diffusion region, breaking the condition of a magnetically confined state, as clearly seen in figure 10.2(a). This condition could induce firehose-type instability in the region ($T_{e\parallel} \gg T_{e\perp}$), although the error bars of the measurement are too large for an exact analysis of the firehose modes.

It is important to note that when the energy deposition rate to electrons $j_e \cdot E$ is decomposed into $j_{e\perp} \cdot E_\perp + j_{e\parallel} E_\parallel$, i.e., separating the inner product of the vectors into that of the perpendicular and parallel components with respect to the local magnetic field lines, $j_{e\perp} \cdot E_\perp$ is measured to be significantly larger than $j_{e\parallel} E_\parallel$, as shown in figure 10.3.

Near the X-point, where energy deposition is maximum, $j_{e\perp} \cdot E_\perp$ is larger than $j_{e\parallel} E_\parallel$ by more than an order of magnitude. This very notable characteristic of energy deposition to electrons was verified by two-dimensional PIC numerical simulation (Yoo thesis, 2013; Yamada et al., 2016a) using the VPIC code (Bowers et al., 2008). While most features of electron flow vectors are reproduced and verified in these simulations, electron flow speeds are much more pronounced near the X-point, as well as on both sides of the separatrices. The thickness of the electron diffusion layer calculated by the two-dimensional numerical code was much smaller than that of the MRX experiment, as discussed extensively before (Dorfman et al., 2008; Ji et al., 2008; Roytershteyn et al., 2010). It was also notable that the energy deposition rate to electrons $j_{e\perp} \cdot E_\perp$ in their simulation was found to be significantly larger than $j_{e\parallel} E_\parallel$ near the X-point, where energy deposition is maximum, although the energy deposition region was much smaller than that observed in MRX.

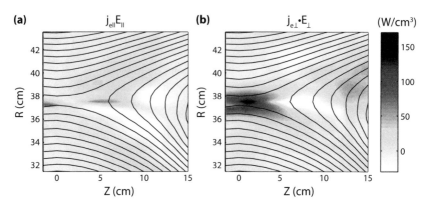

Figure 10.3. Comparison of two compositions of energy deposition rate measured in MRX. (a) $j_{e\parallel}E_{\parallel}$ and (b) $\boldsymbol{j}_{e\perp} \cdot \boldsymbol{E}_{\perp}$. [From Yamada et al. (2015).]

10.5   GENERATION OF AN ELECTRIC POTENTIAL WELL IN THE TWO-FLUID RECONNECTION LAYER

It was experimentally verified in MRX that an inversed-saddle shaped electric potential profile is formed in the reconnection plane to balance the Lorentz force on the electron flows (Yoo et al., 2013). It was found that the flows of magnetized electrons, which cause the Hall effects, produce a strong electric field in the reconnection plane especially across the separatrices as shown in figure 10.4(a), (b). A strong in-plane electric field is generated near the separatrices with a wider and deeper potential well downstream. These MRX potential data are consistent with simulation results (Karimabadi et al., 2007), as well as measurements by the Cluster spacecraft (Wygant et al., 2005), which showed a narrow potential well near the X-point with a half-width in the range 60–100 km (3–$5d_i$), and a deeper and wider well toward the exhaust region.

In the experiment, the electron diffusion region near the X-point was observed to be significantly wider than the electron skin depth (Ren et al., 2008; Ji et al., 2008), in contrast with two-dimensional numerical simulations (Chen et al., 2008a; Karimabadi et al., 2007). The in-plane (Hall) electric field (or potential drop) is mostly perpendicular to the local magnetic field lines, and is strongest near the separatrices. Electric potential is seen to be nearly constant along a poloidal flux contour (or magnetic field line) in the half of the reconnection plane in figure 10.4(a). In this figure, we notice that a large electric field across the separatrices extends to a significantly larger area of the reconnection layer ($L \gg d_i$) than the region in which field-line breaking and reconnection occur. A typical magnitude for the in-plane electric field E_{in} is ~ 700 V/m, which is much larger than the reconnection electric field $E_{\text{rec}} \sim 200$ V/m in MRX.

10.5.1   Observation of a potential well in the magnetotail

During magnetic reconnection in the magnetotail, an explosive release of magnetic energy is considered to occur, as described in chapter 2. It was learned in chapter 8

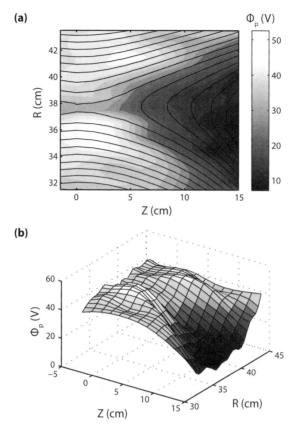

Figure 10.4. See Color Plate 7 for (b). Potential profile of one half of the two-fluid reconnection layer. (a) The inversed-saddle shaped electrostatic potential profile deduced from multiple Langmuir probe measurements and (b) a birds-eye view of the profile. The measured potential profile in MRX is divided into four regions by the separatrices (see figure 5.6). One half of the region is shown in (a). The electric potential tends to be constant along magnetic field lines shown by solid lines. [From Yamada et al. (2015).]

that a current sheet structure caused by Hall effects was identified around the near-earth neutral sheet line during a substorm, based on the data from the Geotail satellite (Asano et al., 2004). A negative potential well, a signature reported by numerical simulations of two-fluid reconnection, was measured. They found a double-peaked current sheet away from the X-line and attributed its cause to Hall current profiles at the separatrices around the neutral sheet.

The detailed characteristics of the neutral sheet in the near-earth magnetotail were measured by the Cluster spacecraft (Wygant et al., 2005). The measurements of electric fields, magnetic fields, and ion energy are used to study the structure and dynamics of the reconnection region in the tail at distances of $18R_E$. They investigated the structure of electric and magnetic fields responsible for the acceleration of ions, and

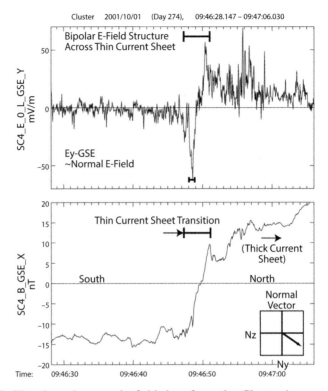

Figure 10.5. Electric and magnetic field data from the Cluster-4 spacecraft from a "thin" ($\delta \sim 4c/\omega_{pe}$) current sheet on 1 October 2001 at 09:46:50 UT. Measurements are (top) E_y-GSE dominated by the normal component of the electric field and (bottom) B_x-GSE showing the current sheet. The distance scale is determined from the normal velocity $V_n \sim 80$ km/s. The geocentric solar ecliptic system (GSE) has its X-axis pointing from the earth towards the sun and its Y-axis is in the ecliptic plane pointing towards dusk. Its Z-axis is parallel to the ecliptic pole (north–south). [From Wygant et al. (2005).]

the formation of the electron current layer during the process of magnetic reconnection in the earth's geomagnetic tail.

Electric field and magnetic field measurements from the thinnest current sheet, obtained between 09:40 and 09:50 UT, are presented in figure 10.5. The electric field has a bipolar signature coinciding with a flip in the direction of the magnetic field. Electric field reversal occurs at the center of the current sheet, suggesting the presence of a strong potential well predicted by many numerical simulations. Wygant focused on measurements of the large amplitude normal component of the electric field observed near the reconnection X-line, the structure of the associated potential drops across the current sheet, and the role of the potential structure in the ballistic acceleration of ions across the current sheet. The measured width of the individual current sheet was often very thin, in the range 60–100 km (3–$5c/\omega_{pe}$). The observed high electric field structure would lead to a large 4–6 kV electric potential well centered in the separatrix region and over 8 kV near the exhaust of the neutral sheet. Measured H^+ velocity space distributions obtained inside the current layers provide evidence that the H^+ ions are accelerated into the potential well along the Z-axis, producing a pair of counter-streaming, double-peaked energetic H^+ beams. These results reveal important signatures of two-fluid reconnection dynamics, described in chapter 5 and in this chapter: the strong potential well and a very thin (electron) current sheet. This is good evidence that the flows of electrons and ions are quite decoupled. Based on their observations, they proposed the following mechanisms for the acceleration of ions in the neutral sheet in the magnetotail:

(1) Incoming field lines bring magnetized electrons to the X-line, compress them, and create a strong negative well near the $z = 0$ line (figure 10.5).
(2) Nonmagnetized ions are accelerated along the Z-axis toward the center, overshoot, and bounce back. During this process, ions are accelerated toward the exit along the x-axis because of the wedge-shaped structure of the potential well; see MRX data in Fig. 10.4. Based on this proposed scenario, the schematic diagrams they developed then are shown in figure 10.6 on page 188. From these dynamics, the generation of the observed counter-streaming ion beams was explained. It was an excellent demonstration of very effective use of cluster satellite data.

Although magnetic measurements by space satellites used to be not as conclusive as laboratory results, for which multiple reproducible plasma experiments can be carried out, the space satellite diagnostics for the particle energy distribution function compensate for this weakness and contribute importantly to the understanding of collisionless reconnection. Multiple satellite observations also greatly improve the reliability and effectiveness of space data analysis.

10.6   ION ACCELERATION AND HEATING IN THE TWO-FLUID RECONNECTION LAYER

Back to MRX, we observe direct acceleration of ions near the separatrices due to the strong electric field mentioned above, whose spatial scale is ~ 2 cm, smaller than the ion gyroradius of ~ 5 cm. Figure 10.7(a) shows a two-dimensional profile of ion flow vectors measured by Mach probes, along with poloidal flux contours (magnetic field lines projected on the reconnection plane) and contours of electric potential Φ_p. It is clearly noticeable that ion flows change their directions at the separatrices and are accelerated in both the Z- and the R-directions. Figure 10.7(b) depicts the spectrum of the 4686 Å line of He II ions measured by the IDSP (Fiksel et al., 1998) at three locations. The spatial resolution of this local spectroscopic measurement is 4 cm. This spectral profile represents the local velocity distributions of ions versus v_Z. Shifted Maxwellian distributions are observed at three typical positions as shown in figure 10.7(b). Notable heating is observed as the ions flow out to the exhaust from the X-region, as demonstrated in (b). The maximum ion outflow of 1.6×10^6 cm/s corresponds to 5 eV of energy per helium ion, which is much smaller than the magnitude of the potential

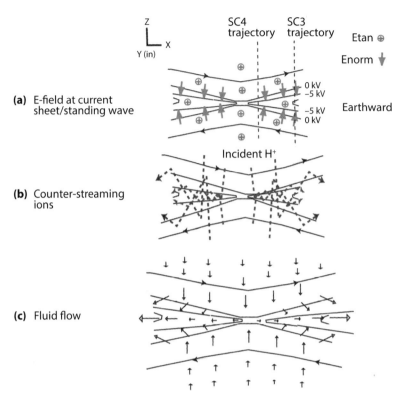

Figure 10.6. Schematic diagrams of the ion acceleration mechanism contemplated by Wygant et al. (2005). (a) The presence of a strong normal component of the electric field was observed with a standing wave/current layer near the X-line. (b) The trajectory of representative ions ballistically accelerated into 4–6 kV potential-well-forming quasi-trapped counter-streaming beams. Counter-streaming beams produce effective pressure, deceleration of ion fluid across a potential drop, and a jet at $1 V_A$ along the outflow direction. (c) Schematic of ion fluid flow vectors consistent with averaging over single particle trajectories.

decrease across the separatrices (~ 30 V). This indicates that ions must lose considerable momentum as they pass through the downstream region.

The cause of this anomalously fast slowdown of ions, together with ion heating, is considered to be "remagnetization" of the outgoing ions. Since it is difficult to verify this mechanism experimentally, two-dimensional fully kinetic simulations have been carried out to verify this remagnetization and understand how ions are heated downstream. In these simulations, realistic MRX global boundary conditions are used in the PIC code VPIC (Bowers et al., 2008). In addition, Coulomb collisions are modeled using the Takizuka–Abe particle-pairing algorithm (Takizuka and Abe, 1977; Daughton et al., 2006), such that ν_{ii}/Ω_i and $\lambda_{i,\mathrm{mfp}}/d_i$ are matched to the experimentally measured values, where ν_{ii} is the ion–ion collision frequency, Ω_i is the upstream ion

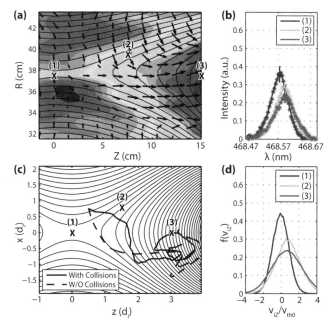

Figure 10.7. See Color Plate 7. Ion dynamics in the ion diffusion region and sample distribution functions. (a) 2D profile of measured flow vectors of ions are shown, together with contours of poloidal flux lines and the electrostatic potential profile shown in figure 10.6. Across the separatrices, ions are accelerated toward the exhaust direction. (b) Normalized spectra of measured He II 4686 Å line (ion) at three different locations specified with crosses in (a). The He II spectral lines are renormalized by local ion temperature. (c) Sample ion trajectories in a PIC-simulated reconnection plane with (thick solid line) and without (thick dashed line) collisions. (d) Data from numerical simulation corresponding to the measurement (b) is shown. [From Yamada et al. (2016b).]

cyclotron frequency, and $\lambda_{i,\mathrm{mfp}}$ is the ion mean free path. As the normal (R) component of reconnected magnetic field becomes stronger further downstream, as shown in figure 10.7(c), the ion trajectory is significantly affected by the magnetic field of the exhaust, and thus ions are remagnetized. With collisions, ions are almost fully thermalized with a higher temperature than the initial value. We note that the ion and electron dynamics are primarily dictated by (collision-free) two-fluid physics, even though some energy loss mechanisms are influenced by collisions.

We obtain good agreement between the observed ion temperature profile and numerical simulation results only with the correct collision frequencies. Figure 10.7(d) shows the ion distribution functions in the simulation at three locations: at the X-point, separatrix, and exhaust. With realistic collisions, ions are almost fully thermalized at the exhaust with a higher T_i than the upstream value. On the other hand, in the collisionless simulation ion distribution is different from Maxwellian, although a broadening in the

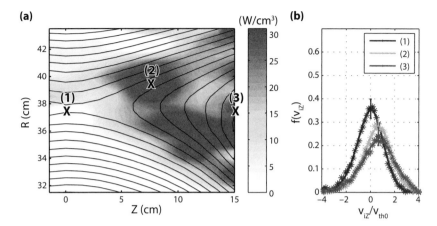

Figure 10.8. See Color Plate 7 for (a). (a) Measured energy deposition rate to ions $j_i \cdot E$. High-energy deposition is primarily due to $j_{i\perp} \cdot E_\perp$, which is concentrated in the ion diffusion region. Energy deposition to ions occurs across the separatrices and in a much wider region than for electrons. (b) Normalized spectra of measured He 4686 Å at three locations marked with crosses in (a). [From Yamada et al. (2015).]

ion distribution exists when it is compared to that at the X-point. These results indicate that ion thermalization is due both to remagnetization and to collisional effects in the downstream region.

An electrostatic acceleration of ions was observed near the separatrices due to the strong electric field mentioned above, whose spatial scale is ~ 2 cm, smaller than the ion gyroradius of ~ 5 cm.

Figure 10.8(a) shows profiles of the energy deposition to ions through $j_i \cdot E$. It is found that this occurs primarily in the exhaust side of the separatrix lines and is concentrated near the separatrices in the exhaust region, as seen in figure 10.8(a), (b).

This acceleration and heating of ions happens in a wide region, extending over an ion skin depth: the ion diffusion region. These accelerated ions are then thermalized by remagnetization through stochastic ion motions and some collisions in the downstream region. When the energy deposition rate to ions $j_i \cdot E$ is decomposed into $j_{i\perp} \cdot E_\perp + j_{i\parallel}E_\parallel$, the perpendicular component $j_{i\perp} \cdot E_\perp$ is again found to be dominant over $j_{i\parallel}E_\parallel$ in the regions where energy deposition to ions is maximum (Yamada et al., 2015).

10.7   EXPERIMENTAL STUDY OF THE DYNAMICS AND THE ENERGETICS OF ASYMMETRIC RECONNECTION

With the recent launch of the MMS (Burch et al., 2016b), the focus of reconnection research has turned to the dynamics of asymmetric reconnection, in which the plasma density of one side of the inflow region is significantly larger than the other (by factor of 10 or more). This is one of the most important features of the magnetopause reconnection, in which the pileup of the solar-wind plasma density is significantly larger than the magnetosphere density by a factor of 10–50. In the reconnection layer in the magnetopause, the solar-wind plasma pressure balances with the magnetic field pressure of the earth dipole field. In the reconnection layer, the thickness of the current sheet becomes comparable to the ion skin depth as well as the ion gyroradius, as we learned in chapter 5. Ions become demagnetized within the reconnection region as the magnetic field becomes small, while electrons are still magnetized and remain frozen to field lines until they reach very close to the X-line. This reconnection region is still in the two-fluid regime. In the vicinity of the X-line, even electrons become demagnetized and diffuse; thus we call this region the electron diffusion region.

Unique features of asymmetric reconnection have been observed in space (Mozer et al., 2008a) and numerical simulations (Mozer et al., 2008b; Tanaka et al., 2008). In asymmetric reconnection, the quadrupole out-of-plane magnetic field (QF) and the bipolar in-plane electric field, which are two signatures of symmetric collisionless reconnection, become almost bipolar and unipolar, respectively. Moreover, strong density gradients form near the low-density-side separatrix, where strong electric field fluctuations are frequently observed (Pritchett et al., 2012; Roytershteyn et al., 2012). The upstream density asymmetry also impacts the ion flow pattern by shifting the ion inflow stagnation point to the low-density side.

Here, let us study the first quantitative analysis of asymmetric reconnection in a laboratory plasma reported recently by Yoo et al. (2013), in which a significant upstream density ratio of up to 8 was systematically created in MRX.

Figure 10.9(a) shows a cross-section of the MRX device in the R–Z plane for asymmetric reconnection experiments. The two gray circles are "flux cores" that each contain two independent coils: a PF coil and a TF coil. As shown in figure 10.9(b), the upstream density asymmetry is generated during the plasma formation period (before driving pull-reconnection) using the inductive electric field from the increasing TF coil current E_{TF}. For this experimental campaign, the direction of E_{TF} is radially outward during the plasma formation, such that ions are pushed to the outboard side, increasing the outer density and thus generating a density asymmetry along the R-direction. Generally, the heavier the ion species is, the longer this process takes. Thus, the density asymmetry during the quasi-steady period can be varied by controlling the TF waveform and using different gas species. We use helium as a filling gas for asymmetric reconnection, and deuterium to create a relatively symmetric plasma. In addition, for the scaling study, the helium fill pressure is varied for further control of the upstream density ratio up to 10.

The main diagnostics for this study are the same as those of symmetric reconnection experiments described earlier. Extensive R–Z scans of Langmuir probes and Mach probes are conducted to obtain two-dimensional profiles of n_e, T_e, Φ_p, and V_i for both asymmetric (4.5 mT helium discharges) and symmetric (4 mT deuterium discharges) cases.

Figure 10.9(c) shows clear differences in the radial density profile at $Z = 0$ between the asymmetric and symmetric cases. For the asymmetric case, the outboard side ($R > 37.5$ cm) has up to 8 times larger density than the inboard side ($R < 37.5$ cm). The main transition from low to high density occurs on the low-density side. The measured

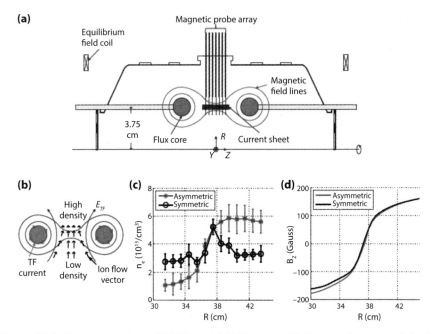

Figure 10.9. A cross-section of MRX. (a) MRX setup. (b) Schematic view of the ion dynamics for generating an asymmetric density profile during the plasma formation period. The blue arrows indicate the direction of the TF coil current. The arrows demonstrate the ion flow patterns before we start the reconnection operation by pulling field lines toward the X-point. (c) Radial electron density profiles at $Z = 0$ for both asymmetric and symmetric cases. (d) Radial profiles of the reconnecting magnetic field component (B_Z) at $Z = 0$. [From Yoo et al. (2013).]

two-dimensional density profile (not shown) shows that a strong density gradient forms at the low-density-side separatrix, which is consistent with numerical simulations (Pritchett, 2008; Tanaka et al., 2008). For the symmetric case, the upstream density is very similar and density peaks at the center of the current sheet ($R = 37.5$ cm).

Reconnecting magnetic field (B_Z) profiles at $Z = 0$ are rather similar, as shown in figure 10.9(d). Despite the large density asymmetry, the low-density side has only about a 15–30% larger B_Z magnitude than the high-density side. The asymmetry in the magnetic field magnitude does not have to be strong to ensure the pressure balance across the current sheet because the magnetic pressure is proportional to B^2, while the particle pressure is proportional to n. In many space observations, the upstream magnetic field strength ratio is also much smaller than the upstream density ratio (Mozer and Pritchett, 2011).

The upstream density asymmetry modifies the quadrupole out-of-plane magnetic field profile, as shown in figure 10.10. For the asymmetric case, the magnitude of the QF on the high-density side (~ 90 G) is about twice the symmetric case (~ 45 G). The

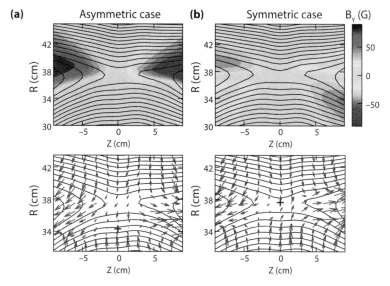

Figure 10.10. See Color Plate 8. 2D profiles of the out-of-plane magnetic field (B_Y) for (left) asymmetric and (right) symmetric cases. Compared to the symmetric case, the quadrupole magnetic field component is enhanced on the high-density side ($R > 37.5$ cm) and suppressed on the low-density side ($R < 37.5$ cm). Plus signs (in red) are stagnation points for poloidal ion flows, showing a strong shift to the low-density side for the asymmetric case. Black lines indicate contours of the poloidal magnetic flux that represent magnetic field lines. [From Yoo et al. (2013).]

low-density side, on the other hand, is only about 10–20 G, which is about one-third of the symmetric case.

This asymmetric quadrupole profile can be explained by the Hall term in the generalized Ohm's law. Since the reconnection electric field, which is balanced by the $J \times B$ Hall term in the upstream region (Ren et al., 2008), should be the same on both sides of the layer, we have

$$E_{rec} = -v_1 B_1 \approx -v_2 B_2, \tag{10.1}$$

where subscript 1 indicates upstream quantities on the high-density side, while the subscript 2 means those on the low-density side. Thus, the Hall current of the high-density side, $J_1 \approx (n_1 v_1)/(n_2 v_2) J_2 \sim (n_1/n_2)(B_2/B_1) J_2 \approx (n_1/n_2)^{1/2} J_2$ is larger by $(n_1/n_2)^{1/2}$ over the lower density side based on a pressure balance between the high and low density plasma systems. Since the QF is generated by the Hall current, the high-density side thus has a larger magnitude of the QF by the square root of density ratio, 3-4.

Using IDSPs to measure the ion temperature, about 300 pieces of magnetic probe to measure the vector magnetic field B, Mach probes to measure ion flows, and triple

Langmuir probes to measure electron temperature and density, the dynamics of the plasma within the reconnection layer were studied without an externally imposed guide field.

The different electron flow patterns generate quite different Hall field patterns in the reconnection plane, as seen in figure 10.10(a), (b). The measured flow profiles of ions clearly demonstrate that two-fluid reconnection is at work in MRX. Ions, which become demagnetized as they enter the ion diffusion region, are accelerated across the separatrices, flowing outward to the exhaust direction, as seen in (c). In contrast to the case of symmetric reconnection, we observe that inflowing ions also form a stagnation point (denoted by a circle) near the X-line (X-point) on the low-density side, with a shift of about 2–3 cm (0.3–$0.5d_i$). This was also verified in the two-dimensional simulation that will be described later.

10.7.1 Energy deposition and structure of the electron diffusion region in the asymmetric reconnection layer

As in the case of symmetric reconnection, we identify a two-scale diffusion layer in which an electron diffusion layer (half-width ~ 5 mm in MRX) resides inside the ion diffusion layer (half-width ~ 6 cm in MRX (He gas)). In this situation, the ion diffusion layer is characterized by the regime where $E + V_e \times B = 0$ with $E + V_i \times B \neq 0$. The electron diffusion layer is the regime $E + V_e \times B = E' \neq 0$, where E' is the electric field in the electron fluid frame. Just outside the electron diffusion layer, $E' = 0$ holds, namely electrons move with magnetic field lines in the reconnection plane (electron-flux freezing; chapter 5), and this relationship was clearly verified by quantitatively evaluating force balance in MRX (Yoo et al., 2013; Yamada et al., 2014). We note that in the case of MRX, the difference between E and E' is relatively small since V_e is much smaller than the electron thermal velocity ($V_e \sim 0.1V_{\text{eth}}$, where V_{eth} is the electron thermal velocity), but this does not apply for MMS data.

In the asymmetric MRX experiments, we observe distinctly different flow patterns compared to the symmetric case which was presented previously. Figure 10.11(a), (b) presents measurements of the electron flows in two- and three-dimensional views within one half of the reconnection plane. As seen in the figure, electrons are flowing out in the Y-direction as well as toward the exhaust in the Z-direction. The out-of-plane electron drift velocity becomes very large at the stagnation point of in-plane electron flows, which is located near the X-line but shifted toward the lower-density side by several electron skin depths ~ 5–$7c/\omega_{pe}$. The out-of-plane magnetic field nominally exhibits a quadrupole pattern during symmetric reconnection, a signature of the Hall effect, but is modified significantly during asymmetric reconnection due to shifted patterns of Hall currents, as seen in the color contours of (b). Due to this nearly bipolar structure seen in the reconnection plane (a), wherein reconnecting field lines move together with electrons, the reconnection plane is tilted to the Y-direction as shown in (b). This plane is orthogonal to the y-axis for symmetric reconnection. It should be noted that the tilt is strong in the high-density inflow region due to the stronger out-of-plane Hall field component. The electrons, which move together with the flow of magnetic field lines in the tilted plane in the high-density side, become demagnetized

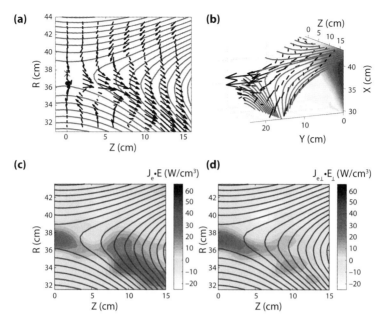

Figure 10.11. Electron dynamics observed in an asymmetric reconnection. (a) In the reconnection plane, electron flows together with reconnecting field lines. The X marker at $(R, Z) = (37.6, 0)$ is the X-line and the black circle denotes the stagnation point of in-plane electron flow. (b) 3D views of electron flow vectors with respect to the reconnecting field lines. Because of the dipole structure of the out-of-plane magnetic field, the reconnection plane is tilted in the Y-axis in the high-density (upper) side. Note the strong electron current in the Y-direction at the stagnation point. (c) Energy deposition to electrons through $j_e \cdot E$ is concentrated in the electron diffusion region around the stagnation point, as well as in the lower-density side of the exhaust. (d) The perpendicular component $j_{e\perp} E_\perp$ is much larger than $j_{e\parallel} E_\parallel$ and is concentrated in the electron stagnation point, while the contribution from $j_{e\parallel} E_\parallel$ is notable on the lower-density side of the exhaust as seen in (c) and (d). [From Yamada et al. (2018).]

in the electron diffusion region and stream out in the Y-direction as well as in the Z-direction.

With respect to the energy deposition rate to electrons $j_e \cdot E$, $j_{e\perp} E_\perp$ is measured to be significantly larger than $j_{e\parallel} E_\parallel$ near the X-line. Near the electron stagnation point, $j_{e\perp} E_\perp$ is larger than $j_{e\parallel} E_\parallel$ by more than an order of magnitude. It was shown earlier in MRX that in symmetric reconnection without a guide field, the energy dissipation to electrons occurs primarily due to $j_{e\perp} E_\perp$ only near the center of the electron diffusion region, the X-line. However, in asymmetric reconnection, it is verified that $j_e \cdot E$ peaks up through $j_{e\perp} E_\perp$ at the *stagnation point* of the electrons' in-plane flow, which is separated from the X-line by ~ 5–$6c/\omega_{pe}$. Recent analysis of data from MMS also verified this key feature by demonstrating that the value of $j_e \cdot E$ peaks when MMS

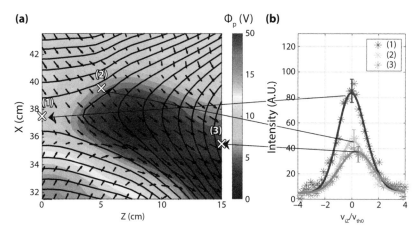

Figure 10.12. Electrostatic potential profile and ion acceleration in MRX in asymmetric reconnection. (a) The inversed-saddle shaped potential well is shifted toward the low-density side and a sharp potential drop occurs in the low-density side of the exhaust region, as is visible in the dark section shifted to the bottom section of reconnection layer. (b) The ion velocity distribution in the Z-direction was measured by the IDSP spectroscopically. The peak of the ion distribution function in V_{iZ} increases to higher values toward the downstream. Measured velocity V_{iZ} is normalized by the ion thermal velocity in the X-line region. Compared to the symmetric case, ions are not significantly accelerated near the separatrix on the high-density side (green line). The ion temperature in the exhaust region (red line) is also slightly lower. The change of the in-plane potential profile is responsible for these differences. See the dark section in the lower bottom part of the figure. Error bars are the square root of the number of photons. [From Yamada et al. (2018).]

encounters (Burch et al., 2016b) the electron stagnation point in the electron diffusion region, as will be discussed in chapter 12.

10.7.2   Ion dynamics and energetics in MRX and the magnetosphere

The large in-plane electric field plays a key role in ion acceleration and heating, as discussed in earlier sections. Recent studies have identified the mechanisms involved in the in-plane electrostatic field being generated by the force balance of the electrons flowing through the center of the reconnection layer. During symmetric reconnection in MRX, it was found that the measured in-plane electric potential profile is an inversed-saddle shape and the resulting electric field is three times larger than the reconnection electric field. However, during asymmetric reconnection, the inversed-saddle shaped potential well is shifted toward the low-density side and a sharp potential drop occurs on the low-density side of the exhaust region, as shown in figure 10.12(a). The unmagnetized ions are accelerated by the in-plane electric field in the exhaust region in both the Z- and X-directions, primarily in the high-density side, and are heated further downstream.

This is in contrast to the symmetric case in which ions are accelerated on both sides of the separatrices. For the case of symmetric reconnection, the value of $\boldsymbol{j}_{i\perp} \cdot \boldsymbol{E}_\perp$, which is the ion energy gain per unit time and unit volume, is about 30–40 W/cm^3 near the separatrices (Yamada et al., 2018). For the asymmetric case it is reduced to about 15 W/cm^3, although the ion acceleration region is notably wider. Figure 10.12(b) presents the velocity space distribution of ions at the three locations denoted in (a). One can see that the ions are drifting downstream with an elevated temperature, which is caused by stochastic motions of ions and some collisional effects.

Chapter Eleven

Analysis of energy flow and partitioning in the reconnection layer

11.1 FORMULATION FOR A QUANTITATIVE STUDY OF ENERGY FLOW IN THE RECONNECTION LAYER

During magnetic reconnection, the conversion of magnetic energy into particle energy in the form of particle heating and acceleration with high nonthermal energies is perhaps the most important aspect. Here, let us investigate the mechanisms of the energy release and conversion processes by formulating an inventory of energy in a prototypical reconnection layer. Keeping in our minds the two types of reconnection layers shown in figure 11.1, we compare the fundamentally different natures of energy conversion analysis. We start with an MHD (magnetohydrodynamic) formulation and then move into the energy dynamics of the magnetic reconnection layer where the two-fluid and kinetic physics mechanisms become dominant. A previous investigation was carried out in an MHD regime using (ad hoc) resistivity, which provides dissipation and enables reconnection (Birn and Hesse, 2005). In more realistic models of collisionless reconnection, such as in the magnetosphere and hot laboratory plasmas, the dissipation from plasma kinetic waves electron inertia (which causes nongyrotropy of the electron and pressure tensor) (Vasyliunas, 1975; Hesse et al., 1999, 2001) becomes more important. We here note that quantitative results on energization and energy conversion can depend on the system size considered. In smaller systems, such as those usually considered in particle simulations of magnetic reconnection or laboratory experiments, the actual dissipation in the reconnection layer is substantial and can be more easily quantified with respect to the inflow magnetic energy. In this chapter, we use the formulation of Yoo thesis (2013) and Yamada et al. (2016a) recently developed. Let us start our investigation by considering a simple Sweet–Parker-type reconnection layer in an MHD formulation, as described by figure 11.1(a), and then move to a two-fluid formulation in (b), for which we learned the mechanisms for energy conversion to electrons and ions in the MRX (Magnetic Reconnection Experiment) in chapter 10. We will try to develop a simple analytical theory for energy flows and will experimentally measure the energy inventory and partitioning in the reconnection layer; we compare the results with PIC (particle-in-cell) simulations.

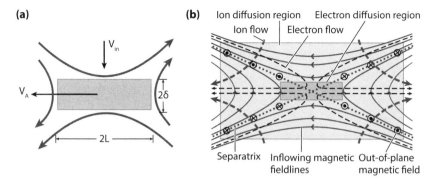

Figure 11.1. (a) Sweet–Parker model based on resistive MHD. Electrons and ions move together throughout the reconnection region. (b) Two-fluid model for a collisionless reconnection layer. Note the different motions of electrons and ions.

11.1.1   Basic MHD formulation for the Sweet–Parker model

In our resistive MHD formulation, magnetic fields move and diffuse in the rectangular reconnection region of figure 11.1(a), where incoming plasma flux is balanced with the outgoing flux, satisfying continuity equations for both plasma fluid and magnetic flux. The reconnection rate depends on the Lundquist number L_q (to avoid confusion with the Poynting vector (S), in this chapter we use L_q instead of the conventional S to denote the Lundquist number), which is usually extremely large: 10^2–10^8 in laboratory fusion plasmas and 10^2–10^{15} in space plasmas. In steady state, the inductive equation introduced in chapter 3 can be simplified to

$$V_{\text{in}}B = (\eta/\delta\mu_0)B, \tag{11.1}$$

where V_{in} is the inflow speed and δ is the half-width of the diffusion region. Using the continuity equation for plasma flows in the reconnection layer, $V_{\text{in}}L = V_{\text{out}}\delta$, where L is the half-length of the diffusion region and V_{out} is the outflow speed. Using pressure balance between the upstream ($B^2/2\mu_0$) and the downstream ($\sim \rho V^2/2$, where ρ is the mass density) regions, we have $V_{\text{out}} = V_A$, which leaves us a very simple formula for the reconnection speed V_{in} as explained in chapter 3:

$$\frac{V_{\text{in}}}{V_A} = \frac{\delta}{L} = \frac{1}{\sqrt{L_q}}, \tag{11.2}$$

where $V_A = B/\sqrt{\mu_0\rho}$ is the Alfvén velocity and $L_q = \mu V_A L/\eta$ is the Lundquist number, the ratio of the Ohmic diffusion time to the crossing time of the Alfvén waves.

We note that the slowness of the Sweet–Parker model comes from the assumption that both plasma and magnetic flux have to go through a narrow rectangular neutral sheet of thickness $\delta = L/\sqrt{L_q}$. However, it can be made much broader if we introduce an effective resistivity η_{eff}, as shown in figure 11.1(a).

11.1.2   MHD description of energy flow

Let us first study an inventory of overall energy flow in the reconnection region by the single-fluid (MHD) model. We use a formulation with the energy transport equation (Birn and Hesse, 2005),

$$\frac{\partial}{\partial t}\left(\frac{B^2}{2\mu_0} + \frac{\epsilon_0 E^2}{2} + u + \frac{\rho}{2}V^2\right) + \nabla \cdot (S + H + K) = 0, \tag{11.3}$$

where $u = (3/2)p$ is the internal energy density, $p = n_e T_e + n_i T_i$ is the pressure, $\rho = m_e n_e + m_i n_i$ is the mass density, V is the single-fluid velocity, $S = (E \times B)/\mu_0$ is the Poynting flux, $H = (u + p)V$ is the enthalpy flux, and $K = (\rho/2)V^2 V$ is the flow energy flux. In nonrelativistic plasmas, the electric field energy ($\epsilon_0 E^2/2$) is usually neglected. The neglect of inductive electric field energy is a valid approximation for the evolution of macroscopic magnetic fields, provided that the characteristic timescale is much longer than the light crossing time. In nonrelativistic plasmas, the electrostatic field energy can also be neglected. The above equation is used to analyze energy conversion in the Sweet–Parker model, which is the most well-known MHD model for magnetic reconnection.

Following the previous description of the Sweet–Parker model in chapter 3, the incoming Poynting flux (S_{in}), flow energy (K_{in}), and enthalpy (H_{in}) flux of reconnection layer are expressed by

$$S_{\text{in}} = \frac{E_{\text{rec}} B_{\text{rec}}}{\mu_0} = \frac{B_{\text{rec}}^2}{\mu_0} V_{\text{in}}, \tag{11.4}$$

$$K_{\text{in}} = \frac{\rho V_{\text{in}}^3}{2} = \frac{1}{2L_q} S_{\text{in}}, \tag{11.5}$$

$$\text{and} \quad H_{\text{in}} = \frac{5}{2} p_{\text{in}} V_{\text{in}} = \frac{5}{4} \beta S_{\text{in}}. \tag{11.6}$$

Here, p_{in} is the upstream pressure and β is the ratio of upstream plasma pressure to reconnecting magnetic pressure. When $L_q \gg 1$ and $\beta \ll 1$, as is typical of magnetized astrophysical plasmas, the total incoming flux is dominated by the Poynting flux.

The outgoing fluxes can also be expressed in terms of the incoming magnetic energy flux S_{in}. Since the reconnection electric field is uniform over the layer due to the steady-state assumption, we have

$$E_{\text{rec}} = V_{\text{in}} B_{\text{rec}} = V_A B_{\text{out}}, \tag{11.7}$$

where B_{out} is the magnetic field strength in the exhaust region. The outgoing Poynting (S_{out}) and flow energy (K_{out}) fluxes are given by

$$S_{\text{out}} = \frac{E_{\text{rec}} B_{\text{out}}}{\mu_0} = \frac{S_{\text{in}}}{\sqrt{L_q}} \tag{11.8}$$

and

$$K_{\text{out}} = \frac{\rho V_A^3}{2} = \frac{\sqrt{L_q}}{2} S_{\text{in}}. \tag{11.9}$$

The outgoing enthalpy flux (H_{out}) can be obtained using eq. (11.3). With the steady-state assumption and the divergence theorem, the relation between the incoming and outgoing fluxes is

$$(S_{\text{in}} + H_{\text{in}} + K_{\text{in}})L = (S_{\text{out}} + H_{\text{out}} + K_{\text{out}})\delta. \tag{11.10}$$

With this relation, and eqs. (11.4)–(11.9), H_{out} is found to be

$$H_{\text{out}} = \left[\left(\frac{1}{2} + \frac{5}{4}\beta \right)\sqrt{L_q} - \frac{1}{2\sqrt{L_q}} \right] S_{\text{in}}. \tag{11.11}$$

The above equations indicate that most of the incoming electromagnetic energy is dissipated within the rectangular-shaped diffusion region and that the energy is equally converted to plasma flow and thermal energy. The change in the magnetic energy flow (ΔW_M) inside the diffusion region per unit time and unit length along the out-of-plane direction is given by

$$\Delta W_M = -4(L S_{\text{in}} - \delta S_{\text{out}}) = -4L S_{\text{in}} \left(1 - \frac{1}{L_q} \right). \tag{11.12}$$

The outgoing magnetic energy is smaller than the incoming energy by a factor of $1/L_q$. Since $L_q \gg 1$ for most astrophysical and large laboratory plasmas, the outgoing magnetic energy is negligible, which means that most of the incoming magnetic energy is dissipated within the diffusion region by resistivity. At the same time, the changes in the flow (ΔW_K) and enthalpy (ΔW_H) energy are written as

$$\Delta W_K = 4(L K_{\text{in}} - \delta K_{\text{out}}) = 2L S_{\text{in}} \left(1 - \frac{1}{L_q} \right) = -\Delta W_M/2, \tag{11.13}$$

$$\Delta W_H = 4(L H_{\text{in}} - \delta H_{\text{out}}) = 2L S_{\text{in}} \left(1 - \frac{1}{L_q} \right) = -\Delta W_M/2. \tag{11.14}$$

Thus, there is an equipartition between the flow and thermal energy gains in the Sweet–Parker model (Birn and Hesse, 2005; Yoo thesis, 2013). This equipartition means that half of the incoming magnetic energy must be converted to flow energy in order to achieve the required Alfvénic outflow.

11.2 ANALYSIS OF ENERGY FLOW IN THE TWO-FLUID FORMULATION

Energy transport and conversion are again governed by conservation laws that can be derived as moments of the Vlasov equation which governs the particle distribution in phase space, in combination with Maxwell's equations.

For two-fluid dynamics, eq. (11.3) is modified to include the microscopic heat flux q and the scalar pressure p, which is generalized to the total pressure tensor P:

$$\frac{\partial}{\partial t}\left[\frac{B^2}{2\mu_0} + \sum_{s=e,i}\left(u_s + \frac{\rho_s}{2}V_s^2\right)\right] + \nabla\cdot\left[S + \sum_{s=e,i}\left(H_s + K_s + q_s\right)\right] = 0. \quad (11.15)$$

Here, u_s, the internal energy of species s, is derived from the pressure tensor as $u_s = \mathrm{Tr}(P_s)/2$, and $H_s = u_s V_s + P_s \cdot V_s$ is the enthalpy flux for species s. In this form, the only term added to the MHD energy transport equation (eq. (11.3)) is the divergence of the microscopic heat flux of each species q_s. If we can assume the heat flux at the boundary is negligible and the diagonal terms of the pressure tensor are dominant for both electrons and ions, the two-fluid energy transport equation should reduce to the MHD energy transport equation.

Because electron and ion dynamics are quite different in the two-fluid regime, a quantitative analysis of the energy partition in a two-fluid reconnection layer is not straightforward to carry out. In fact, this difficulty is closely related to the fact that there is a lack of a full analytical theory of two-fluid reconnection. Such a theory should be able to self-consistently predict key reconnection parameters, such as the reconnection rate, plasma outflow velocity, layer aspect ratio, and energy deposition. It is known that the reconnection mechanisms can depend on many factors, including the boundary conditions, asymmetry in upstream parameters, and the strength of the guide field.

11.3   EXPERIMENTAL STUDY OF THE ENERGY INVENTORY IN TWO-FLUID ANALYSIS

For the energy inventory analysis in a reconnection layer, it is necessary to choose a properly sized volume boundary and boundary conditions, since the energy conversion process occurs not only over the ion diffusion region but also at so-called reconnection fronts where plasma jets originating from an active reconnection site interact with background plasmas (Angelopoulos et al., 2013). The energy conversion process at the reconnection front often depends on the boundary conditions there. To exclude effects from a specific choice of boundary conditions, we set the volume size for the energy inventory analysis such that it covers most of the ion diffusion region (IDR) as seen in figure 11.2. In the MRX, this region was set as a $2L_i \times 2L_i$ box, where L_i is the length of the IDR. In the case of the He plasma used in the experiment (Yamada et al., 2014), L_i is about 12 cm or $2c/\omega_{pi}$ and the half-width of the current sheet δ is roughly 2 cm or $0.3(c/\omega_{pi})$.

The energy inventory during two-fluid reconnection has been carefully examined in a laboratory plasma (Yamada et al., 2014, 2015). Energy flux terms, as well as time-derivative terms, are evaluated within a boundary that covers most of the IDR, as shown in figure 11.2. Including the time-derivative terms is important since the plasma is not perfectly steady state, and these terms represent changes in the energy enclosed in the plasma volume. Details of the energy inventory analysis can be found in Yamada et al. (2015).

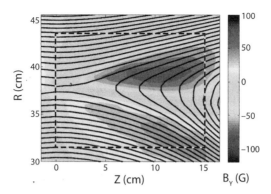

Figure 11.2. Boundary for the energy inventory analysis on MRX. The dashed magenta box of $\sim 2L_i \times 2L_i$ shows the region where the analysis is conducted. The black lines are the poloidal flux contours, representing magnetic field lines. [From Yamada et al. (2015).]

Figure 11.3 shows the overall energy inventory during two-fluid reconnection in MRX. The numbers in this figure are normalized by the incoming magnetic energy per unit time, which is

$$W_{\text{in}} = \int_{\Gamma_b} d^3x \, \nabla \cdot S_{\text{in}}, \tag{11.16}$$

where Γ_b is the volume of the plasma specified by the boundary in figure 11.2 and $S_{\text{in}} = E_Y B_Z / \mu_0 e_R$ is the incoming Poynting flux, with e_R being the unit vector along the R-direction. The actual value of W_{in} is 1.9 MW. There are major differences in the energy inventory between two-fluid reconnection and the Sweet–Parker model. First, there is a significant outgoing Poynting flux, which accounts for 50% of the magnetic energy. Since the aspect ratio of the ion diffusion region is only about 3, the MHD component of the outgoing Poynting flux $S_{\text{MHD}} = -(E_Y B_R / \mu_0) e_Z$ is not so small. Moreover, there is also an outgoing Poynting flux associated with Hall fields, $S_{\text{Hall}} = (E_R B_Y / \mu_0) e_Z - (E_Z B_Y / \mu_0) e_R$, whose contribution is as large as that of S_{MHD}, as shown in figure 11.3. Second, the energy gain is dominated by an increase in thermal energy for both ions and electrons; there is no equipartition. As mentioned in the previous section, it is important to note that the energy deposition to electrons predominantly occurs in the EDR (electron diffusion region) through $j_{e\perp} \cdot E_\perp$, and the energy deposition to ions occurs in the ion diffusion region through $j_{i\perp} \cdot E_\perp$. Accordingly, we have to quantitatively evaluate how magnetic energy is converted to the thermal and flow (kinetic) energy of electrons and ions within a toroidal boundary of minor radius 12 cm and length 15 cm on MRX.

If the system is in a steady state, the time-derivative terms of eq. (11.15) become zero. However, during the quasi-steady period of MRX, the plasma quantities are slowly changing, while the reconnection rate is almost steady. For example, due to the decreasing PF (poloidal field) current, the vacuum component of the magnetic field is

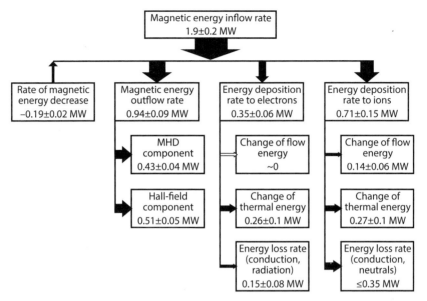

Figure 11.3. Energy inventory during two-fluid reconnection in MRX. Every number is normalized by the incoming magnetic energy per unit time $W_{in} = 1.9$ MW. The electron flow energy increase is not shown because it is extremely small ($\sim 10^{-4}$). [From Yamada et al. (2015).]

decreasing during the quasi-steady period. Thus, the total magnetic energy in Γ_b is also decreasing, and is not negligible due to the large volume over which the integral is conducted. Therefore, the contributions from the time-derivative terms are all included.

The total energy conversion rate to each species (per second) is separately computed by

$$W_s = \int_{\Gamma_b} d^3x \, \boldsymbol{j}_s \cdot \boldsymbol{E}. \tag{11.17}$$

To obtain a change in a specific form of energy, we group associated terms in eq. (11.15). The flow energy change of species s is given by

$$\Delta W_{K,s} = \int_{\Gamma_b} d^3x \left[\frac{\partial}{\partial t} \left(\frac{\rho_s}{2} V_s^2 \right) + \nabla \cdot \left(\frac{\rho_s}{2} V_s^2 \boldsymbol{V}_s \right) \right] \tag{11.18}$$

and the thermal energy change of species s is defined as

$$\Delta W_{H,s} = \int_{\Gamma_b} d^3x \left[\frac{\partial}{\partial t} \left(\frac{3}{2} n_s T_s \right) + \nabla \cdot \left(\frac{5}{2} n_s T_s \boldsymbol{V}_s \right) \right]. \tag{11.19}$$

We note that quantities in the inflow region are taken into account, although those quantities are much smaller than the incoming magnetic energy due to low $\beta \sim 0.2$ in the inflow region.

We estimate the energy loss rate of each species by considering the electron and ion heat flux, electron energy loss by impurity radiation, and ion energy loss to neutrals by charge-exchange collisions. The impurity radiation ($\sim 13\%$ of W_e) is primarily from oxygen ions based on the spectral measurements. The ion energy loss via charge-exchange collisions with neutrals was also evaluated (Yoo thesis, 2013).

In figure 11.3, all quantities are shown as rate of energy flow in and out ($W_{M,\text{in}} = 1.9\,\text{MW}$). The outgoing Poynting flux is sizable in MRX where two-fluid reconnection occurs because the outgoing energy associated with Hall magnetic fields plays a significant role. It was quantitatively evaluated how magnetic energy is converted to thermal and flow (kinetic) energy of electrons and ions within a cylindrical boundary with radial width $12\,\text{cm}$ (from $R = 32$ to $R = 44\,\text{cm}$) and length $15\,\text{cm}$. In our local energy flux inventory, about half of the incoming magnetic energy is converted to particle energy, one-third of which goes to electrons and two-thirds to ions. Our quantitative measurements show that half of the incoming magnetic energy is converted to particle energy with a remarkably fast speed of $\sim 0.1\text{--}0.2(B^2/2\mu_0)V_A$, in comparison with the rate calculated by MHD of $\sim (B^2/2\mu_0)V_A L/S^{1/2} = 0.03(B^2/2\mu_0)$, $S = 900$.

11.4 PARTICLE-IN-CELL SIMULATIONS FOR THE MRX ENERGETICS EXPERIMENTS

To confirm that our idea of analysis is fundamentally valid, let us study the results from the recent two-dimensional PIC simulations with two different boundary conditions. Another reason for comparison with numerical simulations is that there are several possible constraints on the applicability of the MRX experimental setup and the analytical calculation based on a simple model to astrophysical and space plasmas, including the effects of different boundary conditions, system size, and different ratios of ion to electron temperatures that are found in typical space plasmas (e.g., the earth's magnetotail). In order to study these criteria, we can look at the data from kinetic simulations, wherein each of these constraints can be independently relaxed. For these purposes, two sets of two-dimensional simulations with different boundary conditions were performed. The geometry of the first simulation is shown in figure 11.4(a).

For all cases, the study employed the PIC code VPIC (Bowers et al., 2008), but two types of simulations were carried out: (i) open boundary simulation without constraints from the MRX flux core and (ii) PIC simulation reproducing MRX boundary conditions. Length scales are normalized to the ion skin depth and timescales are normalized to the upstream ion cyclotron frequency ω_{ci}. In the Harris equilibrium for the open boundary simulation, the initial magnetic field profile is given by $\boldsymbol{B} = B_0 \tanh(x/\delta)\hat{z}$, and the initial density profile is then $n_e = n_b + n_0 \operatorname{sech}^2(x/\delta)$; see figure 11.4(a). In contrast, in the MRX simulation case in which an initial field is determined by the flux core coil currents, the initial density profile is uniform.

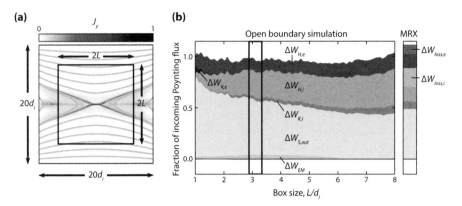

Figure 11.4. (a) In the open boundary simulations, the energy inventory is computed within a square box shown here. Note that here we have normalized the box size L to the upstream ion skin depth d_i, rather than the initial ion skin depth d_{i0}. For this simulation, $d_i \sim 2d_{i0}$. (b) The energy inventory from the open-boundary simulation for MRX is very similar to that of the MRX experiment (right-hand bar), suggesting that downstream boundary conditions do not play a strong role in determining the total energy conversion rate. The flows of the energy are defined here in the same way as in section 11.3; namely $\Delta W_{H,e}$, $\Delta W_{H,i}$, are electron and ion thermal enthalpy flows, $\Delta W_{K,e}$, $\Delta WK, i$, are kinetic energy flows in the exhaust region, $\Delta W_{S,\mathrm{out}}$ represents outflow of magnetic energy through the Poynting vector. In the experiments, heat flux is included in the loss terms $\Delta W_{\mathrm{loss},e}$ by way of radiative losses and energy transfer $\Delta W_{\mathrm{loss},i}$ to neutrals. [From Yamada et al. (2015).]

In the Harris sheet simulation discussed here (Le et al., 2013), the initial sheet thickness is given by $\delta = 0.5d_{i0}$, where d_{i0} is the initial ion skin depth evaluated with n_0. The ion to electron mass ratio is matched to hydrogen, $m_i/m_e = 1,836$, and the sheet temperature ratio is matched to a typical value in the magnetotail, $T_{i0}/T_{e0} = 5$. Due to computational constraints, $\omega_{pe}/\Omega_e = 2$ and the domain is $20d_{i0} \times 20d_{i0}$ with open boundaries (Daughton et al., 2006), and consists of $5,120 \times 5,120$ cells with $\sim 10^{10}$ particles per species. In the MRX simulation reported in Yoo et al. (2014c), the size of the boundary box where all relevant energy fluxes are computed is $L_i \times L_i$ where $L_i = 2d_i$ is the size of the boundary calculated and d_i is the ion skin depth evaluated as defined in MRX.

The effects of the boundary conditions on the overall energy inventory can be studied by comparing the two sets of simulations mentioned above. The open boundaries allow continual quasi-steady reconnection to proceed, while they eliminate any possible effects from downstream boundary conditions, such as the flux cores, in both the experiment and MRX simulation. It also covers higher ion to electron temperature ratios, and the lack of collisions more closely matches space plasma conditions.

In all cases, the energy inventory is quantitatively evaluated following the same procedure as for the experimental data; however, we neither directly compute nor estimate

Table 11.1. Summary of energy conversion during collisionless reconnection. The quantities are normalized to the magnetic energy inflow rate W_B.

	MRX measurement		Simulation		
	$1.5d_i \times 1.5d_i$	$2d_i \times 2d_i$	MRX boundary	Open boundary	Space observation
W_B inflow	1.0	1.0	1.0	1.0	1.0^a
W_B outflow	0.51	0.49	0.6	0.53	0.1–0.3
W_p to ions	0.31	0.37	0.26	0.29	0.39
W_p to electrons	0.21	0.18	0.19	0.11	0.18

[a]The space data has uncertainties in measurements of the total incoming magnetic energy and the exact size of the boundaries. The box size for the simulations is $2d_i$.

the heat flux in the simulations. In the experimental results, the heat flux is estimated and incorporated along with estimates for radiation and energy transfer to neutrals into the total loss terms $\Delta W_{s,\text{loss}}$. As shown in figure 11.4, the obtained energy inventory in the simulation balances quite well (the sum is approximately 1) despite the neglect of the heat flux terms. As a result, we can conclude that heat flux plays a negligible role in the overall energy balance.

Since we are interested in studying the energy inventory during quasi-steady reconnection, the time window over which we compute the energy inventory is carefully chosen. For the direct MRX (ii) simulation case, we choose data from around the time $t \sim 0.5\tau$, where τ is the flux core driving timescale, while in the Harris sheet case we choose the last time-point of the simulation, $t/\Omega_I = 34$, as shown in detail in Yamada (1995). These choices eliminate any transient phenomena associated with the onset of reconnection and allow for a well-developed reconnection layer to be present.

Results from the simulations with different MRX boundary conditions are presented in table 11.1. With a box size of $2d_i \times 2d_i$, the energy inventory of the MRX simulation is qualitatively similar to that of the present experiment. The total outgoing magnetic energy is about 60% of the incoming magnetic energy ($W_{S,\text{in}}$). The contribution of the first term on the left-hand side of eq. (11.15) is about 5% of $W_{S,\text{in}}$. The energy deposition to electrons (W_e) is about 19% of $W_{S,\text{in}}$, and the energy deposition to ions (W_i) is 26% of $W_{S,\text{in}}$.

The results of the energy inventory in the open boundary simulation are shown in figure 11.4. Remarkably, we find that over a broad range of scales, $1.5d_i < L \lesssim 4d_i$, the energy inventory is approximately independent of box size. Furthermore, our simulation results show decent agreement with the experimental results; approximately half of the incoming Poynting flux is converted into particle energy, with most of this energy going to the ion enthalpy. As a result, we can conclude that the experimental constraints outlined above are likely not important in determining the energy conversion efficiency during antiparallel magnetic reconnection.

11.5   A SIMPLE ANALYTICAL MODEL OF ENERGY CONVERSION IN THE TWO-FLUID RECONNECTION LAYER

In this section, let us develop a simple analytical model based on what we learned from experimental results, as well as from numerical simulations. Let us consider how we make a quantitative analysis regarding energy conversion in the reconnection layer using a simplified model. Here we consider a prototypical two-dimensional antiparallel reconnection geometry and describe a simple quantitative analysis of the energy inventory in the IDR. As electrons and ions move into the reconnection layer with different paths, the magnetized electrons penetrate deep into the reconnection layer, generating a strong potential well around the diffusion region. Simultaneously, the magnetic energy is transferred to the electrons. While energy deposition to electrons occurs mostly within the EDR (Pritchett, 2010; Yamada et al., 2015), it also happens along the separatrices. But the amount of the energy gain in the separatrix is much smaller than that in the EDR. In this situation the electric potential $\Phi_p(x, z)$ is constant along the reconnected field lines in the exhaust of the separatrix as shown in figures 10.4 and 11.1(b).

Energy deposition to ions occurs throughout the IDR, which extends to the separatrix regimes, making the estimation a little difficult. We consider first the ion energy gain in a two-dimensional geometry from the in-plane electrostatic field, which we found to be the most dominant energy source for ion acceleration and heating (Yoo et al., 2013). Assuming a simplified model for the analysis of electron current sheet dynamics, with the IDR represented by a $2L_z \times 2L_x$ rectangular box as shown in figure 11.5(a), we can calculate analytically how the magnetic energy is transferred to plasma ions. In this region, magnetic field pressure energy is converted to electric field potential energy through different motions of electrons and ions. The force balance of the Harris sheet described in chapter 4 generates an electron current sheet with thickness δ and B_z-profile of a reconnecting magnetic field as $B = B_0 \tanh(x/\delta)$, where B_0 is the shoulder value of the reconnecting field. The dynamics of the electron current with respect to ions ($V_e \gg V_i$) produce a strong well of electric potential as described by eq. (11.28) (see figure 11.5(b), (c)). The ions are mostly demagnetized in the IDR, and the energy gain for a single ion from the inflow region to the exhaust is $e\Delta\Phi_p$, where $\Delta\Phi_p$ is the plasma potential difference across the separatrices, which can be estimated by the equation of motion for electrons (Yoo et al., 2013; Yamada et al., 2015). In other words, magnetic field pressure energy is converted to electric field potential energy through the motions of magnetized electrons in the background of nonmagnetized ions.

11.5.1   Electron dynamics generate a potential well and ions gain energy in the reconnection layer

The electron equation of motion with respect to the x-direction (inflow direction) in the reconnection plane is written as

$$E_x = V_{ey} B_z - \frac{1}{en_e} \frac{\partial p_e}{\partial x},$$

$$(11.20)$$

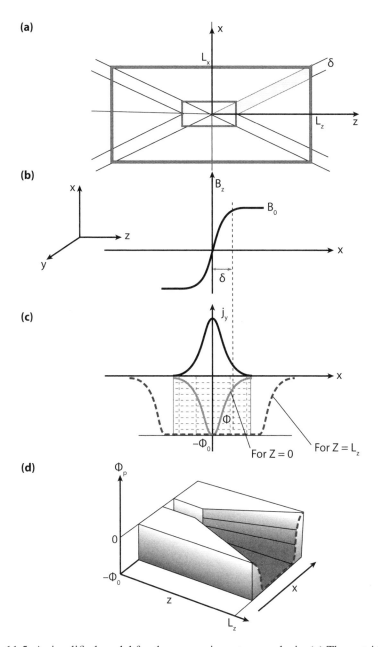

Figure 11.5. A simplified model for the energy inventory analysis. (a) The outside rectangular box of $\sim 2L_z \times 2L_x$ shows the region where the analytical model was constructed. (b) The force balance of the Harris sheet generates an electron current sheet with thickness δ_s and B_z-profile of a reconnecting magnetic field as $B = B_0 \tanh(x/\delta_s)$, where B_0 is the shoulder value of the reconnecting field. (c) The dynamics of the electron current with respect to ions $(V_e \gg V_i)$ produce a strong well of electric potential. (d) 3D presentation of the potential profile Φ_P on the (x, z)-plane. The broken line represents the potential profile at $L = L_Z$. Ions are accelerated toward the lower potential direction (darker gray color) to the exit.

where x and y denote the normal and out-of-plane directions, respectively, and the pressure tensor is simplified to be a scalar pressure. The second term of the right-hand side is less than 10% in most cases and this suggests that the in-plane electric field is primarily determined by the flux freezing principle introduced in chapter 5.

Here, let us employ the descriptions of a generalized Harris sheet (chapter 5, Yamada et al. 2000) by approximating an electron current profile of width δ extending all the way to the exhaust region, as shown in figure 11.5(b), (c):

$$B_z = B_0 \tanh\left(\frac{x}{\delta}\right) \tag{11.21}$$

and

$$j_y = \frac{B_0}{\mu_0 \delta} \operatorname{sech}^2\left(\frac{x}{\delta}\right). \tag{11.22}$$

The current sheet thickness δ is given by

$$\delta = \frac{c}{\omega_{pi}} \frac{\sqrt{2(T_e + T_i)/m_i}}{V_i - V_e} = \frac{c}{\omega_{pi}} \frac{\sqrt{2}V_s}{V_{\text{drift}}}, \tag{11.23}$$

where $V_s \equiv \sqrt{(T_e + T_i)/m_i}$ and $V_{\text{drift}} \equiv V_e - V_i$ is the relative drift velocity between electrons and ions. In the present case, $V_e \gg V_i$. It should be noted that the above solution is more general than the original Harris solution, which is limited to $E_x = \phi = 0$. For the present case, setting $T_e = T_i = T$ and $V_e \gg V_i$ (in the reconnection layer of MRX, electron currents dominate in all directions) yields $\delta = (c/\omega_{pi})(2\sqrt{T/m_i}/V_{ey})$. In the MRX with helium gas, $\langle V_e \rangle \sim 6\sqrt{T/m_i}$, thus we obtain $\delta \sim (1/3)(c/\omega_{pi})$, which is in good agreement with experimental results.

By substituting eqs. (11.21) and (11.22) into eq. (11.20) we obtain

$$E_x = \frac{B_0^2}{\mu_0 \delta} \operatorname{sech}^2\left(\frac{x}{\delta}\right) \tanh\left(\frac{x}{\delta}\right) - \frac{1}{e n_e} \frac{\partial p_e}{\partial x}. \tag{11.24}$$

By partial integration of

$$\int_0^1 \operatorname{sech}^2(x') \tanh(x')\, dx' = \int_0^1 \operatorname{sech}(x')\left(\operatorname{sech}(x') \tanh(x')\right) dx', \tag{11.25}$$

and by using the relationship

$$\int_0^x \operatorname{sech}(x') \tanh(x')\, dx' = -\operatorname{sech}(x), \tag{11.26}$$

we obtain

$$\int_0^1 \operatorname{sech}^2(x') \tanh(x')\, dx' = \frac{1}{2}, \tag{11.27}$$

where $x' = x/\delta$.

Thus, using the above results and by integrating eq. (11.24) along the x-direction, we obtain the potential drop across the current sheet as

$$\Delta \Phi_p = \Phi_0 = \int_0^{\delta} E_x dx \tag{11.28}$$

$$= \frac{B_0{}^2}{2\mu_0 e \langle n_e \rangle} \int_0^1 \text{sech}^2(x')\tanh(x')dx' - \frac{\Delta T_e}{e}, \tag{11.29}$$

$$= \frac{B_0^2}{2\mu_0 e \langle n_e \rangle} - \frac{\Delta T_e}{e}, \tag{11.30}$$

where $\langle n_e \rangle$ is the electron density averaged over the current sheet, B_0 is the shoulder value (the value just outside the current sheet) of the reconnecting magnetic field, and ΔT_e is the electron temperature difference between the center of the current sheet and a point just outside. Based on the high electric conductivity of electrons along the field lines, we can assume that this potential drop through the electron current layer extends to the exhaust region along the separatrix to the edge of the IDR (namely the edge of the boundary box), as seen in figure 11.5(b), (c). [Yamada et al., (2021)]

The temperature difference ΔT_e is related to bulk electron heating during reconnection. This is found to be small with respect to the incoming magnetic energy per electron–ion pair $m_i V_A^2 = B_{\text{rec}}^2/\mu_0 n_e$ ($< 5\%$; Phan et al., 2014; Shay et al., 2014). If we assume $B_0 = B_{\text{rec}}$, and using eqs. (11.4) and (11.28), the ratio of total ion energy gain per unit length of the out-of-plane direction, W_i to W_{in} for each quadrant of the reconnection plane, becomes

$$\frac{W_i}{W_{\text{in}}} = \frac{e\Delta\Phi_p L_i}{S_{\text{in}} L_i} = \frac{B_0^2/2\mu_0}{B_{\text{rec}}^2/\mu_0} = \frac{1}{2}. \tag{11.31}$$

This result has a significant meaning, namely that half of the incoming magnetic energy is converted to electrical potential energy due to the constrained motion of magnetized electrons against ions, and ions are accelerated to the exhaust direction, converting most of the electric potential energy to its flow and ion thermal energy.

11.5.2   Energy deposition to electrons in the reconnection layer

How about energy conversion to electrons? At the moment, the mechanisms of energy conversion to electrons are not known. Here, let us employ a Sweet–Parker-type rectangular-box-shaped energy conversion region for the EDR. Since the EDR is relatively thin (~ 3–$5c/\omega_{ce}$) and of uniform density compared to the ion scale, the incompressible assumption of the Sweet–Parker model is justifiable. Then the electron outflow $V_{e,\text{out}} \sim (L_e/\delta_e)V_{e,\text{in}}$, where L_e and δ_e are the half-length and half-width of the EDR respectively, and $V_{e,\text{in}}$ is the electron inflow speed. From (11.7), one can show that the outgoing magnetic energy is $(L_e/\delta_e)^2$ times smaller than the total incoming magnetic energy to the EDR. Since $(L_e/\delta_e) \gg 1$, the outgoing magnetic energy is negligible, which means that most of the incoming magnetic energy to the EDR is converted to electron energy. Then the ratio of the electron energy gain per unit length along the out-of-plane direction, W_e, to the total incoming magnetic energy per quadrant and per unit length along the out-of-plane direction to the IDR, W_{in}, becomes

$$\frac{W_e}{W_{in}} \approx \frac{S_{in} L_e}{S_{in} L_i} = \frac{L_e}{L_i}, \tag{11.32}$$

where S_{in} is the incoming Poynting flux associated with the reconnecting magnetic field and the reconnection electric field, which is the same as in the Sweet–Parker model (eq. (11.4)).

The length of the EDR is typically of order the ion skin depth (Daughton et al., 2006; Shay et al., 2007; Karimabadi et al., 2007; Ren et al., 2008). The length of the IDR, however, is harder to determine. If, for example, we define the IDR as the region where the Hall effects exist then the IDR can reach several ion skin depths. Due to these characteristics and a lack of theory of two-fluid reconnection, there is no consensus for L_e/L_i, which can also depend on the system size for our modeling or boundary conditions. In MRX, L_e/L_i was measured to be about $1/5$, with $L_e \sim d_i$ and $5L_i \sim L$, where L is the system size; thus we expect that about 20% of the incoming magnetic energy is transferred to electrons. Generally, as we expand the system size the electron partition goes down based on this relationship.

An important feature of energy conversion in the two-fluid reconnection layer is that the outgoing magnetic energy is not negligible, owing to the relatively small aspect ratio. From the above quantitative discussions, $(W_e + W_i)/W_{in}$ is about 50%, which means that we should expect that about 50% of the incoming magnetic energy flows out without conversion. This should be caused by the characteristics of the two-fluid reconnection layer: (1) Since the fast reconnection rate is primarily facilitated by the Hall fields, a large outgoing Poynting flux should be generated by the in-plane E field and the out-of-plane Hall magnetic field. (2) With the fast reconnection speed facilitated by Hall effects, the ratio of the inflow velocity to the outflow velocity would be close to a fraction of unity, resulting in a smaller aspect ratio for the ion diffusion region (L_i/δ_i). As a result, a sizable amount of the outgoing Poynting flux should persist. These observations are generally consistent with recent space observations (Angelopoulos et al., 2013; Eastwood et al., 2013). A comparative study between the energy inventory in the reconnection layer of MRX and space observations (Eastwood et al., 2013) is described in more detail in Yamada et al. (2016b).

11.6   SUMMARY AND DISCUSSIONS ON THE ENERGY INVENTORY OF THE RECONNECTION LAYER

In this chapter we have studied how the magnetic energy is transferred to plasma particles in the reconnection layer using both MHD and two-fluid models. We have found that in the MHD model, the outflowing magnetic energy is very small and most of magnetic energy is tranfered to particles, while the reconnection velocity is slow. In the two-fluid reconnection regime, on the other hand, the reconnection velocity is much faster while 50% of magnetic energy is converted to particle energy, primarily ions and the other 50% flows out of the exhaust region.

In the two-fluid reconnection, there is an outgoing Poynting flux associated with the Hall fields, i.e., the out-of-plane magnetic field and in-plane electric field. For reconnection with a negligible guide field, this outgoing flux (S_{Hall}) is much larger than the MHD-based outgoing Poynting flux associated with B_{out} and E_{rec}, S_{MHD} (Eastwood et al., 2013; Yamada et al., 2014). Overall the quantitative measurements show that half of the incoming magnetic energy is converted to particle (enthalpy) energy with a remarkably fast speed, 0.1–$0.2(B^2/2\mu_0)V_A$ in comparison with the rate calculated by MHD, $\sim (B^2/2\mu_0)V_A L/S^{1/2} = 0.03(B^2/2\mu_0)V_A$, $S = 900$.

The above energy inventory is for the case without a significant asymmetry across the current sheet. Recently, reconnection with a significant (~ 10) density asymmetry has been studied (Yoo et al., 2014b). These laboratory studies of asymmetric reconnection are of importance since reconnection in nature often has large asymmetry in plasma parameters, such as density and temperature, across the current sheet. A typical example is reconnection at the dayside magnetopause, where the solar-wind plasma interacts with magnetospheric plasma (Mozer and Pritchett, 2011). Then, the natural question is how the energy inventory changes in asymmetric reconnection.

It is found that the fraction of magnetic energy converted in the ion diffusion region, which is about 50%, does not notably change, but the detailed energy inventory is different in asymmetric reconnection (Yoo et al., 2017; Yamada et al., 2018). In particular, the ratio of ion energy gain to electron energy gain changes to a smaller value than 2. This change comes mostly from the fact that the density asymmetry changes the Hall field profiles: the Hall electric field on the high-density side, where most of the ions are flowing to the exhaust, becomes much weaker, and thus the ion energy gain becomes smaller. For electrons, there is additional energy gain near the low-density-side separatrices where a strong pressure gradient exists, thereby increasing the electron energy gain.

How the energy inventory changes for different situations, such as reconnection with a guide field, has yet to be studied. The systematic dependence of the energy inventory on various upstream parameters, such as plasma beta and collisionality, should also be studied in the future.

Chapter Twelve

Cross-discipline study of the two-fluid dynamics of magnetic reconnection in laboratory and magnetopause plasmas

12.1 BACKGROUND OF A COLLABORATIVE STUDY OF TWO-FLUID DYNAMICS IN THE RECONNECTION LAYER

As we learned in chapters 8 and 10–12, we now know that two-fluid dynamics, by decoupling ions and electrons, dictate the mechanisms of the reconnection layer in both laboratory plasma and the magnetosphere. Hall effects have been shown to facilitate the fast reconnection observed in collisionless magnetospheric plasmas and nearly collision-free laboratory plasmas.

In the reconnection layer of the magnetopause, the solar-wind plasma pressure balances with the magnetic field pressure of the earth dipole field, thus creating the condition that the plasma pressure is equal to the magnetic pressure of the magnetosphere: $\beta \sim 1$. In the reconnection layer, the thickness of the current sheet becomes comparable to the ion skin depth as well as the ion gyroradius. Ions become demagnetized within the reconnection region as the magnetic field becomes small, while electrons are still magnetized and remain frozen to field lines until they reach very near the X-line. In other words, this reconnection region is dominated by two-fluid dynamics. In the vicinity of the X-line, even electrons become demagnetized and diffuse, and thus we call this region the electron diffusion region.

In this chapter, let us directly compare the dynamics and energetics of an asymmetric reconnection layer observed both in the laboratory plasma of MRX and in the magnetopause by the MMS (Magnetospheric Multiscale Satellite) and discuss our results in the context of two-fluid physics, aided by simulations. As mentioned in chapter 11, a laboratory study on the mechanisms of energy conversion and energy partitioning made significant progress toward understanding these issues in a nearly collision-free environment. Simultaneous measurements by a few hundred magnetic probes can capture global features of field evolution in the reconnection layer in the MRX plasma. On the other hand, coordinated MMS measurements by four satellites can document detailed local properties, including measurements of the velocity–space particle distributions.

Thanks to self-similar scaling, both MRX (Magnetic Reconnection Experiment) and the magnetosphere plasma systems reside in the same regime of magnetic reconnection dynamics in which two-fluid physics dominates. Table 12.1 shows key

Table 12.1. A comparison of representative plasma parameters and scale sizes of MRX and the magnetopause reconnection layer shows amazingly similar characteristics. Note that $S \gg 1$ is satisfied for ideal MHD to be valid globally in both cases. [From Yamada et al. (2018).]

	MRX	Magnetopause	Ratio
System scale size (L) half recon. layer length	0.1–0.2 m	100–200 km	10^6
Ion skin depth ($d_i = c/w_{pi}$)	3–6×10^{-2} m	30–60 km	10^6
Electron skin depth ($d_e = c/w_{pe}$)	1–2 mm	1–2 km	10^6
Normalized scale length ($L^* = L/d_i$)	3	3	~ 1
Plasma pressure high-β side (β_{high})	0.5–1	0.5–1	~ 1
Plasma pressure low-β side (β_{low})	0.05–0.1	0.03–0.1	~ 1
Lundquist number (S)	$> 10^3$	$> 10^{10}$	Ideal MHD is valid globally

parameters of both systems, indicating that the length scale of the reconnection region is about three times the ion skin depth ($d_i = c/\omega_{pi}$). The table demonstrates that the plasma parameters of the MRX and the magnetopause are in the same regime of two-fluid physics, namely different motions of ions and electrons, $S \gg 1$, the same normalized length scale ~ 3, and a similar density asymmetry of a factor of 10. In both systems, the density asymmetry of the inflowing plasma is about 10. Additionally, the Lundquist number S (the ratio of resistive magnetic diffusion time to the Alfvén transit time) is also significantly larger than 1 ($S \gg 1$), which makes it possible to describe global plasma dynamics by ideal MHD (magnetohydrodynamics) except at the reconnection layer. Due to the above self-similar conditions and relationship, reconnection in MRX is expected to share key qualitative and quantitative characteristics with reconnection in the magnetosphere in terms of plasma dynamics and energetics. This allows a high level of cross-discipline examination between laboratory measurements and space observations. Moreover, magnetic reconnection in both MRX and the magnetosphere is driven by external forcing, i.e., flux cores in MRX and the solar wind in the magnetosphere. Using the two-fluid physics analysis, we compare our results on the plasma dynamics as well as on the energy conversion mechanisms during asymmetric, antiparallel reconnection measured in MRX with MMS satellite measurements at the magnetopause. For example, the observational verification of electrons' motion frozen to field lines outside the electron diffusion region described in chapter 8 matches the

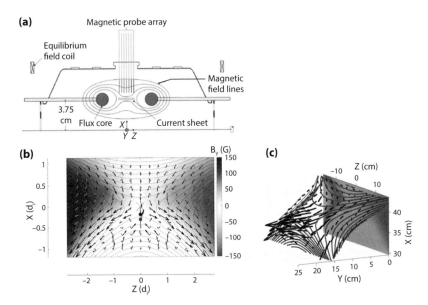

Figure 12.1. MRX apparatus and demonstration of two-fluid effects. (a) MRX apparatus to generate asymmetric reconnection in the current sheet. Each flux core (gray circles) contains two sets of internal coils that are used to create plasma and to drive reconnection. By controlling the sequence of coil currents and the initial plasma flows, asymmetric reconnection is formed with electron density asymmetry up to 10. (b) Measured flow vectors of electrons (long dark arrows) and ions (short thin arrows) in the full reconnection plane, together with poloidal flux contours (gray lines) and contours of the out-of-plane magnetic field. The marker X at $(X, Z) = (37.6, 0)$ cm denotes the location of the X-line where the magnetic field is near zero, the upper filled circle at $(X, Z) = (36.5, 0)$ cm is the stagnation point of in-plane electron flows, and the lower circle at $(X, Z) = (35.8, 0)$ cm is the stagnation point of ion flows. (c) 3D view of magnetic reconnecting field lines. The movement of the field lines in the reconnection plane can be tracked in a supplementary video (attached link for this book) from MRX data. Because of the dipole structure of the out-of-plane magnetic field, the plane where field lines move with electrons is tilted with respect to the Y-axis on the high-density side. [From Yamada et al. (2018) see a color version, Fig. 1. https://mrx.pppl.gov/mrxmovies /Asym.mp4]

earlier measurements in MRX well (described in chapters 5, 10) in which the electron dynamics are analyzed in terms of two-fluid physics in symmetric reconnection. If the same two-fluid mechanisms in two-dimensional analysis should operate in both systems despite vastly different scale sizes and collisional conditions, we expect good agreement between the two data sets.

   Figure 12.1 shows a schematic of MRX (in (a)), together with the measured flow of electrons and ions in the reconnection layer overlaid on contours of the poloidal flux (in (b)). Experiments are carried out in a setup in which two toroidal plasmas, each with an annular cross-section, are formed around two flux cores (gray circles in figure 10.1(a)). Each flux core contains both TF (toroidal field) and PF (poloidal field) coils. As usually done on MRX and explained in detail in chapters 5 and 6, by controlling the currents in the two coils, we can routinely generate the reconnection layer in a controlled manner and detailed plasma parameters are measured by internal probes. In this experimental campaign, by controlling the sequence of coil currents and the initial plasma flows in a special way, asymmetric reconnection is formed with an electron density asymmetry factor of 8–10. Plasma parameters are typically as follows: $n_e{}^I$ (high-density, sheath side) $\sim 4 \times 10^{13}\,\mathrm{cm}^{-3}$, $n_e{}^{II}$ (low-density, magnetosphere side) $\sim 5 \times 10^{12}\,\mathrm{cm}^{-3}$, $T_e = 5\text{–}15\,\mathrm{eV}$, $B = 0.1\text{–}0.2\,\mathrm{kG}$, S (Lundquist number) ≥ 500; the electrons are mostly magnetized (gyroradius $\rho_e \sim 1\,\mathrm{mm} \ll L$, where L is the length of the reconnection layer), while the ions are not. The mean free path for electron–ion Coulomb collisions is in the range 5–30 cm ($\geq$ the layer thickness) and, as a result, the reconnection dynamics are dominated by two-fluid and kinetic effects, despite some collisional effects that were seen. In our coordinate system (X, Y, Z), B_Z is the reconnecting field component and Y is along the out-of-plane axis.

Using a variety of diagnostics, a study of the dynamics of the plasma within the reconnection layer was made without an externally imposed guide field. Figure 12.1(b) depicts flow vectors of ions and electrons across the whole reconnection plane, along with poloidal flux contours and colored contours of the out-of-plane magnetic field component B_Y (see a color version, Fig. 1 of Yamada et al. 2018). It should be noted that the high-density side of the reconnection plane in which magnetic field lines move with electrons, is strongly tilted with respect to the Y-axis, as shown in figure 12.1(c). This unique feature of the asymmetric reconnection is caused by the decoupling of electrons and ions, and Hall effects which generate a dipole field shown in figure 12.1(b). The Hall current is carried by electrons flowing toward (away from) the X-line in the high- (low-) density side, as seen in figure 12.1(b). The electron flows toward the X-line in the separatrix region of the high-density side have been measured by MMS at the magnetopause in agreement with figure 12.1(b). This measurement was also verified in a two-dimensional simulation. The electron flow vectors in the reconnection plane are derived from the electron current profile, which is obtained from the magnetic field profile measured by fine-scale magnetic probes. The measured flow profiles of electrons and ions clearly demonstrate that two-fluid reconnection is at work in MRX. Ions become demagnetized as they enter the ion diffusion region, whose width is about the ion skin depth, and are accelerated across the separatrices, flowing outward to the exhaust direction, as seen in figure 12.1(b). In contrast to the case of symmetric reconnection, we observe that inflowing ions also form a stagnation point (denoted by a blue circle) near the X-line (X-point) on the low-density side with a shift of about 2–3 cm ($0.3\text{–}0.5d_i$).

12.2   DYNAMICS OF THE ELECTRON DIFFUSION REGION AND ENERGY DEPOSITION MEASURED BY MRX

First, the experimental analysis both on MRX and MMS demonstrates that the primary energy deposition on electrons occurs again through $j_{e\perp} \cdot E_\perp$, which is now strong at the stagnation point located near the X-point (Yamada et al., 2018). The potential

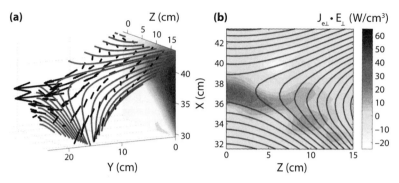

Figure 12.2. Electron dynamics and energy deposition observed in an asymmetric reconnection. (a) 3D views of electron flow vectors with respect to the reconnecting field lines. Because of the dipole structure of the out-of-plane magnetic field, the reconnection plane is tilted in the Y-axis in the high-density (upper) side. Note the strong electron current in the Y-direction at the stagnation point. (b) Perpendicular component $j_{e\perp} \cdot E_{\perp}$ is dominant and concentrated in the electron stagnation point, while the contribution from $j_{e\|} \cdot E_{\|}$ is notable on the lower-density side of the exhaust. [From Yamada et al. (2018).]

well is observed to shift toward the lower-density side of the exhaust region in MRX, as described in chapter 10. As was the case in symmetric reconnection, the accelerated ions are thermalized by remagnetization in the downstream region with some additional collisional effects. Near the electron stagnation point, $j_{e\perp} \cdot E_{\perp}$ is significantly larger than $j_{e\|} \cdot E_{\|}$. It was observed in MRX that in asymmetric reconnection without a guide field, $j_{e\perp} \cdot E_{\perp}$ peaks up through $j_{eY} \cdot E_Y$ at the stagnation point of the electrons' in-plane flow, which is separated from the X-line by ~ 5–8 electron skin depths to the lower density side.

Figure 12.2 shows the profiles of the electron flow vectors and energy deposition rate. The electron flow vectors in the reconnection plane are derived from the electron current profile, which is obtained from the magnetic field profile measured by fine-scale magnetic probes as explain in chapter 10. But in this asymmetric reconnection case, the reconnection plane is tilted in the Y-axis in the high-density (upper) side because of the dipole structure (not a quadrupole) of the out-of-plane magnetic field. It should also be noted that there is a strong electron current in the Y-direction at the stagnation point.

Recent analysis of data from MMS also verified this key feature by demonstrating that the value of $j_{e\perp} \cdot E_{\perp}$ peaks when MMS encounters the electron stagnation point in the electron diffusion region.

12.3   DYNAMICS OF THE ELECTRON DIFFUSION REGION AND ENERGY DEPOSITION MEASURED BY MMS

To investigate where and how energy conversion occurs in magnetopause reconnection, we discuss an encounter with both the electron and ion diffusion regions by the MMS spacecraft on 16 October 2015. The event has a negligible guide field, and the general features are described in Burch et al. (2016b). The plasma density and flow velocity, magnetic field, and electric field measurements shown in figure 12.3 are from the Fast Plasma Investigation, FluxGate magnetometers, and double-probe electric field sensors, respectively. To facilitate the comparison with MRX experimental results, here we use the same coordinate system (X, Y, Z), which corresponds to $(N, -M, L)$ in the boundary normal LMN coordinates determined in Yamada et al. (2018). We note that this (X, Y, Z) coordinate system differs from the geocentric solar magnetospheric (GSM) system.

Figure 12.3 presents energy conversion to electrons in the electron diffusion region during magnetopause reconnection. The kinetic structure of the same diffusion region, including higher-frequency fluctuations, has been discussed previously in chapter 8. Here we filter out electric field fluctuations (such as whistler and lower hybrid waves) with frequencies higher than the ion cyclotron frequency in order to observe the longer-timescale structure of the plasma dynamics and energy conversion. Figure 12.3(a) shows the approximate MMS trajectory through the electron diffusion region on the profile of J_Y. We note here that in MMS data, current density is expressed by J, while lowercase j is used for most of this book. The trajectory is determined based on a comparative study of MMS measurements and PIC (particle-in-cell) predictions of electron distribution functions. The average velocity of the magnetopause plasma (X-line included) along the X-axis during the electron diffusion region crossing is about -30 km/s based on the four-spacecraft magnetic field measurements, and the velocity along the Z-direction is estimated to be about 97 km/s. The energy deposition through the work done by the reconnection electric field in the magnetosphere, E_Y, is about 2 mV/m. This value translates to 0.1–$0.2 V_A^*$, where V_A^* is the hybrid Alfvén velocity as defined by Cassak and Shay. The measured value of the reconnection electric field agrees with the MRX data, which show 140 V/m of E_Y, corresponding to $0.25 V_A^*$ ($B_1 = 175$ G and $B_2 = 200$ G).

One of the most important results of the MRX–MMS collaboration research was to clarify the role of the electron diffusion region, together with the energy deposition to electrons. In both measurements, a two-scale diffusion region was identified as the electron diffusion region ($\delta_e \sim 5$ mm in MRX, 5–10 km in MMS) residing inside the ion diffusion layer ($\delta_i \sim 60$ mm in MRX (He gas) and ~ 50 km in MMS). In this situation, the ion diffusion layer is defined by the regime where $E + V_i \times B \neq 0$ with $E + V_e \times B = 0$. The electron diffusion layer is the regime $E + V_e \times B = E' \neq 0$, where E' is the electric field in the electron fluid frame. Just outside the electron diffusion layer, $E' = 0$ holds, namely electrons move with magnetic field lines in the reconnection plane (electron-flux freezing), and this relationship was clearly verified by Burch et al. (2016b) and by quantitatively evaluating force balance in MRX (Yoo et al., 2013). It should be noted that in the case of MRX, the difference between E and E' is relatively small since U_e is much smaller than the electron thermal velocity ($U_e \sim 0.1 V_{eth}$, where V_{eth} is the electron thermal velocity), but this does not apply for MMS data. In the high-electron-dissipation region defined by high $j_{e\perp} \cdot E_\perp$, a crescent-shaped electron distribution function was detected, showing a strong electron flow along Y, the out-of-reconnection-plane direction (figure 12.4(f)). The distribution

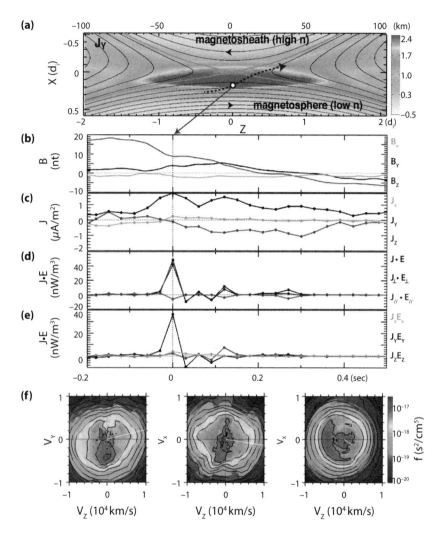

Figure 12.3. See Color Plate 1. Electron dynamics in the magnetophere measured by MMS. (a) Approximate MMS trajectory through the electron diffusion region of the magnetosphere. The trajectory is determined based on a comparative study of MMS data and 2D numerical (PIC) studies. (b)–(f) Time evolution of key components of local plasma parameters showing that $J_\perp \cdot E_\perp$ becomes maximum at the electron diffusion region (d). The electron velocity distributions in (f) show that they predominantly flow in the Y-direction as shown in the MRX data in figure 1.7(a). The documented MMS data are remarkably consistent with the electron dynamics measured by MRX. [Yamada et al. (2018) adapted from MMS data, Burch et al. (2016b).]

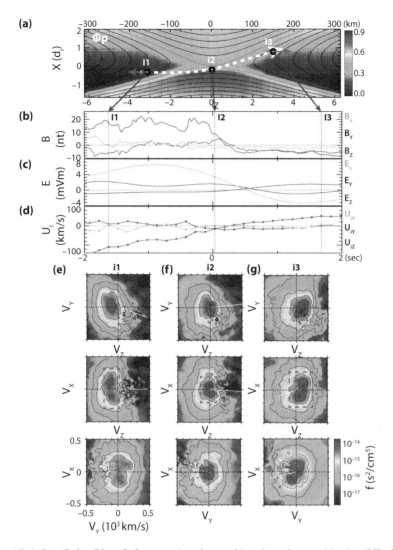

Figure 12.4. See Color Plate 8. Ion acceleration and heating observed in the diffusion of magnetopause reconnection (same MMS event as in figure 12.3). (a) MMS trajectory sketched on the in-plane plasma potential from a PIC simulation, where the directions of the in-plane ion flow vector for times i1 and i3 are indicated by magenta arrows. (b) Magnetic field vector components. (c) Electric field vector components (only for frequencies $f < f_{ci} \sim 0.5$ Hz, using the upstream magnetic field strength). (d) Ion bulk flow velocity vector components. (e), (f), (g) Ion velocity distributions measured by the MMS spacecraft (color contours represent the phase-space density) in three velocity planes for times i1, i2, and i3, which are marked with vertical dashed lines in (b)–(d); corresponding locations marked with white circles in (a). These figures indicate that ions are accelerated toward the exhaust directions in the Z-direction from the X-line. Here we note that the effects of the magnetopause motion have been subtracted from both the electric field and the ion measurements. [Yamada et al. (2018).]

function provides the kinetic view for strong electron out-of-plane flow along the Y-direction in MRX, as seen in figure 10.2(b).

Very recently, Burch et al. (2020) analyzed in detail the electron inflow and outflow velocities surrounding the electron diffusion region. The inflow velocities agree with predicted reconnection rates in which about one-tenth of the incident magnetic field lines break and reconnect. Unlike the ion flows, whose outflow velocities are directed strongly along and against the direction of magnetic fields (north and south) in the plane of reconnection, the strongest electron outflow is generally perpendicular to the plane of reconnection. This result shows that three-dimensional aspects (or more precisely 2.5-dimensional aspects) of reconnection are important, again in agreement with figures 12.2 and 12.1 obtained in MRX.

12.4   ION DYNAMICS AND ENERGETICS IN MRX AND THE MAGNETOSPHERE

As already described in chapter 10, a large in-plane electric field plays a key role in ion acceleration and heating. Recent studies have identified how an in-plane electrostatic field is generated by the force balance of the electrons flowing through the center of the reconnection layer. During asymmetric reconnection, the inversed-saddle shaped potential well is shifted toward the low-density side and a sharp potential drop occurs on the low-density side of the exhaust region. The unmagnetized ions are accelerated by the in-plane electric field in the exhaust region in both the Z- and X-directions, primarily in the high-density side, and are heated further downstream.

A very similar phenomenon is observed inside the ion diffusion region during the same MMS passage. Figure 12.4 shows characteristics of ion acceleration in the magnetopause. While the spacecraft was in the left exhaust region, a strong ion flow at the point i1 toward the $-Z$-direction was observed, and an opposite ion flow to the $+Z$-direction was measured at the point i3 in the right exhaust. The V_Z–V_X velocity distributions (figure 12.4(e)–(g)) show counter-streaming populations along V_X, which is shifted in V_Z, indicating that unmagnetized ions bounce around the B_Z reversal and are being accelerated by the in-plane electric field (gradient of the in-plane potential shown in figure 12.4(a)). The shift in V_Z, which is the same feature as the drift along Z shown in the ion distribution measured in MRX, is the result of acceleration by E_Z. Hence, both MMS and MRX results support that the primary energy deposition to ions occurs due to acceleration by the in-plane electric field in asymmetric reconnection layers, similar to the case for symmetric reconnection described in chapters 11 and 14. Both the V_Z–V_X and V_X–V_Y distributions show that the V_X thermal spread increases toward downstream, suggesting the gradient of in-plane potential contributes to the acceleration of ions. While the observed ion acceleration in MMS is consistent with the results from MRX, the counter-streaming populations detected by MMS are not completely thermalized.

Chapter Thirteen

The dynamo and the role of magnetic reconnection

Magnetic fields are seen everywhere in the universe. They are observed in the majority of astrophysical objects: planets, stars, galaxies, clusters of galaxies, and even in the interstellar medium. The average energy density of a magnetic field is almost equal to the kinetic and thermal energy of plasma, the main component of baryon particles existing in the universe. We could express it as $\beta_p \sim 1$, where β_p represents the ratio of the plasma energy to the magnetic field energy. Dynamo is the opposite plasma process to magnetic reconnection. The latter occurs when and where the magnetic energy of a plasma system dominates over its kinetic energy. When there is abundant kinetic energy in a plasma with respect to magnetic energy, magnetic fields are considered to be generated through a converse process: a dynamo mechanism in plasma.

It is considered that there were no magnetic fields in the universe right after the Big Bang. Where have they come from? It is still uncertain exactly when and how the magnetic fields were generated. However, there is enough evidence that the magnetic fields in the earth, the sun, and other stars are sustained by dynamo activity. This is inferred from the cyclical behavior of the fields in these systems, which occurs on timescales many orders of magnitude less than the time in which magnetic fields diffuse by collisional dissipation. The sun's magnetic field changes its polarity every 11 years. The earth's dipole field changes its polarity in a timescale of a few hundred thousands years, but not by an exact periodic rate. The dynamo generation of galactic magnetic fields is also widely accepted. In most cases, the magnetic field is large in spatial scale, of the same scale as the astrophysical body, and believed to be generated by the combined effect of smaller-scale fluctuations in plasma flow and magnetic field.

Different forms of dynamo action have also been demonstrated and observed in laboratory plasmas. For the most clear example, we observe the generation of a huge magnetic field of 10^6 G when we inject a high-power laser into a small target of hydrogen for inertia fusion experiments in which no magnetic field is applied initially. How is magnetic field created? In RFP (reversed field pinch) and spheromak plasmas for magnetic fusion experiments, a dynamo action was observed in the form of flux conversion.

In this chapter we consider how magnetic field is generated in the universe and how magnetic reconnection plays a role in the dynamo. Until recently, most dynamo theory, including mean field theory, was based on single-fluid MHD (magnetohydrodynamics). Since this book is on magnetic reconnection, only an introductory description of dynamo phenomena is made, particulary in relationship to magnetic reconnection processes. As the primary theme of this monograph is two-fluid physics mechanisms, here we specially consider the two-fluid effects of dynamo action in fusion laboratory plasmas, after a brief introduction to MHD dynamo theory.

13.1   GALACTIC MAGNETIC FIELDS AND BASIC MHD THEORY

13.1.1   Observation of galactic magnetic fields

Astronomical magnetic fields are generally strong enough to influence the dynamics of gas in present-day galaxies, and could have played an important role in the formation and early evolution of galaxies. Yet the origin of these magnetic fields remains unclear. Magnetic fields in galaxies are strong enough to significantly affect interstellar gas dynamics on both global scales characteristic of galactic structure and small scales characteristic of star formation (Zweibel and Brandenburg, 1997). Cosmic rays, which heat, ionize, and supply significant pressure to the interstellar medium, are magnetically confined to galaxies and accelerated electromagnetically.

13.1.2   Observation of galactic magnetic field

Observations of large numbers of external galaxies can reveal correlations between global galactic parameters and magnetic field structure. The best documented example is the correlation of 21 cm microwave continuum intensity with far-infrared luminosity. The former depends on the relativistic electron population and magnetic energy density, and the latter is a measure of the rate at which massive stars are formed. The correlation holds over a wide range of spiral galaxy subtypes and star formation rates. In external galaxies the field lines closely follow the spiral arms seen in figure 13.1. In some cases, the magnetic spirals are better defined than the material spiral arms: two outstanding examples are the galaxy NGC 6946, in which the magnetic arms are particularly well defined in the inter-arm regions, and the flocculent galaxy NGC 5055. A leading theory of spiral structure in galaxies, known as density wave theory, predicts the inclination angle of the field lines to be larger inside spiral arms than in inter-arm regions.

13.1.3   Basic MHD dynamo theory

After Cowling (1934) proved that dynamo action cannot generate a net magnetic field in an axisymmetric system, many physicists tried to build a model to explain geomagnetic dynamo generation in the actual three-dimensional earth geometry. Parker (1955) found an elaborate explanation using a nonaxisymmetric set of velocities in a rotating conductive medium (hot core mantle on the earth). The early efforts led to the well-known α–Ω mean field dynamo theory by Steenbeck et al. (1966), which involved equations for the mean magnetic field called the mean field dynamo equations.

By taking ensemble averages of the induction equations in chapter 3, we obtain

$$\frac{\partial \boldsymbol{B}}{\partial t} = \nabla \times (\boldsymbol{V} \times \boldsymbol{B}) + \nabla \times (\alpha \boldsymbol{B}) + \left(\beta + \frac{\eta}{\mu_0}\right)\nabla^2 \boldsymbol{B}, \tag{13.1}$$

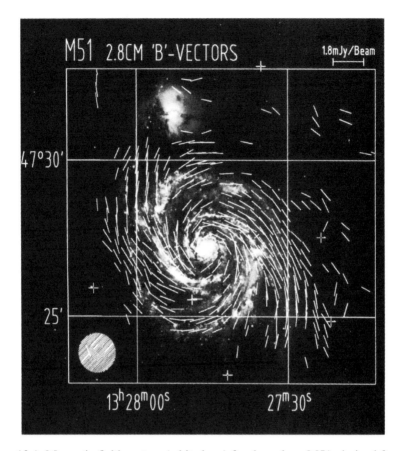

Figure 13.1. Magnetic field vectors (white bars) for the galaxy M51, derived from the polarized continuum intensity at a wavelength of 2.8 cm, which is thought to be synchrotron radiation. The lengths of the vectors are proportional to intensity. The map is superimposed on an optical image. The white striped circle in the lower-left corner indicates the size of the telescope beam. The crosses represent foreground stars. [From Neininger (1992).]

where α is a constant for an eddy motion (Kulsrud 2005; page 394). The β term can be interpreted as a turbulent mixing term and is often called turbulent resistivity. For the fast dynamo case $\beta \gg \eta/\mu_0$, the dynamo fluid is regarded as an ideal fluid, just like in the case of magnetic reconnection. In earlier decades, the astrophysical dynamo problem was addressed theoretically through this type of "kinematic mean-field dynamo" analysis (Kulsrud and Anderson, 1992; Kulsrud, 1998, 1999). In this approach, a certain velocity flow field, such as toroidal rotation or vortex motion, is assumed and the resulting magnetic field is calculated. This simple linear problem does not treat the key aspects of saturation and sustainment of the magnetic field. In recent years, the solution of the nonlinear MHD equations has been worked on for model problems and for the laboratory situation (Zweibel and Brandenburg, 1997). Since it would need an entire book by a dynamo expert to describe this phenomenon, I refer to the above references for this subject. Instead, this chapter discusses the experimental research and its two-fluid physics aspect.

13.2 THE BIERMANN BATTERY DYNAMO

It is considered that there was no magnetic field at the beginning of the universe. Thus, it is important to develop a theory to describe the generation of magnetic field from zero. As described in chapter 3, plasma follows the flux conservation law based on the flux freezing principle.

We consider the resistive MHD equation, with initial condition $B = 0$ everywhere:

$$\frac{\partial B}{\partial t} = \nabla \times (V \times B) + \frac{\eta}{\mu_0} \nabla^2 B. \tag{13.2}$$

The first term on the right-hand side is called the dynamo term and the effects of plasma velocities in changing magnetic fields are called dynamo action. If flux were exactly conserved and magnetic field B were initially zero, then the flux through any region would remain zero and B would always be zero. Even a finite resistivity in the second term would not help with this problem. Thus the validity of eq. (13.1) comes into doubt and the important question of whether magnetic field is generated from zero and sustained by such dynamo action arises.

However, in the two-fluid MHD regime, a more general form of Ohm's law becomes important. Here, as was discussed in chapter 5, the Ohm's law of MHD can be replaced by the generalized Ohm's law in order to describe the force balance of an electron flow, namely

$$E + V \times B = \eta j + \frac{j \times B}{e n_e} - \frac{\nabla \cdot P_e}{e n_e} - \frac{m_e}{e} \frac{d V_e}{dt}. \tag{13.3}$$

By neglecting the inertial and resistive terms, this equation becomes

$$E + V_e \times B = -\frac{\nabla \cdot P_e}{e n_e}, \tag{13.4}$$

where $V = V_i + V_e$ and V, V_i, V_e are the bulk plasma flow vector, ion flow velocity, and electron flow velocity respectively.

If we take the curl of this equation and combine it using Faraday's equation, we obtain

$$\frac{\partial B}{\partial t} = \nabla \times (V_e \times B) + \nabla \times \left(\frac{\nabla p_e}{n_e e}\right) = \nabla \times (V_e \times B) - \frac{(\nabla n_e \times \nabla p_e)}{n_e^2 e}. \tag{13.5}$$

The last term in this equation is the Biermann battery term, which does not go to zero for $B = 0$ because it is associated with nonuniform n_e and p_e. If p_e is a function of

n_e, or $p = p(n_e)$, their cross product is zero. In order to break this constraint in astrophysical plasmas, a rotational motion is key (Kulsrud, 2005). This rotation can come from the protogalaxy's nonuniform chaotic collapse (although it may take millions of years). Also, if a section of an astrophysical plasma is heated by a shock structure or by penetration of a hot plasma regime, it would induce a situation where $\nabla p_e \neq \nabla n_e$.

Kulsrud (2005) calculated how big we could expect the magnetic field strength developed by the Biermann battery to become in astrophysical plasma systems, based on Kolmogorov turbulence theory. He obtained an extremely small field of 2.5×10^{-18} G. Since this field strength is reached in his calculation after one turnover time of the largest eddy, there may be other models for the Biermann battery to initiate dynamo action in astrophysical plasmas.

A Biermann dynamo situation, with $\nabla p_e \neq \nabla n_e$, also happens in the formation process of laser-heated plasma in which a strong local temperature gradient is generated. In this case, the temperature gradients are not in the same direction as the electron density gradient. This situation happens quite often in laboratory fusion device plasmas such as tokamaks or RFPs in which strong local electron heating occurs due to fluctuations or external heating mechanisms (by RF waves or high-energy neutral beam injections).

As an example, Nilson et al. (2006) presented measurements of evidence for strong electron heating at laser-irradiated columns in plasmas, created by injecting two laser beams closely focused on a planar foil target. The two plasmas typically collide and stagnate, and for laser-spot separations of about seven focal spot sizes, two very distinct, highly collimated jets were observed. In addition to a sharp reconnection current sheet between the two columns, as already described in chapter 6 (figure 6.12), they observed that strong azimuthal fields ($\sim 10^6$ G) are created around the two columns. They concluded that the azimuthal magnetic fields are generated by a Biermann battery mechanism of $\nabla n_e \times \nabla T_e$ around each laser spot. In their experiment, ∇T_e is in the radial direction of each column and ∇n_e is in the axial (Z-) direction, thus generated magnetic fields are in the azimuthal direction. Recently, Fox et al. (2018) developed a fully kinetic simulation model for first-principles simulation of the Vulcan experiments (Nilson et al., 2006), including the dynamics of magnetic fields, and verified their magnetic field generation by the Biermann battery effect and Weibel instability.

13.3 RESEARCH ON DYNAMO EFFECTS IN LABORATORY FUSION PLASMAS

The RFP, mentioned many times in earlier chapters, utilizes a steady-state dynamo of large magnitude. Namely, a toroidal electric field is applied to initiate the plasma and drives part of the steady-state current. But a large portion of the current is not driven directly by the electric field. Some part of the edge plasma current flows in a direction even opposite to the electric field. This is illustrated in figure 13.2, which displays the electric field (parallel to the magnetic field) and the parallel plasma current versus minor radius of the toroidal plasma. Note that there is a large mismatch between the two profiles. The remaining current is driven by a dynamo effect, namely the poloidal

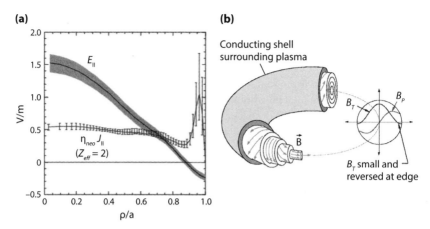

Figure 13.2. (a) Radial profile of the parallel electric field $E_\parallel$ and measured current density terms in Ohm's law for the RFP, where $E_\parallel$ was calculated from nonlinear MHD. The difference in the curves reflects the MHD dynamo. (b) Magnetic profile of RFP plasmas in the MST. [From Prager et al. (2005).]

flux is converted to toroidal flux, thus generating an enhanced toroidal flux. Here we will call it generation of toroidal magnetic flux through flux conversion: a special form of dynamo action, amplification of toroidal flux through a magnetic reorganization process. Profiles in experiments show that this dynamo current is large, up to several hundred kiloamperes, more than half of the total current.

The dynamo is also evident through spontaneous generation of toroidal magnetic flux. In the MST (Madison Symmetric Torus), toroidal magnetic flux is generated in periodic dynamo events, as shown in figure 13.3(a). The sudden jumps in the toroidal flux are produced by a dynamo mechanism. Between the sudden events, the dynamo is mostly absent and the toroidal flux decays. A key parameter for dynamo studies is the Lundquist number S (the ratio of the resistive diffusion time to the Alfvén transit time, from about 10^3 to 10^7 in RFPs). One of the challenges for the dynamo problem is to understand the physics mechanisms of its saturation at the high-S regime. (The geo-dynamo is somewhat distinguished by the low-S value associated with liquid metals.)

The major approach to understanding the dynamo, both natural and laboratory, has been the MHD model. The dynamo effect enters into the mean-field Ohm's law

$$\langle \boldsymbol{E} \rangle + \langle \boldsymbol{V} \times \boldsymbol{B} \rangle = \langle \eta \boldsymbol{j} \rangle, \tag{13.6}$$

where the vector variable includes fluctuating quantities, and $\langle\ \rangle$ denotes the mean field (an average over the fluctuations). The second term of the left-hand side represents the MHD dynamo. It is an electromotive force arising from fluctuations in the plasma flow velocity and magnetic field. This term is often referred to as the alpha effect. In single MHD theory, all dynamo effects enter through this term. It has been measured in laboratory plasmas, in which the flow is measured spectroscopically or by inferring

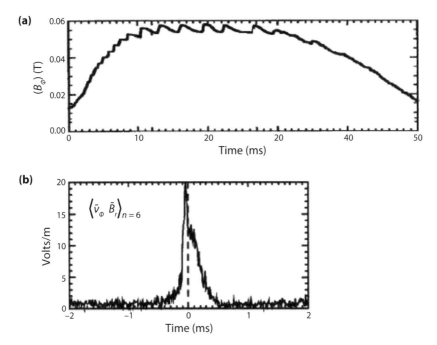

Figure 13.3. (a) Toroidal magnetic flux versus time in MST. Discrete jumps are dynamo events. (b) Measured MHD dynamo versus time. Flow velocity fluctuations are measured spectroscopically; magnetic field is measured with magnetic probes. Data are obtained from an ensemble of the impulsive dynamo events shown in (a). [From Den Hartog et al. (1999).]

it from the measured $E \times B$ drift. For example, it is seen in the RFP that under some conditions the alpha effect term, shown in figure 13.3(b), accounts for driven current.

The difference between the various dynamo situations is the origin of the fluctuations or local vorticities. In some natural dynamos, the flow is generated by thermal convection or rotation of plasmas. The magneto-rotational instabilities are typical examples. In magnetic fusion laboratory plasmas, the flow is often generated by magnetic instabilities. There are many types of dynamo problems, with differing boundary conditions, plasma parameters, and geometry. However, there is a strong physics commonality to all the problems.

13.4   EFFECTS OF A TWO-FLUID DYNAMO IN AN RFP PLASMA

Until recently, most dynamo theory, including mean field theory, was based on single-fluid MHD. Here we consider the two-fluid effects of a dynamo along the lines of thought in RFP and spheromak fusion plasma research. There are two ways in which the effects of two fluids (electrons and ions) can enter into dynamo action, altering the physics of flux conversion or magnetic reconnection and tearing instabilities in these plasmas. They introduce new spatial scales into the problem, such as the ion skin depth c/ω_{pi}, arising from the Hall term. These effects are discussed in dealing with two-fluid magnetic reconnection. The key two-fluid dynamo mechanisms come from the generalized Ohm's law. If we include the electron pressure and Hall terms, we can express the parallel mean-field Ohm's law in a form averaging over the time range of fluctuations,

$$\langle E \rangle + \langle V \times B \rangle - \left\langle \frac{j \times B}{ne} \right\rangle + \left\langle \frac{1}{ne} \nabla \cdot P_e \right\rangle = \langle \eta j \rangle. \tag{13.7}$$

Here we neglected the electron inertia term, and all vectors include fluctuating components. The combined effect from the second and third terms of the left-hand side of eq. (13.7) (or the second term of eq. (13.6)) represents the Hall dynamo effects which influence $E_{\parallel}$, and the fourth term is for pressure-driven dynamo effects, which are of opposite sign to the magnetic reconnection case. These are additional terms that represent the two-fluid dynamo mechanisms. A linear theory of the two-fluid tearing instability was investigated by Mirnov et al. (2003), concluding that the dynamo effect caused by the $v_e \times B$ Hall term dominates over the contribution from the alpha effect term in the narrow electron layer, while in the broader ion layer these terms are comparable.

Ji et al. (1995) measured fluctuation-induced dynamo electric fields over a wide range of electron collisionality in the outer edge of an RFP called TPE-1RM20. The fluctuation fields are primarily from tearing modes that exist at the rational surfaces mentioned in chapter 9 (figures 9.8 and 9.10). They concluded that in the collisionless regime where the mean free paths of plasma particles are longer than the fluctuation wavelength, the usual dynamo effects from the $v_e \times B$ Hall term can sustain the parallel current, but in the collisional regime a dynamo mechanism associated with the fluctuations directly related to the fourth term of eq. (13.7) is responsible for dynamo action.

It is important to summarize the sustaining mechanism of RFP plasma as follows. The external inductive drive provides an extra poloidal flux to drive the toroidal plasma current. Then excessive poloidal field generates a dynamo action by $j_e \times B_p$ or by the diamagnetic currents, which would generate the electric field $E_{\parallel}$ to sustain the poloidal current at the edge, which internally generates additional toroidal flux, thus maintaining the toroidal flux as mentioned in the earlier section. These types of mechanisms are thus called flux conversion because the poloidal flux is converted to toroidal flux through the magnetic reconnection process. Thus magnetic reconnection plays a key role in the dynamo process. However, the magnetic energy of the whole system should decay without an external supply of flux. So this type of dynamo action in a low-β plasma should be classified as a flux-conversion dynamo, different from the conventional dynamo in which magnetic energy is generated from zero or a minute initial value. As mentioned in chapter 9, magnetic helicity is conserved during this global magnetic reconnection process, while magnetic energy decreases. In a low-β plasma, there is not enough free kinetic energy to drive dynamo mechanisms to amplify the total magnetic energy. As mentioned in chapter 1, if there is abundant kinetic or thermal energy in a plasma with respect to magnetic energy, just like in the early universe ($W_p \gg W_B$), magnetic fields are generated through a dynamo process.

There are arguments that two-fluid effects may be active in protostellar accretion disks, in white dwarfs, and in neutron stars. In protostellar disks, ion inertia is increased by collisions with neutrals. In addition, the disks are thin. These two effects imply that Hall terms and electron MHD effects may be important in these systems. In crystallized neutron star crusts, solid-state stresses can act to inhibit ion motion, thus introducing similar two-fluid dynamics.

Chapter Fourteen

Magnetic reconnection in large systems

In astrophysical plasmas, the ratio of global to kinetic scales is very large and the ratio of mean free path to plasma scales is small, thus MHD (magnetohydrodynamic) models are considered to be practical for treating space astrophysical phenomena. Except in magnetosphere plasmas, in which two-fluid physics dominates, MHD has been the most common theory for treating astrophysical plasma physics problems. Is it possible to have fast reconnection in MHD, without invoking kinetic processes? The earlier argument based on MHD was that fast reconnection is explained by an "open scissor"-shaped X-point, which Petschek introduced. To employ the MHD model, a short current sheet, through which not all field lines pass, has been assumed, while local enhanced resistivity is considered to be enhanced as a function of local, high current density. But the exact cause of the enhanced ("anomalous") resistivity has yet to be conclusively determined by laboratory experiments or by theory.

In recent decades, many numerical theories and laboratory experiments have investigated reconnection in relatively small plasma systems of 10–100 ion skin depths. In much larger systems, however, it was found in recent numerical simulations that multiple current sheets or reconnection layers develop in the reconnection region, which can increase the reconnection rate in both collisional and collisionless regimes. For example, in simple two-dimensional resistive MHD simulations for a large plasma with a large Lundquist number, a group of theoretical works (Shibata and Tanuma, 2001; Bhattacharjee et al., 2009; Huang and Bhattacharjee, 2010; Loureiro et al., 2012) found that a laminar Sweet–Parker layer is transformed to a chain of secondary magnetic islands and the reconnection process becomes inherently impulsive, and thus the reconnection rate averaged over time becomes faster. The appearance of multiple layers would become dominant, particularly in three-dimensional systems. This process can invoke turbulence in the layer, and new approaches are required to properly describe this turbulent layer. This type of multilayer reconnection can occur in solar flares as well as in fusion plasmas. Lazarian and Opher (2009) found that driven MHD turbulence can enhance the reconnection rate significantly. In this chapter, let us study the relatively recent development of this new approach to studying magnetic reconnection in a large system, theoretically and experimentally.

14.1 DEVELOPMENT OF PLASMOID THEORY

Two decades ago, Tanuma et al. (2001) found in their MHD simulation study for reconnection in solar flares that the reconnection current sheet changes shape as a result of

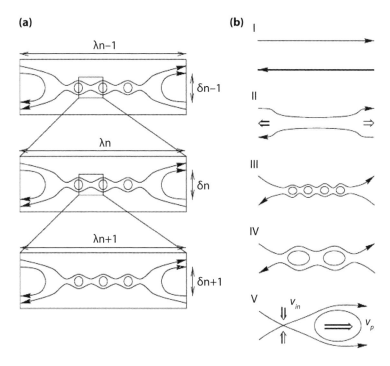

Figure 14.1. (a) Schematic view of the fractal reconnection. (b) A scenario for fast reconnection: (I) The initial current sheet. (II) Current sheet thinning in the nonlinear stage of the tearing instability or global resistive MHD instability. The current sheet thinning stops when the sheet evolves to the Sweet–Parker sheet. (III) Secondary tearing in the Sweet–Parker sheet. The current sheet becomes fractal because of further secondary tearing as shown in (a). (IV) The magnetic islands coalesce with each other to form bigger magnetic islands. The coalescence itself proceeds in a fractal hierarchy manner. During phases (III) and (IV), a microscopic plasma scale (ion gyroradius or ion inertial length) is reached, so that fast reconnection becomes possible at small scales. (V) The greatest energy release occurs when the largest plasmoid (magnetic island or flux rope) is ejected. The maximum inflow speed (V_{inflow} = reconnection rate) is determined by the velocity of the plasmoid (V_p). Hence this reconnection is termed a plasmoid-induced reconnection. [From Shibata and Tanuma (2001).]

a local tearing instability. They found that a localized current sheet becomes thinner as it breaks up and the thinning is saturated when the sheet thickness becomes comparable to that of the Sweet–Parker sheet thickness. A secondary tearing instability occurs, and further local current-sheet thinning follows. If the sheet becomes sufficiently thin to produce anomalous resistivity, a Petschek-type reconnection starts with locally enhanced resistivity. On the basis of nonlinear MHD simulations, Shibata and Tanuma (2001) proposed that the current sheet eventually has a fractal structure, consisting of multiple layers of magnetic islands (plasmoids) with different sizes, as shown in figure 14.1.

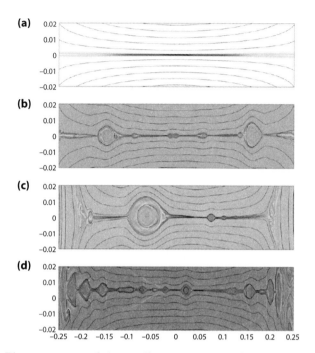

Figure 14.2. Time sequence of the nonlinear evolution of the current density j_Y of a Sweet–Parker current sheet in a large system of Lundquist number $S \sim 6 \times 10^5$. The black lines represent surfaces of constant ψ: (a) $t = 3.0$, (b) $t = 6.0$, (c) $t = 9.1$, (d) $t = 12.0$. [From Bhattacharjee et al. (2009).]

Once the current sheet has a fractal hierarchy structure, it becomes possible to connect macroscale dynamics (with a flare size of 10^9 cm) and microplasma-scale dynamics (with ion gyroradius or ion skin depth of 10^2 cm). Then anomalous resistivity could be invoked, they thought.

Shibata and Tanuma (2001) presented a scenario for fast reconnection in the solar corona as shown in figure 14.1. By applying their scenario to an actual solar coronal condition, they found that a fractal hierarchy number of $n \sim 6$ is necessary to connect the solar flare reconnection scale from the ion gyroradius or ion skin depth. Their current sheet becomes a fractal sheet consisting of many plasmoids with different sizes. The plasmoids tend to coalesce with each other, as theorized by Tajima et al. (1987), to form bigger plasmoids. When the biggest island (i.e., a monster plasmoid) is ejected out of the sheet, we have the most violent energy release, which may correspond to the impulsive phase of flares. The tearing mode instability in the Sweet–Parker current sheet has been studied by Loureiro et al. (2007), and is sometimes called a plasmoid instability.

Nonlinear evolution of plasmoid-dominated reconnection has been extensively studied in more recent years using MHD simulations; see figure 14.2. The plasmoid distribution in the nonlinear regime, knowing the features of which is essential for

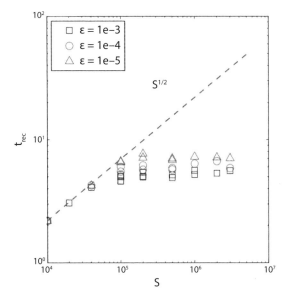

Figure 14.3. The reconnection time t_{rec} for various S and ε, where ε is a small parameter that controls the noise level of the plasma density variation in a single step of system evolution. The dashed line is the Sweet–Parker scaling. Above S_c, a critical value of S, the reconnection time t_{rec} deviates from the Sweet–Parker scaling and becomes shorter. In the plasmoid unstable regime, the reconnection time is nearly independent of S and has a weak dependence on the noise level ε throughout the large-S range they tested. The plateaued values of t_{rec} in the high-S regimes for $\varepsilon = 10^{-3}$, 10^{-4}, 10^{-5} are 5.30 ± 0.27, 6.10 ± 0.41, 7.05 ± 0.16, respectively. [From Huang and Bhattacharjee (2010).]

understanding the current sheet thinning process, has been discussed. Cassak et al. (2009), Bhattacharjee et al. (2009), and Huang and Bhattacharjee (2010) found that once plasmoid instability sets in, the reconnection rate becomes nearly independent of the Lundquist number for $S > 10^4$.

The Sweet–Parker layer in a system that exceeds a critical value of the Lundquist number (S) is unstable to the plasmoid instability. Huang and Bhattacharjee (2010) carried out a two-dimensional numerical study with an island-coalescing system driven by a low level of random noise. The primary Sweet–Parker layer breaks into multiple plasmoids and even thinner current sheets through multiple levels of cascading if the Lundquist number is greater than a critical value. To quantify the speed of reconnection, they measured the time it takes to reconnect a significant portion of the magnetic flux within two merging islands. They denote the time as t_{rec} for a certain designated significant portion of flux to reconnect. The range corresponds to reconnecting 25% of the initial flux so that the layer does not shorten too much compared with that in the initial condition. Figure 14.3 shows the reconnection time t_{rec} for various S and ε. For lower S, the reconnection time $t_{rec} \sim S^{1/2}$, as expected from the Sweet–Parker theory (see chapter 3). The critical Lundquist number for plasmoid instability is about

$S_c \leq 4 \times 10^4$. Above S_c, the reconnection time t_{rec} deviates from the Sweet–Parker scaling and becomes shorter. In the plasmoid unstable regime, the reconnection time is nearly independent of S. However, the reconnection time has a weak dependence on the noise level ε throughout the S range they tested (ε is a small parameter that controls the noise level of the plasma density variation in a single step of system evolution). They tested the convergence of their numerical results by varying the resolution, the time step, and the random seed for selected runs. In the figure, these are represented by multiple data points with the same parameters. The results are fairly consistent, with fluctuations of no more than a few percent. As a result of the plasmoid instability, the system realizes a fast nonlinear reconnection rate that is nearly independent of S, and is only weakly dependent on the level of noise. By considering stochastic generation, growth, coalescence, and ejections of plasmoids, Uzdensky et al. (2010) predicted the dependence of the plasmoid distribution function $f(\phi)$ on the flux quantity ϕ contained by plasmoids as $f(\phi) \sim \phi^{-1}$, and Loureiro et al. (2012) carried out two-dimensional MHD simulations to investigate the plasmoid distribution, obtaining double-power-law-like distributions.

It should be noted, however, that all of these studies were carried out in two dimensions. It is not obvious that two-dimensional results will remain unchanged for three-dimensional astrophysical situations with a high Lundquist number ($S > 10^{12}$ for solar flares).

14.2   EFFECTS OF MHD TURBULENCE ON MAGNETIC RECONNECTION

It has been asked by many whether small-scale driven turbulence in a broad reconnection layer can simply be treated as some form of enhanced resistivity, albeit from MHD fluctuations rather than micro-instabilities. The basic idea is that the flow would tangle the magnetic field, producing small-scale magnetic structures that resistively decay. Turbulent diffusion rates are generally set by the eddy turnover time l/v_l and depend only weakly on the microscopic diffusion rate. We might therefore expect the turbulent diffusivity η_t to be of the eddy form $l v_l$, where l is the outer scale of the turbulent cascade, where most of the power is, and v_l is the turbulent velocity at that scale.

14.2.1   Lazarian–Vishniac model for turbulent MHD reconnection

Lazarian and Vishniac (1999) considered that the Sweet–Parker model (described in chapter 3) can be generalized by considering the effects of MHD turbulence. In their concept, two regions of the reconnecting magnetic fields are separated into a laminar flow region and a turbulent reconnection layer region, shown in figure 14.4. Here, let us follow their arguments. They considered that the outflow in the turbulent flow is not limited by the relatively thin region determined by resistivity as modeled by Sweet and Parker, but is rather broad and determined by magnetic field wandering. Assuming that the field wandering is the cause of the reconnection zone opening up, they at first

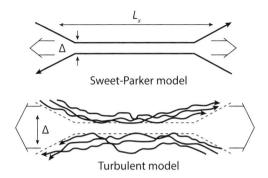

Sweet-Parker model

Turbulent model

Figure 14.4. (Top) Sweet–Parker model of reconnection. The outflow is limited to the thin width of Sweet–Parker. (Bottom) The thickness Δ is an astrophysical scale of L_{LV}.

calculated the layer thickness, Δ for the case where the turbulence injection scale L_i is less than the total size L_X. One finds that the perpendicular extent of the eddy at the injection scale is $L_i M_A^2$, where M_A is the Alfvén Mach number. The transverse contributions from different eddies at the injection scale are not correlated and therefore are the result of random walk with a step of $L_i M_A^2$. The number of steps along L_x is L_x/L_i and thus

$$\Delta \simeq \left(\frac{L_x}{L_i}\right)^{1/2} L_i M_A^2 \quad \text{for } L_i < L_x. \tag{14.1}$$

In the Sweet–Parker laminar model, the magnetic field lines are subject to Ohmic diffusion, in which magnetic field-lines diffuse by resistivity in a time t given by

$$\langle y^2(t) \rangle \sim \lambda t, \tag{14.2}$$

where λ is the magnetic diffusivity. The field lines are advected out of the sides of the reconnection layer of length L_x at a velocity of order V_A. They find the time that the lines can spend in the resistive layer is the Alfvén transit time $t_A \sim L_x/V_A$. Thus for the Sweet–Parker model, the thickness of the diffusion region should be

$$\Delta \simeq \sqrt{\langle y^2(t_A) \rangle} \sim \sqrt{\lambda t_A} = L_x/\sqrt{S}, \tag{14.3}$$

where S is the Lundquist number. Using relationships obtained in chapter 3, we obtain $V_{\text{rec}} \sim V_A/\sqrt{S}$.

For the case of turbulent reconnection, Lazarian and Vishniac (1999) used the Richardson diffusion coefficients instead of Ohmic diffusion. In this case, the mean-squared separation of particles is $\langle |x_1(t) - x_2(t)|^2 \rangle \sim \epsilon t^3$, where ϵ is the energy cascade rate:

$$\Delta \simeq \sqrt{\epsilon t_A^3} \simeq L_x \left(\frac{L_x}{L_i}\right)^{1/2} M_A^2 \quad \text{for } L_x < L_i \tag{14.4}$$

and

$$v_{\text{rec}} \simeq V_A \left(\frac{L_x}{L_i}\right)^{1/2} M_A^2 \quad \text{for } L_i > L_x. \tag{14.5}$$

Thus, eq. (14.5) together with (14.1) provides a description of fast reconnection for turbulent reconnection in the presence of sub-Alfvénic turbulence.

The scaling predicted in Lazarian and Vishniac (1999) was later tested by their numerical simulation and reported in Kowal et al. (2009) as $v_{\text{in}} \propto \mathcal{P}^{1/2} l_i$. With some discrepancy between observed power dependence in two- and three-dimensional cases, more comprehensive studies are being carried out.

14.2.2   Observation of multiple reconnection layers by kinetic simulations

Daughton et al. (2009) found that a collisionless reconnection layer breaks up into many islands and current layers, generating a highly turbulent reconnection region even in their two-dimensional PIC (particle-in-cell) simulations. Using fully kinetic simulations with a Fokker–Planck collision operator, they demonstrated that Sweet–Parker reconnection layers are unstable to plasmoids (secondary islands) for Lundquist numbers beyond $S > 1,000$. The instability is increasingly violent at higher Lundquist numbers, both in terms of the number of plasmoids produced and the super-Alfvénic growth rate. They observed dramatic enhancement in the reconnection rate when the half-thickness of the current sheet between two plasmoids approaches the ion inertial length; see figure 14.5.

This study has been extended to three dimensions (Daughton et al., 2011) to simulate the MRX (Magnetic Reconnection Experiment) experiment carried out in the near collision-free condition to discover the three-dimensional features of breakup shown in figure 14.6. This result suggests that turbulence can significantly broaden the electron diffusion region, as well as the ion diffusion region. Three-dimensional simulations were carried out on two peta-scale supercomputers, using the kinetic PIC code VPIC (Bowers et al., 2008), which solves the relativistic Vlasov–Maxwell system of equations. The simulation results in the figure reveal some dramatic differences from the basic ideas discussed before. First, the concept of magnetic islands in real three-dimensional systems corresponds to extended flux ropes, which can interact in a variety of complex ways not possible in two-dimensional models. In contrast, the flux ropes shown in figure 14.6 are generated by the tearing instability propagating with wide oblique angles. These results demonstrate a few dominant modes that are localized near the center of the initial layer. Some of the interesting three-dimensional features in the figure arise from the interaction of these flux ropes, leading to a complex connectivity of magnetic field lines. However, the observed dynamics at this time was simpler than expected from nonlinear theories. A more detailed description is given in Daughton et al. (2011). A numerical simulation of MRX experiments was also carried out.

In three dimensions, these islands form extended current sheets or flux ropes at oblique angles above and below the reconnection layer. Over longer timescales, these flux ropes coalesce and intertwine in complex ways to produce new, highly elongated

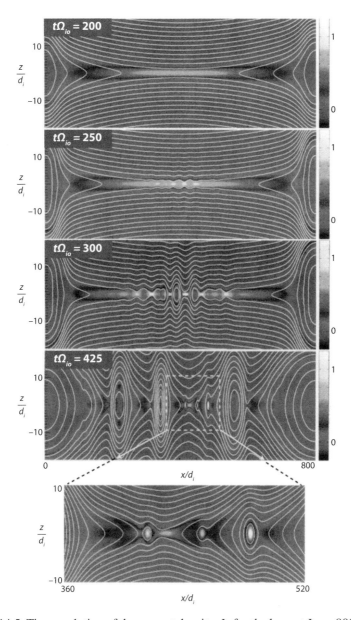

Figure 14.5. Time evolution of the current density J_y for the largest $L_x = 800d_i$ simulation. The white lines are the magnetic flux surfaces and the bottom panel is a close-up of the region indicated at $t = 425(1/\Omega_i)$ to illustrate the repeated formation of new plasmoids within the electron layer. The current density is normalized to the initial peak value J_0. [From Daughton et al. (2009).]

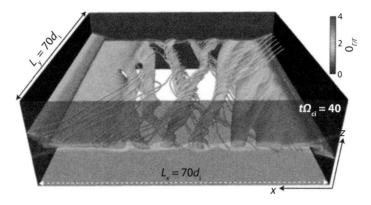

Figure 14.6. See Color Plate 8. Formation of primary flux ropes in a current sheet where electron flows dominate. Tearing instability gives rise to flux ropes as illustrated by an iso-surface of the particle density colored by the magnitude of the normalized current density, along with sample magnetic field lines (yellow). [From Daughton et al. (2011).]

current sheets, which are also unstable to new magnetic island (flux rope) formation. This situation can cause high resistivity and thus fast reconnection, as previously mentioned. The motion of flux ropes would induce impulsive reconnection. Thus we could say that a generalized Sweet–Parker model with enhanced resistivity and viscosity may be able to qualitatively describe fast reconnection, as described in chapter 3.

On the other hand, it should be noted that a remarkable experimental result has very recently been obtained in a super-high-density plasma, generated by laser radiation, with a long reconnection layer of aspect ratio of 100 (Fox et al., 2020). In this high-β plasma of $\beta > 10$, a stable 2 mm-long distinct neutral sheet (reconnection layer) was identified with the width of the ion skin depth, $20\,\mu$m. If normalized by the ion skin depth, this plasma is considered to be a large system, since the ratio of the plasma size ($L = 2$–4 mm) to the ion skin depth is 200. Now this type of laser-generated plasma has become an important medium in which two-fluid physics can be studied in a large high-β plasma condition.

14.3   EXPERIMENTAL STATUS OF MAGNETIC RECONNECTION RESEARCH FOR A LARGE SYSTEM

Thus, reconnection research has recently been moving toward covering dynamics beyond the idealized, classical, single quasi-stationary X-line geometry described in the earlier part of this book, to explore more realistic highly dynamic reconnection characteristics of large systems, such as those found in most space and astrophysical environments. Also, as the fusion research devices have become larger, the ratios of global plasma sizes have become significantly large with respect to the characteristic scale of kinetic processes, which include the ion skin depth and the ion gyroradius (a few

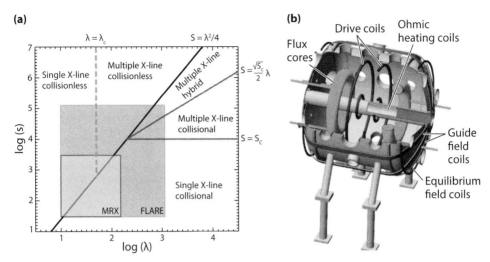

Figure 14.7. (a) A phase diagram for magnetic reconnection in 2D. If either S or the normalized size λ $(= L/\rho)$ is small, reconnection with a single X-line occurs in collisional or in collisionless phases. When both S and λ are sufficiently large, three new multiple X-line phases appear with magnetic islands. The dynamics of new current sheets between these islands are determined either by collisional physics or by collisionless physics, as described in this chapter. The conditions for electron runaway are shown as a broken vertical gray line. The existing experiments, such as MRX, do not have access to these new phases. (b) The FLARE device at PPPL, Princeton University, is accessible to these new phases, which are relevant to reconnection in heliophysical and some astrophysical plasmas. [From Ji and Daughton (2011).]

centimeters). These large reconnection systems will often contain multiple X-lines, plasmoids, and flux rope formations due to secondary instabilities. They will lead to the self-consistent emergence of turbulence and accompanying coherent structures under a variety of plasma conditions.

An experimental study of magnetic reconnection in a large system has started at PPPL (Princeton Plasma Physics Laboratory), Princeton University, utilizing a newly constructed device FLARE (Facility of LArge Reconnection Experiment), with improved diagnostics. The experimentally accessible regimes in the phase diagram of reconnection (Ji and Daughton, 2011) are shown in figure 14.7(a). The size of the device is roughly twice that of MRX. FLARE has initiated a major operation recently.

In figure 14.7(a), let us look at the reconnection phase diagram of Ji and Daughton (2011), with the physics issues of each regime explained by Loureiro and Uzdensky (2016). Different reconnection regimes are indicated as a function of Lundquist number (vertical axis) and of the ratio between the system size L and the ion sound Larmor radius ρ_s (horizontal axis), where $\rho_s = \sqrt{(T_i + T_e)m_i}/q_i B_T$. The solid black diagonal line is yielded by comparing the Sweet–Parker reconnection width δ_{SP} with the kinetic skin depth (designated by ρ_s) and indicates whether we are in a collisional Sweet–Parker regime or in a collisionless regime (respectively, below or above the solid black line). Without a guide field, this criterion corresponds to the criterion for transition from a collisional regime (represented by δ_{SP}) to a collisionless two-fluid reconnection regime (c/ω_{pi}) as expressed by eq. (5.21), which is equivalent to $\delta_{SP} = c/\omega_{pi}$, as described by Yamada et al. (2006). The vertical gray broken line labeled $(L/\rho_s)_c$ is an empirical line that separates the regimes where simulations of collisionless reconnection have a tendency to exhibit multiple plasmoids from those with a tendency to show a single X-point, although there is no analytical theory to back this threshold. The horizontal (solid) gray line labeled S_c indicates the resistive MHD plasmoid instability threshold, which is described in this chapter: the Sweet–Parker current sheet is plasmoid unstable in the region above this line, and stable below it (another threshold that exists, but that we omit for simplicity, results from the case when the MHD plasmoid instability, in its nonlinear stage, triggers a transition to the kinetic scales). The regions of operational space roughly covered by a selection of reconnection experiments (both existing) are also indicated.

Understanding the generation and influence of secondary reconnection instabilities is one of the primary goals of a new reconnection experimental project, the FLARE device shown in figure 14.4(b). Accordingly, a key part of our new kinetic simulation efforts will be directed at modeling these devices in order to validate numerical simulation codes and test theoretical ideas. These efforts will build upon our previous fully kinetic simulations in the MRX, mentioned above, which over the past several years have evolved to include realistic boundary conditions, Coulomb collisions, and full three-dimensional dynamics.

Comparisons with past experiments have resolved some issues regarding the role of electromagnetic fluctuations (Ji et al., 2004; Dorfman et al., 2013) in MRX. At the moment, simulations and experiment are addressing whether this instability plays an important role in energy dissipation. A study will also be carried out on the formation of small-scale flux ropes that have recently been observed and the influence of a guide field on the reconnection dynamics. In the presence of sufficient guide magnetic field, can we form large-scale current layers by extending the system size to that of kinetic simulations by Daughton et al. (2009) or Daughton et al. (2011). Uncovering physical mechanisms for fast dissipation of magnetic energy, through emerging three-dimensional mode structures and turbulence, will be a focus for the upcoming research campaign. Experimental efforts will span a wide range of parameters in collisionality, guide field, and external driving conditions and global boundary conditions from a broad suite of experiments. In three-dimensional geometry, a broad spectrum of oblique tearing modes can become unstable.

14.4   MAGNETIC RECONNECTION IN A LARGE SYSTEM OF ELECTRON–POSITRON PAIR PLASMA

The physics of collisionless magnetic reconnection in electron–positron plasmas is of significant interest in a number of high-energy astrophysical phenomena, including the jets from active galactic nuclei, pulsar winds, and models of gamma-ray bursts, as discussed in chapter 7. Furthermore, a better understanding of the pair plasma limit may shed light on the Hall effects, which should be quite different due to the equal masses of electrons and positrons. The fundamental question of how the dynamical evolution will scale to macroscopic systems of physical relevance was examined through numerical simulation by Daughton et al. (2011) using two-dimensional fully kinetic simulations with both open and periodic boundary conditions. They found that repeated formations and ejections of plasmoids play a key role in controlling the average structure of a diffusion region and preventing the further elongation of the layer. The reconnection rate is modulated in time as the current layers expand and new plasmoids are formed. The observed averaged reconnection rate is very fast and is remarkably insensitive to the system size for sufficiently large systems. This dynamic scenario could lead to a different explanation for fast reconnection in large-scale systems.

14.5 IMPULSIVE RECONNECTION IN A LARGE SYSTEM

It is important to study the key mechanisms of impulsive reconnection that happens in a large system with a large Lundquist number. Energy transport is fast during three-dimensional self-organization in large astrophysical and fusion plasmas. A general feature of reconnection in a large plasma system is its impulsive, bursty nature, leading to the powerful release of stored magnetic energy. In magnetospheric substorms, in solar and stellar flares and coronal mass ejections, in various astrophysical events such as in gamma-ray flares in the Crab Nebula and magnetar magnetospheres, and also in the self-organization of magnetic fusion plasmas, reconnection starts suddenly in time. There has been no clear consensus nor established universal explanation for what triggers the onset of a reconnection event in these large systems. But some ideas could be proposed in specific cases mentioned above, as well as in chapters 7, 8, and 9.

When external force is gradually applied to a plasma, the global conditions of magnetic field configuration and plasma pressure profile can be kept stable against MHD instability because of a low magnetic shear geometry or boundary conditions. However, once an instability criterion is reached, MHD modes grow quite fast, with growth time being as fast as Alfvén transit time. Then the evolution of the plasma configuration would create a region of strong magnetic field shear, which drives reconnection. One important lesson from the two- and three-dimensional numerical studies of reconnection in large systems mentioned in this chapter is that the reconnection inflow velocity cannot be significantly slower than about $0.01 V_A$, due to the occurrence of plasmoid instabilities, and is therefore always quite fast.

As mentioned in chapter 7, global reconnection happens fast in solar flares as well as in tokamaks. In both cases we have learned that a gradual change of plasma configurations can make them unstable against a global MHD instability, which then drives fast reconnection by forming a current layer or reconnection layer: "driven reconnection." Current layer disruptions are identified in three-dimensional analysis in MRX as a local way to quickly release magnetic energy. These disruptions are due to the ejection of three-dimensional, high-current-density flux rope (plasmoid) structures. A three-dimensional, two-fluid model consistent with the observations is proposed as a possible disruption mechanism. The onset of fast reconnection in some systems may have unique features that are not universal. In RFP (reversed field pinch) fusion research plasma, which has a periodic boundary condition as mentioned in chapter 9, the magnetic geometry permits the simultaneous appearance of many tearing modes, some of which are linearly unstable due to the gradient of the current profile in the core. Impulsive reconnection, referred to as a sawtooth event, occurs when resonant modes in the core drive modes at the edge through nonlinear interaction of linearly unstable resonant modes. This nonlinear, multimode process underlies the RFP's global magnetic self-organization processes, such as electron and ion momentum transport and noncollisional ion heating and particle energization. In the context of solar flares, models have been proposed for the triggering process involving the detailed magnetic structure in loops and arcades and the nonlinear evolution of multiple reconnection sites (Kusano et al., 2004, 2020).

Chapter Fifteen

Summary and future prospects

A breakthrough in physics research often happens when somebody proposes a simple persuasive picture of complex physical processes. A typical example is the Feynmann diagram for quantum physics. Magnetic reconnection research made a big quantum jump when the Parker–Sweet reconnection picture was created to describe a prototypical layer in which magnetic reconnection takes place, even when nobody had observed such a layer in natural or laboratory plasmas. This simplified, beautiful local model drew much attention from researchers, even though this resistive MHD (magnetohydrodynamics) model based on particle collisions did not provide a quantitatively correct answer to describe the observed reconnection rate. Even a modified MHD model proposed by Petschek was not sufficient to properly describe the key physics of the reconnection layer. Particularly in collisionless plasmas, where the mean free path of current-carrying electrons exceeds the length scale of the system and the MHD condition breaks down, the reconnection rate was found to be very large.

Throughout this book, we have reached a perspective that magnetic reconnection is determined by both local and global plasma dynamics. In this chapter, let us summarize first the major findings for local reconnection layer dynamics and then the global characteristics of magnetic reconnection in the most recent decades.

15.1 MAJOR FINDINGS FROM LOCAL ANALYSIS

As fast reconnection was observed in multiple cases of magnetic self-organization in magnetic fusion plasmas, as well as in space and astrophysical plasmas, the local dynamics of the reconnection layer became one of the central issues of magnetic reconnection research, as mentioned in chapter 1, and is a major problem discussed in this monograph. The major findings of local analysis are as follows:

(1) It has been realized that magnetic reconnection is often driven by external forcing, such as seen in the magnetopause reconnection layer where incoming solar-wind fields compress the reconnection layer so thinly that the motions of ions differ greatly from those of electrons, invoking two-fluid physics mechanisms. Recent controlled laboratory experiments have created such a reconnection layer in the two-fluid regime, leading to a better understanding of the dynamics of collisionless reconnection.

(2) Hall effects were observed in the two-fluid regime, with the typical signature of an out-of-reconnection-plane quadrupole magnetic field structure, in space satellite data, numerical simulations, and laboratory experiments. These Hall effects also generate an enhanced reconnection electric field to induce fast motion of the reconnecting flux line, providing a decisive verification of the enhanced reconnection rate in collisionless plasmas. While this Hall effect itself does not generate energy dissipation, the electron pressure tensor term in the generalized Ohm's equation becomes very large and generates enhanced dissipation near the X-point of the reconnection plane.

(3) In dedicated reconnection experiments, the reconnection rate is found to increase rapidly as the ratio of the electron mean free path to the length scale increases. This result is attributed to the large Hall electric field in the reconnection plane just outside the electron diffusion layer near the X-point. MMS (Magnetospheric Multiscale Satellite) and MRX (Magnetic Reconnection Experiment) observations have verified that in the electron frame just outside the X-region, electrons are frozen to magnetic field lines, namely, "flux freezing" is working for the electron fluid as discussed in chapters 5 and 8 ($E_M \approx (V_e \times B)_M$).

(4) The data from MMS are remarkably in agreement with laboratory experimental results obtained in MRX. Through advances achieved by space technology and analysis, MMS has demonstrated that the actual three-dimensional magnetic reconnection phenomena occurring in space plasmas can be well described by two-dimensional physics analysis made by numerical simulation and laboratory studies on MRX. Further cross-discipline study should lead to a more accurate picture of collisionless magnetic reconnection.

(5) With respect to the energy deposition rate to electrons, $j_{e\perp}E'_{\perp}$ measured in the electron frame is found to be significantly larger than $j_{e\parallel}E'_{\parallel}$ near the X-line, where $E' = E + V_e \times B$, along with its parallel and perpendicular components. In asymmetric reconnection, it is verified in both MRX (Yamada et al., 2014) and MMS data (Burch et al., 2016b) that $\boldsymbol{j}_e \cdot \boldsymbol{E'}$ peaks up through $j_{e\perp}E'_{\perp}$ at the *stagnation point* of the electrons' in-plane flows, which is separated from the X-line by $\sim 5(c/\omega_{pe})$.

(6) Regarding the energetics of magnetic reconnection, studies of the two-fluid reconnection layer in laboratory experiments and in space data have shown that conversion of magnetic energy occurs in a region significantly larger than the narrow electron diffusion region that was previously assumed to be the site of electron energization. An inversed-saddle shaped electrostatic potential profile was measured in the reconnection plane and the resulting electric field was found to accelerate ions toward the exhaust region of the reconnection region. MRX verified that accelerated ions are thermalized by remagnetization in the downstream region. Evidence of the same potential profile and fast ions has been observed in space plasmas.

(7) A quantitative inventory of magnetic energy conversion during reconnection was carried out in a laboratory reconnection layer with a specific well-defined boundary. This study concluded that about half of the inflowing magnetic energy is converted to particle energy, about two-thirds of which is ultimately transferred to ions and one-third to electrons. A local analytical theory is developed for this energy

conversion in chapter 11. While these features of energy conversion and partitioning apparently do not strongly depend on the size of the analysis region over the
tested range of scales, the question still remains whether there is a universal principle for the partitioning of converted energy.

(8) Electrostatic and electromagnetic fluctuations are observed in reconnection layers
in both laboratory and space plasmas, with notable similarities in their characteristics. Although a correlation is found between the reconnection rate and the
amplitude of electromagnetic waves in laboratory experiments, neither conclusive
quantitative correlation nor a causal relationship has yet been identified.

Most recent theories and experiments have focused on local reconnection and, in
general, have not addressed the question of how local reconnection generally connects
to global reconnection. This is understandable since local reconnection is more generic
and considered adaptable to most global situations. But for a proper application of
reconnection mechanisms this connection must be found. As a working hypothesis, it
was often assumed that the field outside the layer is of the same size as the global field
that reverses across the layer. This situation does not always hold. There is often an
additional field (*guide field*) perpendicular to the reconnection plane which is given by
the global picture. In local theory it is arbitrarily included as a guide field, but the current sheet length becomes of global size and the boundary condition has to be adjusted
and realistically considered.

15.2 MAJOR FINDINGS FROM GLOBAL ANALYSIS

In the area of global reconnection research, notable progress has been made in char-
acterizing key features of magnetic self-organization or the relaxation phenomena of
plasmas in solar flares, as well as in laboratory fusion devices, as described in chapters
7 and 9. It has been recognized that global reconnection phenomena occur impulsively
quite often after a very slow buildup phase of magnetic configuration. A change of
global configuration can drive a sharp current layer and can induce fast local reconnection there. Sometimes, fast local reconnection leads to an impulsive global topology
change or global magnetic self-organization phenomena. Thus we find that magnetic
self-organization is influenced and determined both by local plasma dynamics in the
reconnection region and by global evolution of the topology of plasma configurations
(often to an unstable state) and/or external forcing. The major findings from global
studies of magnetic reconnection are as follows:

(1) In solar flares, reconnection sites were observed through topological changes of
plasmas by soft-X-ray emissions. Reconnection sites are also identified with hard-
X-ray emissions near the top of solar flare arcades during coronal eruptions and
CMEs (coronal mass ejections). Reconnection speed was measured to be much
faster than the Sweet–Parker rate. For possible driving mechanisms of the fast
reconnection, MHD instabilities such as the kink instability, torus instability, and
double ark instability have been investigated. In some cases, the predictability of
CME has been discussed based on a quantitative analysis of a physics-based model.

(2) Seeking a general criterion or reason why magnetic energy is stored for a long period and then suddenly released, globally driving the plasma to a relaxed state, we have found that MHD instabilities play a key role in laboratory fusion experimental plasmas. In tokamak plasmas, magnetic reconnection is often driven by an ideal kink-type MHD instability, excited after a gradual change of tokamak equilibrium with a reconnection time much shorter than the Sweet–Parker time. With the recent understanding of two-fluid physics in collisionless plasmas, this is not surprising, since the Sweet–Parker model is only applicable to collisional one-fluid MHD plasmas, while tokamak plasmas are collisionless ($\lambda_{mfp} \gg L$) and we expect fast two-fluid dynamics to be in play during reconnection.

(3) In RFP (reversed-field-pinch) experiments, reconnection occurs in the plasma core and, under some conditions, at the edge. It is observed that two unstable tearing modes in the core region can nonlinearly couple to produce a driven reconnection at a third location in the plasma edge region. It is conjectured that a similar phenomenon can occur in active solar arcade flares, where a spontaneous reconnection at one location can drive reconnection at other locations, leading to eruptions. Most recently, a reconnection model by Kusano et al. (2020), based on the merging of two larger coronal arcs, has been proposed to predict the onset of a major CME.

(4) One of the major questions has been how global systems generate local reconnection structures through the formation of one or multiple current sheets, either spontaneously or forced by boundary conditions. We have found a hint for this question: it depends on the size of the system. In addressing the global issues, we have learned that all classical models fail when particularly long global lengths are assumed for the current layers. This problem was discussed in chapter 14 by addressing the multiple magnetic reconnection layers or plasmoids in a large system. In this respect, we learned that in a large system with a large Lundquist number $S > 10^4$, multiple current layers appear whose lengths are closer to the shorter scale on which the equilibrium settles. *The concomitant occurrence of multiple reconnection layers can provide a key to resolving fast magnetic self-organization or global reconnection phenomena in a large system.*

(5) With respect to the buildup of global stored energy for reconnection, heat transport can influence global reconnection phenomena. The fast transport of electrons through the newly reconnected field lines can change the plasma stability condition, either causing impulsive self-organization or, sometimes, suppressing reconnection by reducing the strong pressure gradients across the flux surfaces that are driving MHD instability and fast reconnection.

We can hypothesize that global magnetic self-organization phenomena in both tokamak sawtooth crashes and solar flares share a common feature. When reconnection occurs in a certain region of the globally connected plasma, a topology change results. A sudden change of magnetic flux is induced in a short time (large $d\Psi/dt$) in a directly connected part of the global plasma. This leads to a large electric field along the magnetic field lines and acceleration of electrons to superthermal energy. Indeed, in reconnection events in both solar flares and tokamak sawteeth, we observe a significant amount of high-energy tail (runaway) electrons. A further, careful, systematic study of tokamak sawteeth and RFP relaxation events can illuminate this important flow channel, revealing the essence of the relationship between the plasma transport and reconnection phenomena and can provide a key to understanding magnetic self-organization in large plasma systems.

15.3   OUTSTANDING ISSUES AND FUTURE RESEARCH

In the past few decades, significant progress has been made to understand the dynamics of the magnetic reconnection layer, particularly in collision-free plasmas. For example, we now understand in the context of two-fluid physics how ions are accelerated and heated in the layer in the presence of a strong electric field induced by the motion of magnetized electrons. This progress has been made by effective cross-discipline collaboration between numerical simulations, laboratory experiments, and space observations.

The understanding of energy conversion and partitioning has moved forward notably. I anticipate further progress will be made by continuing the present, very effective collaborative effort between many researchers around the world. At the moment, the collaboration between magnetosphere research and laboratory experiments is especially strong and successful, as described in chapters 8 and 12.

While it is not straightforward to apply MHD analysis for reconnection in collisionless astrophysical systems, it is important to utilize MHD theory to understand the global nature of magnetic reconnection in large solar and astrophysical plasmas, after clarifying the role of kinetic processes in plasmas. This is an important aspect for future research into reconnection in astrophysical plasmas, in which experiment, simulation, and theory could mutually motivate and reinforce one another. Certain regimes of astrophysical interest are being accessed by current high-energy-density plasma reconnection experiments as briefly mentioned in this book. This is a promising area for future research. Further study of electron–positron pairs should be able to assess the two-fluid kinetic process, while Hall effects due to the large ion–electron mass ratio should diminish in a pair plasma. Also, major progress in the means of research, such as advanced supercomputing, upgraded laboratory facilities, new advanced diagnostics, and multiple clusters of satellites, will help our research progress significantly.

As I mentioned at the beginning of this chapter, magnetic reconnection research made substantial progress when the beautiful Sweet–Parker reconnection picture was proposed. But many reconnection problems discussed in this monograph have demonstrated the necessity of investigating reconnection dynamics beyond the idealized, classical, single quasi-stationary X-line MHD model of Sweet–Parker. We have explored the recently discovered, more realistic, highly dynamic reconnection characteristics of large systems, such as those found in most space and astrophysical environments, as well as in some fusion plasmas. These complex large plasmas often generate multiple X-lines, plasmoids, and flux ropes in their reconnection region. The presence of large amplitude fluctuations would lead to turbulent reconnection. This theme has emerged in the last several years as the new paradigm of how magnetic reconnection really happens in natural plasmas. Understanding the generation and influence of secondary reconnection instabilities is one of the primary goals of new reconnection experimental devices. As mentioned in the previous section, further careful study of the criterion for the simultaneous occurrence of multiple reconnection layers in RFP and tokamak plasmas can provide a key to resolving fast magnetic self-organization in large systems.

While there is ample evidence for the existence of waves and microturbulence in the reconnection layer, no consensus has yet been obtained regarding which waves should affect the reconnection rate directly or indirectly. Extensive study has been carried out to determine how the electron diffusion layer is affected by the presence of wave turbulence, as well as how the profiles of the electron diffusion layer affect overall reconnection dynamics, including energy dissipation. As described in detail in chapter 5, lower hybrid frequency ranges of waves have been observed and identified both in MRX and in the magnetopause. Electron flows in the electron diffusion region were often observed to fluctuate on a variety of timescales, causing turbulent reconnection. The electron current channel becomes unstable due to a sharp radial gradient in the current density, making the local flux transfer rate fluctuate and generating impulsive reconnection. The reconnection rate measured by the flux transfer rate at the diffusion layer was compared with the global rate of flux inflow by Ren et al. (2008) and an experimental campaign has been carried out on this topic in more detail on MRX (Ji et al., 2004; Yoo et al., 2019). It appears to me that the presence of waves of the lower hybrid frequency range should contribute to the enhanced resistivity, as well as the enhanced diffusion of electrons. But more conclusive research is needed to find an important quantitative relationship between the wave amplitudes and observed fast reconnection rate. This issue is one of the central problems to be resolved in the next several years.

Another key problem in recent reconnection research has been the flux and energy transport between the local reconnection layer and the global plasma and its boundaries. How is magnetic helicity conserved during magnetic self-organization in a plasma and how is the magnetic energy conserved or transported between the local region and the global plasma? As we have learned in laboratory flux rope experiments described in chapter 9, magnetic reconnection plays a key role in helicity-conserving magnetic self-organization toward the Taylor state. Perhaps we should revisit and investigate more deeply the fundamental principle and essential physics of the *Taylor state* (Taylor, 1986), to which magnetized plasmas tend to relax.

15.4   CLOSING REMARKS

While this book comes to an end, my three-decade-long journey searching for a solution to the magnetic reconnection problem does not appear to end here. It will continue into the future. The journey has given me the joy of learning many different areas of space astrophysics, which is especially highlighted by the intellectual excitement associated with engaging in the "grand challenge" nature of the magnetic reconnection problem through cross-field, cross-discipline interaction. I am thankful to my colleagues who have shared the challenges and excitement with me. I am convinced that the future of this field is bright, with many new exciting diagnostics, new space satellites, and twenty-first-century technology, including powerful quantum computing.

Many colleagues and my graduate students have made significant contributions to the research results cited in this book, especially Russell Kulsrud, Hantao Ji, Yasushi Ono, Troy Carter, Scott Hsu, Clayton Myers, Yang Ren, Dmitri Uzdensky, and Jong-soo Yoo. Also, the recent progress in magnetic reconnection research has much benefited from collaborations with members of the Center for Magnetic Self-Organization (CMSO),[1] a Physics Frontier Center, and from cross-discipline collaborations with many space physics researchers including Spiro Antiochos, Jim Burch, Li-Jen Chen, William Daughton, Jim Drake, Michel Hesse, Forrest Mozer, Kazunari Shibata, and Saku Tsuneta. Finally, I appreciate support for writing this book from the Princeton Plasma Physics Laboratory colleagues and the Office of Fusion Energy Science of the US Department of Energy, which has helped me to disseminate the large amount of information obtained over the years in the course of my research on magnetic reconnection.

---

[1] Jointly supported by the National Science Foundation and Department of Energy between 2003 and 2014. Principle investigators: Amitava Bhattacharjee, Faust Cattaneo, Cary Forest, Hantao Ji, Russell Kulsrud, Stewart Prager, Robert Rosner, John Sarff, and Ellen Zweibel.

Appendix A

Basic description of waves by dispersion relationship equations

A.1 BASIC DESCRIPTION OF WAVES IN COLD PLASMAS

Waves and associated wave–particle interactions play a key role in the magnetic reconnection process. Since there are so many kinds of waves in plasma, we introduce here the basic description of waves, emphasizing those that are involved in the magnetic reconnection layer. To describe waves, we employ the formalism described in the textbook Stix (1992). The properties of the propagation of waves are characterized by the "dispersion relation" by which the relationship between wave frequency and inverse wavelength k is described. In this chapter, all descriptions of equations and waves are made using CGS units, while the main text is written using MKSA units, as traditionally done in experimental and observational research fields.

The dispersion relation for a plasma is generally obtained using Maxwell's equations and from the condition for a "nontrivial" solution of a homogeneous set of field equations. For the formulation with Maxwell's equations, it is necessary to express the plasma current density j in terms of the electric field E, leading to the use of the dielectric tensor. We employ Fourier analysis in space variables where the first-order quantities to work with are

$$n_1, \quad v_{x1} \sim e^{-i\omega t + ikx}. \tag{A.1}$$

We assume that the zero-order quantities are uniform in pace and constant in time. Wave oscillation can be described by

$$\mathrm{Re}(n_1) = \bar{n}_1 \cos(\boldsymbol{k} \cdot \boldsymbol{r} - \omega t). \tag{A.2}$$

The electric displacement D includes the vacuum displacement and plasma response current, with the relation

$$D = \widehat{K} \cdot E = E + \frac{4\pi i}{\omega} j, \tag{A.3}$$

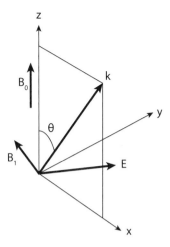

Figure A.1. The propagation vector k is defined in (x, y, z) coordinates with B_0 assumed to be in the z-direction.

where $\widehat{K}$ is the dielectric tensor. The plasma current j is made of fluid (macroscopic) flow velocities v_σ of multiple species as denoted by

$$j = \sum_\sigma n_\sigma q_\sigma e_\sigma v_\sigma, \tag{A.4}$$

where n_σ is the number density of particles of type σ with charge $q_\sigma e_\sigma$.

The equation of motion for a plasma particle type σ is written as

$$m_\sigma \frac{dv_\sigma}{dt} = q_\sigma \left(E + \frac{v_\sigma}{c} \times B \right). \tag{A.5}$$

In Fourier analysis, this equation becomes

$$-i\omega m_\sigma v_\sigma = q_\sigma \left(E + \frac{v_\sigma}{c} \times B_0 \hat{z} \right). \tag{A.6}$$

To simplify the geometry, we assume a zero-order steady magnetic field B_0 along the z-axis with $B_0 = \hat{z} B_0$ as shown in figure A.1 and with

$$B_0 = \begin{bmatrix} 0 \\ 0 \\ B_0 \end{bmatrix}, \quad E = \begin{bmatrix} E_x \\ E_y \\ E_z \end{bmatrix}, \quad v_\sigma = \begin{bmatrix} v_{\sigma x} \\ v_{\sigma y} \\ v_{\sigma z} \end{bmatrix}, \tag{A.7}$$

and we use the relationship

$$v_\sigma \times \hat{z}_\sigma = \begin{bmatrix} 0 & 1 & 0 \\ -1 & 0 & 0 \\ 0 & 0 & 0 \end{bmatrix} v_\sigma \tag{A.8}$$

to derive solutions for v in terms of E_x, E_y, E_z. The solutions are also used to determine the dielectric tensor K. We obtain

$$\begin{bmatrix} \omega & -i\frac{q_\sigma B_0}{m_\sigma c} & 0 \\ i\frac{q_\sigma B_0}{m_\sigma c} & \omega & 0 \\ 0 & 0 & \omega \end{bmatrix} v_\sigma = i\frac{q_\sigma E}{m_\sigma c}, \tag{A.9}$$

which can be written as

$$\begin{bmatrix} \omega & -i\omega_{c\sigma} & 0 \\ i\omega_{c\sigma} & \omega & 0 \\ 0 & 0 & \omega \end{bmatrix} v_\sigma = i\omega_{c\sigma}\frac{E}{B_0}, \tag{A.10}$$

where $\omega_{c\sigma} \equiv \frac{q_\sigma B_0}{m_\sigma c}$ is the cyclotron angular frequency for particle species σ. By using the above relationship between v_σ and E, we can calculate the dielectric tensor $\widehat{K}$.

A.2   THE DISPERSION RELATION

Having obtained the dielectric tensor $\widehat{K}$, we can now solve Maxwell's equations for waves:

$$\nabla \times E = -\frac{1}{c}\frac{\partial B}{\partial t}, \tag{A.11}$$

$$\nabla \times B = -\frac{4\pi j}{c} + \frac{1}{c}\frac{\partial E}{\partial t} = -\frac{1}{c}\frac{\partial D}{\partial t}. \tag{A.12}$$

Using Fourier analysis, we obtain

$$vk \times (k \times E) + \frac{\omega^2}{c^2}\widehat{K} \cdot E = 0. \tag{A.13}$$

To normalize the propagation vector, we can introduce the refractive index as the ratio of the speed of light to the phase velocity vector of waves,

$$n = \frac{kc}{\omega}. \tag{A.14}$$

Thus the reciprocal of n represents the wave phase velocity vector normalized by c, and it is often utilized to characterize the propagation feature of waves by conforming to a "wave normal surface."

Using n, we can write

$$n \times (n \times E) + \widehat{K} \cdot E = 0. \tag{A.15}$$

With our geometrical condition that we have a stationary $\boldsymbol{B}_0$ in the z-axis, we denote the angle of wave propagation $\boldsymbol{n}$ with respect to $\boldsymbol{B}_0 = \hat{z}B_0$ by θ, and the above equation is converted to

$$\begin{bmatrix} S - n^2 \cos^2\theta & -iD & n^2 \cos\theta \sin\theta \\ iD & S^2 - n^2 & 0 \\ n^2 \cos\theta \sin\theta & 0 & P - n^2 \sin^2\theta \end{bmatrix} \begin{bmatrix} E_x \\ E_y \\ E_z \end{bmatrix} = 0. \tag{A.16}$$

Here S, D, P are defined as

$$S \equiv \frac{1}{2}(R+L), \tag{A.17}$$

$$D \equiv \frac{1}{2}(R-L), \tag{A.18}$$

$$L \equiv 1 - \sum_\sigma \frac{\Pi_\sigma^2}{\omega^2}\left(\frac{\omega}{\omega + \omega_{c\sigma}}\right), \tag{A.19}$$

$$R \equiv 1 - \sum_\sigma \frac{\Pi_\sigma^2}{\omega^2}\left(\frac{\omega}{\omega - \omega_{c\sigma}}\right), \tag{A.20}$$

$$P \equiv 1 - \sum_\sigma \frac{\Pi_\sigma^2}{\omega^2} \qquad \Pi_\sigma^2 \equiv \frac{4\pi n_\sigma q_\sigma^2 e^2}{m_\sigma}. \tag{A.21}$$

The dispersion relation is obtained by finding a "trivial" condition, namely for any values of E_x, E_y, E_z, this equation has to hold. The condition for a trivial solution is that the determinant of the left-hand-side matrix is zero. This condition provides the wanted dispersion relationship between $\boldsymbol{k}$ and ω, or equivalently $\boldsymbol{n}$ and ω. So the dispersion equation is

$$An^4 - Bn^2 + C = 0, \tag{A.22}$$

where

$$A = S\sin^2\theta + P\cos^2\theta, \tag{A.23}$$

$$B = RL\sin^2\theta + PS(1 + \cos^2\theta), \tag{A.24}$$

$$C = PRL. \tag{A.25}$$

We obtain the solution as

$$n^2 = \frac{B \pm \sqrt{F}}{2A}, \tag{A.26}$$

where F is written as

$$F = (RL - PS)^2 \sin^4\theta + 4P^2D^2\cos^2\theta. \tag{A.27}$$

Table A.1. Table for basic waves characterized by dispersion relations.

Longitudinal Waves in Homogeneous Plasma

Electron Waves $\omega_{pi}, \omega_{ci} \ll \omega \sim \omega_{pe}, \omega_{ce}$

① Plasma Oscillation ($B_0 = 0$ or $k // B_0$) Bohm-Gross mode

$$\omega^2 = \omega_{pe}^2 + \frac{3}{2}k^2 v_e^2$$

② Upper Hybrid Wave ($k \perp B_0$)

$$\omega^2 = \omega_{pe}^2 + \omega_{ce}^2$$

③ Electron Bernstein mode ($k \perp B_0$)

$$\omega = n\omega_{ce}\left(1+\Delta_n(k)\right)$$

Ion Waves $\omega_{pi}, \omega_{ci} \sim \omega \ll \omega_{pe}, \omega_{ce}$

① Ion Acoustic Wave ($B_0 = 0$ or $k // B_0$)

$$\omega^2 = \gamma_i \frac{k^2 T_i}{m_i} + \frac{\omega_{pi}^2}{1 + k_{pe}^2/k^2}$$

② Electrostatic Ion Cyclotron Wave ($k \perp B_0$)

$$\omega = n\omega_{ci}\left(1+\Delta_n(k)\right)$$

③ Ion Bernstein Wave ($k \perp B_0$)

$$\omega = n\omega_{ci}\left(1+\Delta_n(k)\right)$$

④ Lower Hybrid Wave ($k \perp B_0$)

$$\frac{1}{\omega_{LH}^2} = \frac{1}{\omega_{ci}^2 + \omega_{pi}^2} + \frac{1}{|\omega_{ci}\omega_{ce}|}$$

Electromagnetic Waves in Homogeneous Plasma

Electromagnetic Electron Waves $\omega_{pi}, \omega_{ci} \ll \omega \sim \omega_{pe}, \omega_{ce}$

① Light Wave ($B_0 = 0$)

$$\omega^2 = \omega_{pe}^2 + k^2 c^2$$

② O Wave ($k \perp B_0$, $E_1 // B_0$)

$$\frac{c^2 k^2}{\omega^2} = 1 - \frac{\omega_{pe}^2}{\omega^2}$$

③ X Wave ($k \perp B_0$, $E_1 \perp B_0$)

$$\frac{k^2 c^2}{\omega^2} = 1 - \frac{\omega_{pe}^2}{\omega^2}\frac{\omega^2 - \omega_{pe}^2}{\omega^2 - (\omega_{ce}^2 + \omega_{pe}^2)}$$

④ R Wave ($k // B_0$, whistler)

$$\frac{k^2 c^2}{\omega^2} = 1 - \frac{\omega_{pe}^2}{\omega^2}\frac{\omega}{\omega - \omega_{ce}}$$

⑤ L Wave ($k // B_0$)

$$\frac{k^2 c^2}{\omega^2} = 1 - \frac{\omega_{pe}^2}{\omega^2}\frac{\omega}{\omega + \omega_{ce}}$$

Electromagnetic Ion Waves $\omega < \omega_{pi}, \omega_{ci}$

① Shear Alfvén Wave ($k // B_0$)

$$\omega^2 = k^2 v_A^2 \qquad v_A^2 = \frac{c^2 B_0^2}{4\pi n m_i}$$

② Magnetosonic Wave ($k \perp B_0$)

$$\frac{\omega^2}{k^2 c^2} = \frac{c_s^2 + v_A^2}{c^2 + v_A^2}, \quad \left(c_s^2 = \frac{\gamma_e T_e + \gamma_i T_i}{m_i}\right)$$

This dispersion relationship equation can be written in another form:

$$\tan^2 \theta = \frac{-P(n^2 - R)(n^2 - L)}{(Sn^2 - RL)(n^2 - P)},$$
(A.28)

and in this way we obtain the dispersion relations for waves propagating parallel ($\theta = 0$) and perpendicular ($\theta = \pi/2$) to the base magnetic field B_0 as

$$P = 0, \quad n^2 = R, \quad n^2 = L \quad \text{for } \theta = 0,$$
(A.29)

$$\text{and} \quad n^2 = \frac{RL}{S}, \quad n^2 = P \quad \quad \text{for } \theta = \pi/2.$$
(A.30)

Table A.1 describes the basic dispersion relationship equations for waves propagating parallel ($\theta = 0$) and perpendicular ($\theta = \pi/2$) to the base magnetic field B_0. While the above derivations of dispersion relations are made for cold plasmas, in the figure thermal plasma effects are added. So, in table A.1, the dispersion relations for waves are categorized based on their propagation direction with respect to the stationary background magnetic field, fluctuating vector components with respect to the propagation vector k, and polarization of waves. While all listed waves in the figure can be involved in the reconnection layer, electron whistler waves and lower hybrid waves are more often investigated because the former is expected to cause electron heating and acceleration and the latter is considered to generate strong dissipation as well as anomalous resistivity in the reconnection layer. Ion acoustic waves are basically sound waves in plasma whose dispersion characteristics can also be derived in terms of plasma pressure p divided by the plasma density, or $\omega/k = \sqrt{p/\rho} = \sqrt{(T_e + T_i)/m_i}$. They are strongly damped modes in the reconnection layer plasma because of the strong effects of the ion Landau damping. Another common type of wave is the Alfvén wave, whose dispersion can be similarly derived in terms of magnetic pressure divided by the plasma density, or $\omega/k = \sqrt{p/\rho} = \sqrt{B^2/4\pi\rho}$.

In order to generate enhanced resistivity against the electron current flow in the reconnection layer, waves have to invoke ion dynamics so that the electron momentum flows are resisted by heavier ions. Thus lower hybrid waves (LHWs), which are usually represented by magnetized electrons and demagnetized ions, have been considered the most likely candidates for generating the enhanced resistivity. The dispersion relation for LHWs in the high-β reconnection layer becomes rather complex, as described by Ji et al. (2005). It is expected that it can generate enhanced resistivity as well as dissipation. Recently, a more comprehensive study was carried out by Yoo et al. (2019) by verifying the dispersion characteristics of LHWs, both in the magnetosphere (Yoo et al., 2020) and in MRX (Magnetic Reconnection Experiment) plasmas (Yoo et al., 2019).

Appendix B

Plasma parameters for typical laboratory and natural plasmas

B.1 PLASMA PARAMETER DIAGRAM

Figure B.1 shows a parameter diagram for typical laboratory and natural plasmas.

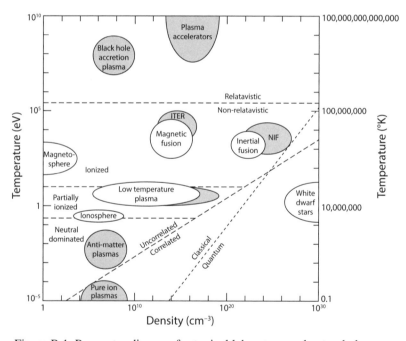

Figure B.1. Parameter diagram for typical laboratory and natural plasmas.

B.2 TYPICAL PLASMA PARAMETERS AND FORMULAE

Table B.1 lists parameters for typical laboratory and natural plasmas, while table B.2 gives typical plasma parameters and formulae in CGS units. Note that the main text is written in MKSA units.

Table B.1. Approximate magnitudes in some typical plasmas.

Plasma type	n cm^{-3}	T eV	ω_{pe} s^{-1}	λ_D cm	$n\lambda_D^3$	ν_{ei} s^{-1}
Interstellar gas	1	1	6×10^4	7×10^2	4×10^8	7×10^{-5}
Gaseous nebula	10^3	1	2×10^6	20	8×10^6	6×10^{-2}
Solar corona	10^9	10^2	2×10^9	2×10^{-1}	8×10^6	60
Diffuse hot plasma	10^{12}	10^2	6×10^{10}	7×10^{-3}	4×10^5	40
Solar atmosphere, gas discharge	10^{14}	1	6×10^{11}	7×10^{-5}	40	2×10^9
Warm plasma	10^{14}	10	6×10^{11}	2×10^{-4}	8×10^2	10^7
Hot plasma	10^{14}	10^2	6×10^{11}	7×10^{-4}	4×10^4	4×10^6
Thermonuclear plasma	10^{15}	10^4	2×10^{12}	2×10^{-3}	8×10^6	5×10^4
Theta pinch	10^{16}	10^2	6×10^{12}	7×10^{-5}	4×10^3	3×10^8
Dense hot plasma	10^{18}	10^2	6×10^{13}	7×10^{-6}	4×10^2	2×10^{10}
Laser plasma	10^{20}	10^2	6×10^{14}	7×10^{-7}	40	2×10^{12}

Table B.2. Formulae for primary plasma parameters expressed in CGS units. In space plasmas, ions are primarily made of protons. In the lab plasma, protons are replaced by ions of different mass and charges, and proton/ion parameters are thus modified accordingly. The quantity λ is the Coulomb logarithm, given by Spitzer (1962).

Symbol	Quantity	Formula	Value
τ_e	Electron collision time	$\frac{3}{\sqrt{32\pi}} \frac{m_e^{1/2}(kT_e)^{3/2}}{\lambda e^4 n_i}$	$\frac{2.9 \times 10^{-2}}{\lambda/10} T_e^{3/2} n_e^{-1}$ s
ω_{ce}	Electron gyrofrequency	$\frac{eB}{m_e c}$	$1.8 \times 10^7 B$ s^{-1}
ω_{cp}	Proton gyrofrequency	$\frac{eB}{m_p c}$	$9.6 \times 10^3 B$ s^{-1}
ω_{pe}	Electron plasma frequency	$\left(\frac{4\pi n_e e^2}{m_e}\right)^{1/2}$	$5.6 \times 10^4 n_e^{1/2}$ s
ω_{pp}	Proton plasma frequency	$\left(\frac{4\pi n_p e^2}{m_p}\right)^{1/2}$	$1.3 \times 10^3 n_p^{1/2}$ s
d_e	Electron skin depth	$\frac{c}{\omega_{pe}}$	$5.4 \times 10^5 n_e^{-1/2}$ cm
d_p	Proton skin depth	$\frac{c}{\omega_{pp}}$	$2.3 \times 10^7 n_p^{-1/2}$ cm
σ	Electrical conductivity	$\frac{\omega_p e^2 \tau_e}{4\pi}$	$7.3 \times 10^6 T_e^{3/2}(10/\lambda)$ s
η	Magnetic diffusivity	$\frac{c^2}{4\pi\sigma} = \frac{\delta_e^2}{\tau_e}$	$\frac{9.9 \times 10^{12}}{T_e^{3/2}}(\lambda/10)$ cm^2 s^{-1}
V_A	Alfvén speed	$\frac{B}{\sqrt{4\pi m_p n_p}} = \omega_{ci}\delta_i$	$2.2 \times 10^{11} B n_p^{-1/2}$ cm s^{-1}
S	Lundquist number	$\frac{Lv_A}{\eta} = \frac{L}{\delta_i}(\omega_{ce}\tau_e)$	$2.3 \times 10^{-2} LBT_e^{3/2}$ $n_e^{-1/2}(\lambda/10)^{-1}$

Appendix C

Common notation

Table C.1. Notation.

A	Vector potential
B	Magnetic induction
c	Speed of light in a vacuum
e	Elementary charge
E	Electric field
η	Electrical resistivity
ε_0	Permittivity of free space
γ	Ratio of specific heats; adiabatic constant
j	Electric current density
J_0, J_1	Bessel functions (of order 0 and 1)
m_e, m_i	Electron, ion mass
M_A	Alfvén Mach number; V/V_A
μ_0	Permeability of free space
n	Number density
m/n	Mode numbers (poloidal/toroidal) in toroidal plasma
ν_{ei}	Electron/ion collision frequency
ω_{ce}, ω_{ci}	Electron/ion gyrofrequency
ω_{pe}, ω_{pi}	Electron/ion plasma frequency
P	Pressure tensor
p	Pressure
$q, q(r)$	Safety factor for toroidal fusion plasma
Ψ	Poloidal flux
Φ	Toroidal flux
ρ	Mass density
S	Lundquist number (in chapter 11, L_q is used)
t	Time
$\tanh(x)$	Hyperbolic tangent function $(\sinh(x)/\cosh(x))$
T	Temperature
V	Fluid flow velocity
v	Particle velocity
V_A	Alfvén speed

Bibliography

Amari, T., J. Luciani, Z. Mikic, and J. Linker, 1999, *Astrophys. J. Lett.* **529**(1), L49.

Anderson, J. K., C. B. Forest, T. M. Biewer, J. S. Sarff, and J. C. Wright, 2004, *Nucl. Fusion* **44**, 162.

Angelopoulos, V., A. Runov, X. Z. Zhou, D. L. Turner, S. A. Kiehas, S. S. Li, and I. Shinohara, 2013, *Science* **341**(6153), 1478.

Antiochos, S., C. DeVore, and J. Klimchuk, 1999, *Astrophys. J.* **510**, 485.

Antiochos, S. K., J. T. Karpen, and C. R. DeVore, 2002, *Astrophys. J.* **575**, 578.

Asano, Y., T. Mukai, M. Hoshino, Y. Saito, H. Hayakawa, and T. Nagai, 2004, *J. Geophys. Res.* **109**, A02212.

Bale, S., F. Mozer, and T. Phan, 2002, *Geophys. Res. Lett.* **29**, 2180.

Bateman, G., 1978, *MHD Instabilities* (MIT Press, Cambridge, MA).

Baum, P., and A. Bratenahl, 1980, *Adv. Electron Phys* **54**, 1.

Becker, U., T. Neukirch, and K. Schindler, 2001, *J. Geophys. Res.* **106**, 3811.

Bellan, P., 2000, *Spheromaks* (Imperial College Press, London).

Belova, E. V., S. C. Jardin, H. Ji, M. Yamada, and R. Kulsrud, 2000, *Phys. Plasmas* **7**, 4996.

Berger, M. A., and G. B. Field, 1984, *J. Fluid Mech.* **147**, 133.

Bergerson, W., C. Forest, G. Fiksel, D. Hannum, R. Kendrick, J. Sarff, and S. Stambler, 2006, *Phys. Rev. Lett.* **96**, 015004.

Bhattacharjee, A., and E. Hameiri, 1986, *Phys. Rev. Lett.* **57**, 206.

Bhattacharjee, A., Y. Huang, H. Yang, and B. Rogers, 2009, *Phys. Plasmas* **16**, 112102.

Bhattacharjee, A., Z. W. Ma, and X. Wang, 2001, *Phys. Plasmas* **8**, 1829.

Binderbauer, M., T. Tajima, L. Steinhauer, E. Garate, M. Tuszewski, L. Schmitz, H. Guo, A. Smirnov, H. Gota, D. Barnes, et al., 2015, *Phys. Plasmas* **22**(5), 056110.

Birdsall, C., and A. Langdon, 1985, *Plasma Physics via Computer Simulation* (McGraw-Hill, New York).

Birn, J., J. Drake, M. Shay, B. Rogers, R. Denton, M. Hesse, M. Kuznetsova, Z. Ma, A. Bhattachargee, A. Otto, and P. Pritchett, 2001, *J. Geophys. Res.* **106**(A3), 3715.

Birn, J., and M. Hesse, 2005, *Ann. Geophys.* **23**(10), 3365.

Birn, J., and E. R. Priest, 2007, *Reconnection of Magnetic Fields: Magnetohydrodynamics and Collisionless Theory and Observations* (Cambridge University Press, Cambridge, UK).

Birn, J., and E. R. Priest, 2007, in *Reconnection of Magnetic Fields: Magnetohydrodynamics and Collisionless Theory and Observations*, edited by J. Birn and E. R. Priest (Cambridge University Press, Cambridge, UK), p. 3.

Biskamp, D., 1986, *Phys. Fluids* **29**(5), 1520.

Biskamp, D., 2000, *Magnetic Reconnection in Plasmas* (Cambridge University Press, Cambridge).

Biskamp, D., E. Schwarz, and J. F. Drake, 1997, *Phys. Plasmas* **4**, 1002.

Bodin, H. A. B., 1990, *Nucl. Fusion* **30**, 1717.

Bowers, K. J., B. J. Albright, L. Yin, B. Bergen, and T. J. T. Kwan, 2008, *Phys. Plasmas* **15**, 055703.

Bratenahl, A., and C. M. Yeates, 1970, *Phys. Fluids* **13**, 2696.

Breslau, J. A., and S. Jardin, 2003, *Phys. Plasmas* **10**, 1291.

Brown, M., 1999, *Phys. Plasmas* **6**, 1717.

Brown, M., C. Cothran, M. Landreman, D. Schlossberg, and W. Matthaeus, 2002, *Astrophys. J.* **577**, L63.

Brown, M. R., C. D. Cothran, and J. Fung, 2006, *Phys. Plasmas* **13**(5), 056503.

Burch, J. L., T. E. Moore, R. B. Torbert, and B. L. Giles, 2016a, *Space Science Rev.* **199**, 5.

Burch, J. L., R. B. Torbert, T. D. Phan, L.-J. Chen, T. E. Moore, R. E. Ergun, J. P. Eastwood, D. J. Gershman, P. A. Cassak, M. R. Argall, S. Wang, M. Hesse, et al., 2016b, *Science* **352**, aaf2939.

Burch, J., J. Webster, M. Hesse, K. Genestreti, R. Denton, T. Phan, H. Hasegawa, P. Cassak, R. Torbert, B. Giles, et al., 2020, *Geophys. Res. Lett.* **47**(17), e2020GL089082.

Cao, D., H. S. Fu, J. B. Cao, T. Y. Wang, D. B. Graham, Z. Z. Chen, F. Z. Peng, S. Y. Huang, Y. V. Khotyaintsev, M. André, C. T. Russell, B. L. Giles, et al., 2017, *Geophys. Res. Lett.* **44**(9), 3954.

Carmichael, H., 1964, in *The Physics of Solar Flares*, edited by W. N. Hess, p. 451.

Carrington, R. C., 1859, *Mon. Not. Roy. Astron. Soc.* **20**(1), 13, ISSN 0035-8711, https://doi.org/10.1093/mnras/20.1.13.

Carter, T., H. Ji, F. Trintchouk, M. Yamada, and R. Kulsrud, 2002a, *Phys. Rev. Lett.* **88**, 015001.

Carter, T., M. Yamada, H. Ji, R. Kulsrud, and F. Trintchouk, 2002b, *Phys. Plasmas* **9**, 3272.

Cassak, P., and S. Fuselier, 2016, in *Magnetic Reconnection* (Springer), pp. 213–276.

Cassak, P., M. Shay, and J. Drake, 2009, *Phys. Plasmas* **16**, 120702.

Cassak, P. A., M. A. Shay, and J. F. Drake, 2005, *Phys. Rev. Lett.* **95**(23), 235002.

Cattell, C., J. Dombeck, J. Wygant, J. Drake, M. Swisdak, M. Goldstein, W. Keith, A. Fazakerley, M. Andre, E. Lucek, and A. Balogh, 2005, *J. Geophys. Res.* **110**, A01211.

Chapman, S. C., and C. G. Mouikis, 1996, *Geophys. Res. Lett.* **23**(22), 3251, https://agupubs.onlinelibrary.wiley.com/doi/abs/10.1029/96GL02845.

Chen, J., 1989, *Astrophys. J.* **338**, 453.

Chen, J., 2017, *Phys. Plasmas* **24**(9), 090501.

Chen, L.-J., N. Bessho, B. Lefebvre, H. Vaith, A. Fazakerley, A. Bhattacharjee, P. A. Puhl-Quinn, A. Runov, Y. Khotyaintsev, A. Vaivads, E. Georgescu, and R. Torbert, 2008a, *J. Geophys. Res.* **113**, A12213.

Chen, L. J., A. Bhattacharjee, P. A. Puhl-Quinn, H. Yang, N. Bessho, S. Imada, S. Muhlbachler, P. W. Daly, B. Lefebvre, Y. Khotyaintsev, A. Vaivads, A. Fazakerley, et al., 2008b, *Nature Phys.* **4**, 19.

Chen, L.-J., M. Hesse, S. Wang, D. Gershman, R. Ergun, C. Pollock, R. Torbert, N. Bessho, W. Daughton, J. Dorelli, B. Giles, R. Strangeway, et al., 2016a, *Geophys. Res. Lett.* **43**(1), 6036.

Chen, Y., G. Du, D. Zhao, Z. Wu, W. Liu, B. Wang, G. Ruan, S. Feng, and H. Song, 2016b, *Astrophys. J. Lett.* **820**(2), L37.

Close, R. M., C. E. Parnell, D. W. Longcope, and E. R. Priest, 2004, *Astrophys. J. Lett.* **612**, L81.

Cothran, C. D., J. Fung, M. R. Brown, and M. J. Schaffer, 2006, *Rev. Sci. Instru.* **77**, 063504.

Cothran, C. D., M. Landreman, M. R. Brown, and W. H. Matthaeus, 2003, *Geophys. Res. Lett.* **30**, 1213.

Cowling, T. G., 1934, *Monthly Not.* **94**, 768.

Daughton, W., 1999, *J. Geophys. Res.* **104**, 28701.

Daughton, W., 2003, *Phys. Plasmas* **10**, 3103.

Daughton, W., G. Lapenta, and P. Ricci, 2004, *Phys. Rev. Lett.* **93**(10), 105004.

Daughton, W., V. Roytershteyn, B. Albright, H. Karimabadi, L. Yin, and K. J. Bowers, 2009, *Phys. Rev. Lett.* **103**(6), 065004.

Daughton, W., V. Roytershteyn, H. Karimabadi, L. Yin, B. J. Albright, B. Bergen, and K. J. Bowers, 2011, *Nat. Phys.* **7**(7), 539.

Daughton, W., J. Scudder, and H. Karimabadi, 2006, *Phys. Plasmas* **13**(7), 072101 (pages 15).

Dawson, J., 1962, *Phys. Fluids* **5**(4), 445.

De Pontieu, B., S. W. McIntosh, M. Carlsson, V. H. Hansteen, T. D. Tarbell, C. J. Schrijver, A. M. Title, R. A. Shine, S. Tsuneta, Y. Katsukawa, K. Ichimoto, Y. Suematsu, et al., 2007, *Science* **318**, 1574.

Démoulin, P., and G. Aulanier, 2010, *Astrophys. J.* **718**, 1388.

Den Hartog, D., J. Chapman, D. Craig, G. Fiksel, P. Fontana, S. Prager, and J. Sarff, 1999, *Phys. Plasmas* **6**(5), 1813.

Den Hartog, D. J., J.-W. Ahn, A. F. Almagri, J. K. Anderson, A. D. Beklemishev, A. P. Blair, F. Bonomo, M. T. Borchardt, D. L. Brower, D. R. Burke, M. Cengher, B. E. Chapman, et al., 2007, *Nucl. Fusion* **47**, L17.

Deng, X. H., and H. Matsumoto, 2001, *Nature* **410**, 557.

Ding, W. X., D. L. Brower, D. Craig, B. H. Deng, G. Fiksel, V. Mirnov, S. C. Prager, J. S. Sarff, and V. Svidzinski, 2004, *Phys. Rev. Lett.* **93**(4), 045002.

Ding, W. X., D. L. Brower, D. Craig, B. H. Deng, S. C. Prager, J. S. Sarff, and V. Svidzinski, 2007, *Phys. Rev. Lett.* **99**(5), 055004.

Dorfman, S., W. Daughton, V. Roytershteyn, H. Ji, Y. Ren, and M. Yamada, 2008, *Phys. Plasmas* **15**(10), 102107.

Dorfman, S., H. Ji, M. Yamada, J. Yoo, E. Lawrence, C. Myers, and T. Tharp, 2013, *Geophys. Res. Lett.* **40**, 233.

Drake, J., and M. Shay, 2007, in *Reconnection of Magnetic Fields: Magnetohydrodynamics and Collisionless Theory and Observations*, edited by J. Birn and E. R. Priest (Cambridge University Press, Cambridge, UK), p. 87.

Drake, J., M. Swisdak, C. Cattell, M. Shay, B. Rogers, and A. Zeiler, 2003, *Science* **299**, 873.

Drake, J. F., P. A. Cassak, M. A. Shay, M. Swisdak, and E. Quataert, 2009, *Astrophys. J.* **700**, L16.

Drake, J. F., M. Swisdak, H. Che, and M. A. Shay, 2006, *Nature* **443**, 553.

Dungey, J., 1953, *Phil. Mag.* **44**, 725.

Dungey, J., 1961, *Phys. Rev. Lett.* **6**(2), 47.

Dungey, J., 1995, in *Physics of the Magnetopause*, edited by P. Song, B. Sonnerup, and M. Thomsen (AGU Monograph 90), p. 81.

Eastwood, J. P., T. D. Phan, J. F. Drake, M. A. Shay, A. L. Borg, B. Lavraud, and M. G. G. T. Taylor, 2013, *Phys. Rev. Lett.* **110**(22), 225001.

Edwards, A. W., D. J. Campbell, W. W. Engelhardt, H.-U. Fahrbach, R. D. Gill, R. S. Granetz, S. Tsuji, B. J. D. Tubbing, A. Weller, J. Wesson, and D. Zasche, 1986, *Phys. Rev. Lett.* **57**, 210.

Egedal, J., W. Daughton, J. F. Drake, N. Katz, and A. Le, 2009, *Phys. Plasmas* **16**(5), 050701.

Egedal, J., W. Daughton, and A. Le, 2012, *Nature Phys.* **8**, 321.

Egedal, J., and A. Fasoli, 2001, *Phys. Plasmas* **8**(5),1935.

Egedal, J., A. Fasoli, and J. Nazemi, 2003, *Phys. Rev. Lett.* **90**, 135003.

Egedal, J., A. Fasoli, M. Porkolab, and D. Tarkowski, 2000, *Rev. Sci. Instr.* **71**, 3351.

Egedal, J., W. Fox, N. Katz, M. Porkolab, M. Øieroset, R. P. Lin, W. Daughton, and D. J. F., 2008, *J. Geophys. Res.* **113**, A12207.

Egedal, J., W. Fox, N. Katz, M. Porkolab, K. Reim, and E. Zhang, 2007, *Phys. Rev. Lett.* **98**(1), 15003.

Egedal, J., A. Le, W. Daughton, B. Wetherton, P. A. Cassak, L.-J. Chen, B. Lavraud, R. B. Torbert, J. Dorelli, D. J. Gershman, and L. A. Avanov, 2016, *Phys. Rev. Lett.* **117**(18), 185101.

Egedal, J., M. Øieroset, W. Fox, and R. P. Lin, 2005, *Phys. Rev. Lett.* **94**(2), 025006.

Ergun, R., L.-J. Chen, F. Wilder, N. Ahmadi, S. Eriksson, M. Usanova, K. Goodrich, J. Holmes, A. Sturner, D. Malaspina, et al., 2017, *Geophys. Res. Lett.* **44**(7), 2978.

Fairfield, D. H., E. W. Hones, Jr., and C.-I. Meng, 1981, *J. Geophys. Res.* **86**, 11189.

Fan, Y., and S. E. Gibson, 2007, *Astrophys. J.* **668**(2), 1232.

Fiksel, G., A. F. Almagri, B. E. Chapman, V. V. Mirnov, Y. Ren, J. S. Sarff, and P. W. Terry, 2009, *Phys. Rev. Lett.* **103**(14), 145002.

Fiksel, G., D. J. den Hartog, and P. W. Fontana, 1998, *Rev. Sci. Instru.* **69**, 2024.

Fontana, P. W., D. J. den Hartog, G. Fiksel, and S. C. Prager, 2000, *Phys. Rev. Lett.* **85**, 566.

Forbes, T. G., and E. R. Priest, 1995, *Astrophys. J.* **446**, 377.

Fox, W., J. Matteucci, C. Moissard, D. Schaeffer, A. Bhattacharjee, K. Germaschewski, and S. Hu, 2018, *Phys. Plasmas* **25**(10), 102106.

Fox, W., M. Porkolab, J. Egedal, N. Katz, and A. Le, 2010, *Phys. Plasmas* **17**, 072303.

Fox, W., M. Porkolab, J. Egedal, N. Katz, and A. Le, 2008, *Phys. Rev. Lett.* **101**, 255003.

Fox, W., D. Schaeffer, M. Rosenberg, G. Fiksel, J. Matteucci, and et al., 2020, *submitted to Phy. Rev. Letts.*

Frank, A. G., 1974, in *Neutral Current Layers in a Plasma* (Izdatel'stvo Nauka, Moscow), pp. 108–166.

Frank, A. G., S. Y. Bogdanov, G. V. Dreiden, V. S. Markov, and G. V. Ostrovskaya, 2006, *Phys. Lett. A* **348**, 318.

Frank, A. G., S. Y. Bogdanov, V. S. Markov, G. V. Ostrovskaya, and G. V. Dreiden, 2005, *Phys. Plasmas* **12**, 052316.

Furno, I., T. P. Intrator, G. Lapenta, L. Dorf, S. Abbate, and D. D. Ryutov, 2007, *Phys. Plasmas* **14**, 022103.

Furth, H., J. Killeen, and M. Rosenbluth, 1963, *Phys. Fluids* **6**, 459.

Fuselier, S., and W. Lewis, 2011, *Space Sci. Rev.* **160**(1-4), 95.

Gabriel, A. H., J. Charra, G. Grec, J.-M. Robillot, T. Roca Cortés, S. Turck-Chièze, R. Ulrich, S. Basu, F. Baudin, L. Bertello, P. Boumier, M. Charra, et al., 1997, *Solar Phys.* **175**, 207, and other articles in the same issue.

Gekelman, W., A. Collette, and S. Vincena, 2007, *Phys. Plasmas* **14**(6), 062109 (pages 13).

Gekelman, W., and H. Pfister, 1988, *Phys. Fluids* **31**, 2017.

Gekelman, W., R. Stenzel, and N. Wild, 1982, *J. Geophys. Res. Space Phys.* **87**, 101.

Giovanelli, R., 1946, *Nature* **158**, 81.

Gold, T., and F. Hoyle, 1960, *Mon. Not. Roy. Astron. Soc.* **120**, 89.

Golub, L., J. Bookbinder, E. Deluca, M. Karovska, H. Warren, C. J. Schrijver, R. Shine, T. Tarbell, A. Title, J. Wolfson, B. Handy, and C. Kankelborg, 1999, *Phys. Plasmas* **6**, 2205.

Gonzalez, W., and E. Parker (eds.), 2016, *Magnetic Reconnection: Concept and Application* (Springer).

Gosling, J. T., R. M. Skoug, D. J. McComas, and C. W. Smith, 2005, *Geophys. Res. Lett.* **32**, 5105.

Graham, D. B., Y. V. Khotyaintsev, C. Norgren, A. Vaivads, M. André, S. Toledo-Redondo, P. A. Lindqvist, G. T. Marklund, R. E. Ergun, W. R. Paterson, D. J. Gershman, B. L. Giles, et al., 2017, *J. Geophys. Res.* **122**(1), 517.

Gray, T., M. Brown, and D. Dandurand, 2013, *Phys. Rev. Lett.* **110**(8), 085002.

Gray, T., V. S. Lukin, M. R. Brown, and C. D. Cothran, 2010, *Phys. Plasmas* **17**, 102106.

Hagenaar, H. J., 2001, *Astrophys. J.* **555**, 448.

Hansen, A. K., A. F. Almagri, D. Craig, D. J. den Hartog, C. C. Hegna, S. C. Prager, and J. S. Sarff, 2000, *Phys. Rev. Lett.* **85**, 3408.

Hansen, J., and P. Bellan, 2001, *Astrophys. J.* **563**, L183.

Harris, E., 1962, *Il Nuovo Cimento* **23**, 115.

Hart, G. W., A. Janos, D. D. Meyerhofer, and M. Yamada, 1986, *Phys. Fluids* **29**(6), 1994.

Harvey, K. L., and F. Recely, 1984, *Solar Phys.* **91**, 127.

Hawkins, J. G., L. C. Lee, M. Yan, Y. Lin, F. W. Perkins, and M. Yamada, 1994, *J. Geophys. Res.* **99**, 5869.

Hesse, M., N. Aunai, D. Sibeck, and J. Birn, 2014, *Geophys. Res. Lett.* **41**(24), 8673.

Hesse, M., M. Kuznetsova, and J. Birn, 2001, *J. Geophys. Res.* **106**, 2983.

Hesse, M., K. Schindler, J. Birn, and M. Kuznetsova, 1999, *Phys. Plasmas* **6**, 1781.

Heyvaerts, J., E. Priest, and D. Rust, 1977, *Astrophys. J.* **216**, 123.

Hibschman, J. A., and J. Arons, 2001, *Astrophys. J.* **560**(2), 871.

Hirayama, T., 1974, *Solar Phys.* **34**, 323.

Hones Jr, E., 1977, *J. Geophys. Res.* **82**(35), 5633.

Hood, A. W., and E. R. Priest, 1981, *Geophys. Astrophys. Fluid Dynamics* **17**, 297.

Horiuchi, R., and T. Sato, 1994, *Phys. Plasmas* **1**, 3587.

Horiuchi, R., and T. Sato, 1999, *Phys. Plasmas* **6**(12), 4565.

Hoyle, F., 1949, *Some Recent Researches in Solar Physics* (Cambridge University Press, Cambridge).

Hsu, S., G. Fiksel, T. Carter, H. Ji, R. Kulsrud, and M. Yamada, 2000, *Phys. Rev. Lett.* **84**, 3859.

Huang, Y.-M., and A. Bhattacharjee, 2010, *Phys. Plasmas* **17**(6), 062104.

Huba, J., J. Drake, and N. Gladd, 1980, *Phys. Fluids* **23**(3), 552.

Hughes, W., 1995, in *Introduction to Space Physics*, edited by M. Kivelson and C. Russell (Cambridge University Press, London, UK), p. 227.

Jemella, B. D., J. F. Drake, and M. A. Shay, 2004, *Phys. Plasmas* **11**(12), 5668.

Ji, H., 1999, in *Magnetic Helicity in Space and Laboratory Plasmas*, edited by M. Brown, R. Canfield, and A. Pevtsov (AGU Monograph 111), p. 167.

Ji, H., A. Almagri, S. Prager, and J. Sarff, 1994, *Phys. Rev. Lett.* **73**, 668.

Ji, H., T. Carter, S. Hsu, and M. Yamada, 2001, *Earth Planets Space* **53**, 539.

Ji, H., and W. Daughton, 2011, *Phys. Plasmas* **18**(11), 111207 (pages 10).

Ji, H., W. Daughton, et al. 2022, "Magnetic reconnection in the era of exascale computing and multiscale experiments," submitted to *Nat. Rev. Phys.*

Ji, H., R. Kulsrud, W. Fox, and M. Yamada, 2005, *J. Geophys. Res.* **110**, A08212.

Ji, H., and S. Prager, 2002, *Magnetohydrodynamics* **38**, 191.

Ji, H., S. C. Prager, and J. S. Sarff, 1995, *Phys. Rev. Lett.* **74**, 2945.

Ji, H., Y. Ren, M. Yamada, S. Dorfman, W. Daughton, and S. P. Gerhardt, 2008, *Geophys. Res. Lett.* **35**, L13106.

Ji, H., S. Terry, M. Yamada, R. Kulsrud, A. Kuritsyn, and Y. Ren, 2004, *Phys. Rev. Lett.* **92**, 115001.

Ji, H., Y. Yagi, K. Hattori, A. F. Almagri, S. C. Prager, Y. Hirano, J. S. Sarff, T. Shimada, Y. Maejima, and K. Hayase, 1995, *Phys. Rev. Lett.* **75**(6), 1086.

Ji, H., M. Yamada, S. Hsu, and R. Kulsrud, 1998, *Phys. Rev. Lett.* **80**, 3256.

Ji, H., M. Yamada, S. Hsu, R. Kulsrud, T. Carter, and S. Zaharia, 1999, *Phys. Plasmas* **6**, 1743.

Kadomtsev, B., 1975, *Sov. J. Plasma Phys.* **1**, 389.

Karimabadi, H., W. Daughton, and J. Scudder, 2007, *Geophys. Res. Lett.* **34**, L13104.

al Karkhy, A., P. Browning, G. Cunningham, S. Gee, and M. Rusbridge, 1993, *Phys. Rev. Lett.* **70**, 1814.

Kennel, C. F., and F. V. Coroniti, 1984a, *Astrophys. J.* **283**, 694.

Kennel, C. F., and F. V. Coroniti, 1984b, *Astrophys. J.* **283**, 710.

Kivelson, M., and C. Russell, 1995, *Introduction to Space Physics* (Cambridge University Press, London, UK).

Klassen, A., H. Aurass, and G. Mann, 2001, *Astron. Astrophys.* **370**(3), L41.

Kliem, B., and T. Török, 2006, *Phys. Rev. Lett.* **96**, 255002.

Kopp, R. A., and G. W. Pneuman, 1976, *Solar Phys.* **50**, 85.

Kornack, T. W., P. K. Sollins, and M. R. Brown, 1998, *Phys. Rev. E* **58**, R36.

Kowal, G., A. Lazarian, E. Vishniac, and K. Otmianowska-Mazur, 2009, *Astrophys. J.* **700**(1), 63.

Krall, N., and P. Liewer, 1971, *Phys. Rev. A* **4**(5), 2094.

Krause, F., and K.-H. Rä, 1980, *Mean-Field Magnetohydrodynamics and Dynamo Theory* (Akademie, Berlin).

Krucker, S., H. S. Hudson, L. Glesener, S. M. White, S. Masuda, J.-P. Wuelser, and R. P. Lin, 2010, *Astrophys. J.* **714**, 1108.

Kruskal, M., and R. Kulsrud, 1958, *Phys. Fluids* **1**, 265.

Kruskal, M., and M. Schwarzschild, 1954, *Proc. R. Soc. Lond. A* **223**(1154), 348, ISSN 0080-4630.

Kulsrud, R., 1998, *Phys. Plasmas* **5**, 1599.

Kulsrud, R., 2001, *Earth Planets Space* **53**, 417.

Kulsrud, R., H. Ji, W. Fox, and M. Yamada, 2005, *Phys. Plasmas* **12**, 082301.

Kulsrud, R. M., 1999, *Ann. Rev. Astron. Astrophys.* **37**(1), 37.

Kulsrud, R. M., 2005, *Plasma Physics for Astrophysics* (Princeton University Press, Princeton).

Kulsrud, R. M., and S. W. Anderson, 1992, *Astrophys. J.* **396**, 606.

Kuperus, M., and M. Raadu, 1974, *Astron. Astrophys.* **31**, 189.

Kuritsyn, A., M. Yamada, S. Gerhardt, H. Ji, R. Kulsrud, and Y. Ren, 2006, *Phys. Plasmas* **13**, 055703.

Kusano, K., T. Iju, Y. Bamba, and S. Inoue, 2020, *Science* **369**(6503), 587.

Kusano, K., T. Maeshiro, T. Yokoyama, and T. Sakurai, 2004, *Astrophys. J.* **610**, 537.

Lapenta, G., and J. Brackbill, 2002, *Phys. Plasmas* **9**, 1544.

Latham, J., E. Belova, and M. Yamada, 2021, *Phys. Plasmas* **28**(1), 012901.

Lazarian, A., and M. Opher, 2009, *Astrophys. J.* **703**(1), 8.

Lazarian, A., and E. T. Vishniac, 1999, *Astrophys. J.* **517**, 700.

Le, A., J. Egedal, O. Ohia, W. Daughton, H. Karimabadi, and V. S. Lukin, 2013, *Phys. Rev. Lett.* **110**, 135004.

Lee, G. S., P. H. Diamond, and Z. G. An, 1989, *Phys. Fluids B* **1**, 99.

Lemons, D., and S. Gary, 1977, *J. Geophys. Res.* **82**(16), 2337.

Levinton, F. M., S. H. Batha, M. Yamada, and M. C. Zarnstorff, 1993, *Phys. Fluids B* **5**, 2554.

Lichtenberg, A., 1984, *Nucl. Fusion* **24**, 1277.

Lin, R. P., 2006, *Space Sci. Rev.* **124**, 233.

Lin, R. P., S. Krucker, G. J. Hurford, D. M. Smith, H. S. Hudson, G. D. Holman, R. A. Schwartz, B. R. Dennis, G. H. Share, R. J. Murphy, A. G. Emslie, C. Johns-Krull, et al., 2003, *Astrophys. J. Lett* **595**, L69.

Liu, Y., 2008, *Astrophys. J. Lett.* **679**(2), L151.

Loureiro, N., R. Samtaney, A. Schekochihin, and D. Uzdensky, 2012, *Phys. Plasmas* **19**(4), 042303.

Loureiro, N., A. Schekochihin, and S. Cowley, 2007, *Phys. Plasmas* **14**, 100703.

Loureiro, N. F., and D. A. Uzdensky, 2016, *Plasma Phys. Control. Fusion* **58**(1), 014021.

Lu, E. T., 1995, *Phys. Rev. Lett.* **74**, 2511.

Lu, E. T., and R. J. Hamilton, 1991, *Astrophys. J. Lett.* **380**, L89.

Ma, Z. W., and A. Bhattacharjee, 1996, *Geophys. Res. Lett.* **23**, 1673.

Ma, Z.-W., T. Chen, H. Zhang, and M. Yu, 2018, *Sci. Rep.* **8**(1), 1.

Mahajan, S. M., 1989, *Phys. Fluids B* **1**, 43.

Mandt, M. E., R. E. Denton, and J. F. Drake, 1994, *Geophys. Res. Lett.* **21**, 73.

Masuda, S., T. Kosugi, H. Hara, S. Tsuneta, and Y. Ogawara, 1994, *Nature* **371**, 495.

Matsumoto, R., T. Tajima, K. Shibata, and M. Kaisig, 1993, *Astrophys. J.* **414**, 357.

McBride, J., E. Ott, J. Boris, and J. Orens, 1972, *Phys. Fluids* **15**(12), 2367.

McComas, D. J., S. J. Bame, C. T. Russell, and R. C. Elphic, 1986, *J. Geophys. Res.* **91**, 4287.

Mikić, Z., D. D. Schnack, and G. van Hoven, 1990, *Astrophys. J.* **361**, 690.

Mirnov, V. V., C. C. Hegna, and S. C. Prager, 2003, *Plasma Phys. Rep.* **29**, 566.

Mirnov, V. V., C. C. Hegna, and S. C. Prager, 2004, *Phys. Plasmas* **11**, 4468.

Moore, R. L., A. C. Sterling, G. A. Gary, J. W. Cirtain, and D. A. Falconer, 2011, *Space Sci. Rev.* **160**(1-4), 73.

Moritaka, T., R. Horiuchi, and H. Ohtani, 2007, *Phys. Plasmas* **14**, 102109.

Mozer, F. S., V. Angelopoulos, J. Bonnell, K. H. Glassmeier, and J. P. McFadden, 2008a, *Geophys. Res. Lett.* **35**, L17S04.

Mozer, F. S., S. Bale, and T. D. Phan, 2002, *Phys. Rev. Lett.* **89**, 015002.

Mozer, F. S., and P. L. Pritchett, 2011, *Space Sci. Rev.* **158**(1), 119, ISSN 0038-6308.

Mozer, F. S., P. L. Pritchett, J. Bonnell, D. Sundkvist, and M. T. Chang, 2008b, *J. Geophys. Res.* **113**, A00C03.

Mozer, F. S., and A. Retinò, 2007, *J. Geophys. Res.* **112**, 10206.

Myers, C. E., 2015, *Laboratory Study of the Equilibrium and Eruption of Line-Tied Magnetic Flux Ropes in the Solar Corona*, Ph.D. thesis, Princeton University.

Myers, C. E., M. Yamada, H. Ji, J. Yoo, W. Fox, J. Jara-Almonte, A. Savcheva, and E. E. DeLuca, 2015, *Nature* **528**(7583), 526.

Myers, C. E., M. Yamada, H. Ji, J. Yoo, J. Jara-Almonte, and W. Fox, 2016, *Phys. Plasmas* **23**, 112102.

Nagayama, Y., K. M. McGuire, M. Bitter, A. Cavallo, E. D. Fredrickson, K. W. Hill, H. Hsuan, A. Janos, W. Park, G. Taylor, and M. Yamada, 1991, *Phys. Rev. Lett.* **67**, 3527.

Nagayama, Y., M. Yamada, W. Park, E. D. Fredrickson, A. C. Janos, K. M. McGuire, and G. Taylor, 1996, *Phys. Plasmas* **3**, 1647.

Neininger, N., 1992, *Astron. Astrophys.* **263**, 30.

Nilson, P. M., L. Willingale, M. C. Kaluza, C. Kamperidis, S. Minardi, M. S. Wei, P. Fernandes, M. Notley, S. Bandyopadhyay, M. Sherlock, R. J. Kingham, M. Tatarakis, et al., 2006, *Phys. Rev. Lett.* **97**(25), 255001.

Ohyabu, N., S. Okamura, and N. Kawashima, 1974, *J. Geophys. Res.* **79**, 1977.

Øieroset, M., R. P. Lin, T. D. Phan, D. E. Larson, and S. D. Bale, 2002, *Phys. Rev. Lett.* **89**(19), 195001.

Øieroset, M., T. D. Phan, M. Fujimoto, R. P. Lin, and R. P. Lepping, 2001, *Nature* **412**, 414.

Ono, Y., R. A. Ellis, A. C. Janos, F. M. Levinton, R. M. Mayo, R. W. Motley, Y. Ueda, and M. Yamada, 1988, *Phys. Rev. Lett.* **61**(25), 2847.

Ono, Y., M. Inomoto, T. Okazaki, and Y. Ueda, 1997, *Phys. Plasmas* **4**, 1953.

Ono, Y., A. Morita, M. Katsurai, and M. Yamada, 1993, *Phys. Fluids B* **5**, 3691.

Ono, Y., M. Yamada, T. Akao, T. Tajima, and R. Matsumoto, 1996, *Phys. Rev. Lett.* **76**, 3328.

Ortolani, S., and D. Schnack, 1993, *Magnetohydrodynamics of Plasma Relaxation* (World Scientific Publishing, Singapore).

Osborne, T., R. Dexter, and S. Prager, 1982, *Phys. Rev. Lett.* **49**(10), 734.

Park, H. K., 2019, *Adv. Phys.: X* **4**(1), 1633956.

Park, H. K., A. J. H. Donné, J. N. C. Luhmann, I. G. J. Classen, C. W. Domier, E. Mazzucato, T. Munsat, M. J. van de Pol, and Z. Xia (TEXTOR Team), 2006a, *Phys. Rev. Lett.* **96**(19), 195004 (pages 4).

Park, H. K., J. N. C. Luhmann, A. J. H. Donné, I. G. J. Classen, C. W. Domier, E. Mazzucato, T. Munsat, M. J. van de Pol, and Z. Xia (TEXTOR Team), 2006b, *Phys. Rev. Lett.* **96**(19), 195003 (pages 4).

Parker, E., 1957, *J. Geophys. Res.* **62**, 509.

Parker, E., 1963, *Astrophys. J. Suppl.* **8**, 177.

Parker, E., 1979, *Cosmical Magnetic Fields* (Clarendon Press, Oxford, UK).

Parker, E. N., 1955, *Astrophys. J.* **122**, 293.

Parker, E. N., and M. Krook, 1956, *Astrophys. J.* **124**, 214.

Parker, E. N., and A. F. Rappazzo, 2016, in *Magnetic Reconnection* (Springer), p. 181.

Paschmann, G., B. Ö. Sonnerup, I. Papamastorakis, N. Sckopke, G. Haerendel, S. Bame, J. Asbridge, J. Gosling, C. Russell, and R. Elphic, 1979, *Nature* **282**(5736), 243.

Petschek, H., 1964, *NASA Spec. Pub.* **50**, 425.

Phan, T., S. Bale, J. Eastwood, B. Lavraud, J. Drake, M. Øieroset, M. Shay, M. Pulupa, M. Stevens, R. MacDowall, et al., 2020, *Astrophys. J. Supp. Ser.* **246**(2), 34.

Phan, T., J. Drake, M. Shay, J. Gosling, G. Paschmann, J. Eastwood, M. Øieroset, M. Fujimoto, and V. Angelopoulos, 2014, *Geophys. Res. Lett.* **41**(20), 7002.

Phan, T., et al., 2000, *Nature* **404**, 848.

Phan, T. D., J. T. Gosling, M. S. Davis, R. M. Skoug, M. Øieroset, R. P. Lin, R. P. Lepping, D. J. McComas, C. W. Smith, H. Reme, and A. Balogh, 2006, *Nature* **439**, 175.

Pneuman, G. W., 1984, in *Physics of Solar Prominences*, edited by E. Jensen, P. Maltby, and F. Orall (IAU Colloq. **44**).

Prager, S. C., J. Adney, A. Almagri, J. Anderson, A. Blair, D. L. Brower, M. Cengher, B. E. Chapman, S. Choi, D. Craig, S. Combs, D. R. Demers, et al., 2005, *Nucl. Fusion* **45**, 276.

Priest, E., and T. Forbes, 1986, *J. Geophys. Res.* **91**, 5579.

Priest, E., and T. Forbes, 2000, *Magnetic Reconnection* (Cambridge University Press).

Pritchett, P. L., 2001, *J. Geophys. Res.* **106**, 3783.

Pritchett, P. L., 2003, in *Space Plasma Simulation* (Springer), p. 1.

P. L. Pritchett, 2008, *J. Geophys. Res. Space Phys.* **113**, A06210.

Pritchett, P. L., 2010, *J. Geophys. Res.* **115**, A10208.

Pritchett, P. L., and F. V. Coroniti, 2004, *J. Geophys. Res.* **109**, 1220.

Pritchett, P. L., F. S. Mozer, and M. Wilber, 2012, *J. Geophys. Res.* **117**, 06212.

Rees, M. J., and J. E. Gunn, 1974, *Mon. Not. Roy. Astron. Soc.* **167**, 1.

Ren, Y., 2007, *Studies of Non-MHD Effects during Magnetic Reconnection in a Laboratory Plasma*, Ph.D. thesis, Princeton University.

Ren, Y., M. Yamada, S. Gerhardt, H. Ji, R. Kulsrud, and A. Kuritsyn, 2005, *Phys. Rev. Lett.* **95**(5), 055003.

Ren, Y., M. Yamada, H. Ji, S. Gerhardt, and R. Kulsrud, 2008, *Phys. Rev. Lett.* **101**, 085003.

Ricci, P., J. Blackbill, W. Daughton, and G. Lapenta, 2004, *Phys. Plasmas* **11**, 4489.

Ricci, P., J. U. Brackbill, W. Daughton, and G. Lapenta, 2005, *Phys. Plasmas* **12**, 055901.

Rosenbluth, M. N., R. Y. Dagazian, and P. H. Rutherford, 1973, *Phys. Fluids* **16**, 1894.

Roytershteyn, V., W. Daughton, S. Dorfman, Y. Ren, H. Ji, M. Yamada, H. Karimabadi, L. Yin, B. Albright, and K. Bowers, 2010, *Phys. Plasmas* **17**, 055706.

Roytershteyn, V., W. Daughton, H. Karimabadi, and F. S. Mozer, 2012, *Phys. Rev. Lett.* **108**, 185001.

Runov, A., R. Nakamura, W. Baumjohann, T. L. Zhang, M. Volwerk, H.-U. Eichel-berger, and A. Balogh, 2003, *Geophys. Res. Lett.* **30**, 8.

Russell, C. T., and R. C. Elphic, 1979, *Geophys. Res. Lett.* **6**, 33.

Sakurai, T., 1976, *Pub. Astron. Soc. Japan* **28**, 177.

Sarff, J., A. Almagri, J. Anderson, D. Brower, D. Craig, B. Deng, D. den Hartog, W. Ding, G. Fiksel, C. Forest, V. Mirnov, S. Prager, et al., 2005, in *The Magnetized Plasma in Galaxy Evolution*, edited by K. T. Chyzy, K. Otmianowska-Mazur, M. Soida, and R.-J. Dettmar, pp. 48–55.

Sato, T., and T. Hayashi, 1979, *Phys. Fluids* **22**, 1189.

Savcheva, A., E. Pariat, A. van Ballegooijen, G. Aulanier, and E. DeLuca, 2012, *Astrophys. J.* **750**(1), 15.

Scholer, M., 1989, *J. Geophys. Res.* **94**, 8805.

Scholer, M., I. Sidorenko, C. Jaroschek, R. Treumann, and A. Zeiler, 2003, *Phys. Plasmas* **10**, 3521.

Scime, E., S. Hokin, N. Mattor, and C. Watts, 1992, *Phys. Rev. Lett.* **68**, 2165.

Sergeev, V., A. Runov, W. Baumjohann, R. Nakamura, T. L. Zhang, M. Volwerk, A. Balogh, H. Rème, J. A. Sauvaud, M. André, and B. Klecker, 2003, *Geophys. Res. Lett.* **30**, 60.

Sergeev, V. A., D. G. Mitchell, C. T. Russell, and D. J. Williams, 1993, *J. Geophys. Res.* **98**, 17345.

Shafranov, V., 1956, *At. Energy* **5**, 38.

Shay, M., J. Drake, R. Denton, and D. Biskamp, 1998, *J. Geophys. Res.* **103**, 9165.

Shay, M., C. Haggerty, T. Phan, J. Drake, P. Cassak, P. Wu, M. Øieroset, M. Swisdak, and K. Malakit, 2014, *Phys. Plasmas* **21**(12), 122902.

Shay, M. A., and J. F. Drake, 1998, *Geophys. Res. Lett.* **25**, 3759.

Shay, M. A., J. F. Drake, B. N. Rogers, and R. E. Denton, 2001, *J. Geophys. Res.* **106**, 3759.

Shay, M. A., J. F. Drake, and M. Swisdak, 2007, *Phys. Rev. Lett.* **99**(15), 155002.

Shibata, K., and T. Magara, 2011, *Living Rev. Solar Phys.* **8**(6).

Shibata, K., S. Masuda, M. Shimojo, H. Hara, T. Yokoyama, S. Tsuneta, T. Kosugi, and Y. Ogawara, 1995, *Astrophys. J.* **451**, L83.

Shibata, K., T. Nakamura, T. Matsumoto, K. Otsuji, T. J. Okamoto, N. Nishizuka, T. Kawate, H. Watanabe, S. Nagata, S. Ueno, R. Kitai, S. Nozawa, et al., 2007, *Science* **318**, 1591.

Shibata, K., S. Nozawa, and R. Matsumoto, 1992, *Publ. Astron. Soc. Japan* **44**, 265.

Shibata, K., and S. Tanuma, 2001, *Earth Planets Space* **53**(6), 473.

Shibata, K. 2016, in *Magnetic Reconnection* (Springer), edited by W. Gonzalez and E. Parker, pp. 378–381.

Shinohara, I., T. Nagai, M. Fujimoto, T. Terasawa, T. Mukai, K. Tsuruda, and T. Yamamoto, 1998, *J. Geophys. Res.* **103**(A9), 20365.

Silin, I., J. Büchner, and A. Vaivads, 2005, *Phys. Plasmas* **12**, 062902.

Soltwisch, H., 1988, *Rev. Sci. Instr.* **59**, 1599.

Soltwisch, H., P. Kempkes, F. Mackel, H. Stein, J. Tenfelde, L. Arnold, J. Dreher, and R. Grauer, 2010, *Plasma Phys. Control. Fusion* **52**(12), 124030.

Sonnerup, B., 1981, *J. Geophys. Res.* **86**, 10049.

Sonnerup, B. U. Ö, 1979, in *Solar System Plasma Physics*, edited by L. Lanzerotti, C. Kennel, and E. Parker (Cambridge University Press, North-Holland New York), p. 45.

Spitzer, L., 1962, *Physics of Fully Ionized Gases* (Interscience Publishers, New York).

Stark, A., W. Fox, J. Egedal, O. Grulke, and T. Klinger, 2005, *Phys. Rev. Lett.* **95**, 235005.

Steenbeck, M., F. Krause, and K.-H. Rädler, 1966, *Z. Naturforsch. A* **21**(4), 369.

Stenzel, R., and W. Gekelman, 1979, *Phys. Rev. Lett.* **42**, 1055.

Stenzel, R., and W. Gekelman, 1981, *J. Geophys. Res. Space Phys.* **86**, 649.

Stenzel, R., W. Gekelman, and N. Wild, 1982, *J. Geophys. Res.* **87**, 101.

Stenzel, R. L., M. C. Griskey, J. M. Urrutia, and K. D. Strohmaier, 2003, *Phys. Plasmas* **10**(7), 2780.

Stix, T. H., 1992, *Waves in Plasmas* (American Institute of Physics, New York).

Strauss, H. R., 1985, *Phys. Fluids* **28**, 2786.

Sturrock, P. A., 1966, *Nature* **211**, 695.

Sturrock, P. A., and S. M. Smith, 1968, *Solar Phys.*, **5**, 87.

Sweet, P., 1958, in *Electromagnetic Phenomena in Cosmical Physics*, edited by B. Lehnert (Cambridge Univ. Press, New York), p. 123.

Syrovatskii, S., 1971, *Sov. Phys. JETP* **33**, 933.

Syrovatskii, S., 1981, *Ann. Rev. Astron. Astrophys.* **19**, 163.

Syrovatskii, S. I., A. G. Frank, and A. Z. Khodzhaev, 1973, *Sov. Phys. Tech. Phys.* **18**, 580.

Szabo, A., D. Larson, P. Whittlesey, M. L. Stevens, B. Lavraud, T. Phan, S. Wallace, S. I. Jones-Mecholsky, C. N. Arge, S. T. Badman, et al., 2020, *Astrophys. J. Supp. Ser.* **246**(2), 47.

Tajima, T., J. Sakai, H. Nakajima, T. Kosugi, F. Brunel, and M. Kundu, 1987, *Astrophys. J.* **321**, 1031.

Takizuka, T., and H. Abe, 1977, *J. Comput. Phys.* **25**, 205.

Tanaka, K. G., A. Retinò, Y. Asano, M. Fujimoto, I. Shinohara, A. Vaivads, Y. Khotyaintsev, M. André, M. B. Bavassano-Cattaneo, S. C. Buchert, and C. J. Owen, 2008, *Ann. Geophys.* **26**(8), 2471.

Tanuma, S., T. Yokoyama, T. Kudoh, and K. Shibata, 2001, *Astrophys. J.* **551**(1), 312.

Tavani, M., A. Bulgarelli, V. Vittorini, A. Pellizzoni, E. Striani, P. Caraveo, M. Weisskopf, A. Tennant, G. Pucella, and A. Trois, 2011, *Science* **331**(6018), 736.

Taylor, J. B., 1974, *Phys. Rev. Lett.* **33**, 1139.

Taylor, J. B., 1986, *Rev. Mod. Phys.* **58**, 741.

Terasawa, T., 1983, *Geophys. Res. Lett.* **10**, 475.

Titov, V., and P. Dé, 1999, *Astron. Astrophys.* **351**, 707.

Torbert, R. B., J. L. Burch, T. D. Phan, M. Hesse, M. R. Argall, J. Shuster, R. E. Ergun, L. Alm, R. Nakamura, K. J. Genestreti, D. J. Gershman, W. R. Paterson, et al., 2018, *Science* **362**(6421), 1391.

Török, T., B. Kliem, and V. S. Titov, 2004, *Astron. Astrophys.* **413**(3), L27.

Török, T., and B. Kliem, 2005, *Astrophys. J.* **630**, L97.

Trintchouk, F., M. Yamada, H. Ji, R. Kulsrud, and T. Carter, 2003, *Phys. Plasmas* **10**, 319.

Tripathi, S. K. P., and W. Gekelman, 2010, *Phys. Rev. Lett.* **105**(7), 075005.

Tsuneta, S., 1996, *Astrophys. J.* **456**, 840.

Ugai, M., and T. Tsuda, 1977, *J. Plasma Phys.* **17**, 337.

Uzdensky, D., 2011, *Space Sci. Rev.* **160**(1-4), 45.

Uzdensky, D., and R. Kulsrud, 1998, *Phys. Plasmas* **5**, 3249.

Uzdensky, D., and R. Kulsrud, 2000, *Phys. Plasmas* **7**(10), 4018.

Uzdensky, D., and R. Kulsrud, 2006, *Phys. Plasmas* **13**, 062305.

Uzdensky, D. A., B. Cerutti, and M. C. Begelman, 2011, *Astrophys. J.* **737**(2), L40.

Uzdensky, D. A., N. F. Loureiro, and A. Schekochihin, 2010, *Phys. Rev. Lett.* **105**, 235002.

Vasyliunas, V., 1975, *Rev. Geophys. Space Phys.* **13**, 303.

von Goeler, S., W. Stodiek, and N. Sauthoff, 1974, *Phys. Rev. Lett.* **33**, 1201.

Waelbroeck, F. L., 1989, *Phys. Fluids B* **1**(12), 2372.

Wang, Y., R. Kulsrud, and H. Ji, 2008, *Phys. Plasmas* **15**, 122105.

Wesson, J., 1987, *Tokamaks* (Clarendon Press, Oxford).

Wilder, F., R. Ergun, D. Newman, K. Goodrich, K. Trattner, M. Goldman, S. Eriksson, A. Jaynes, T. Leonard, D. Malaspina, et al., 2017, *J. Geophys. Res. Space Phys.* **122**(5), 5487.

Winske, D., and N. Omidi, 1996, *J. Geophys. Res. Space Phys.* **101**(A8), 17287.

Woltjer, L., 1958, *Proc. Natl. Acad. Sci. USA* **44**, 489.

Wygant, J., C. Cattell, R. Lysak, Y. Song, J. Dombeck, J. McFadden, F. Mozer, C. Carlson, G. Parks, E. Lucek, A. Balogh, M. Andre, et al., 2005, *J. Geophys. Res.* **110**, A09206.

Wyper, P., S. Antiochos, and C. DeVore, 2017, *Nature* **544**, 452.

Yamada, M., 1995, in *Physics of the Magnetopause*, edited by P. Song, B. Sonnerup, and M. Thomsen (American Geophysical Union Monograph 90), pp. 215.

Yamada, M., 1999, in *Study of Magnetic Helicity and Relaxation Phenomena in Laboratory Plasmas*, edited by M. Brown, R. Canfield, and A. Pevtsov (AGU Monograph 111), p. 129.

Yamada, M., 2007, *Phys. Plasmas* **14**, 058102.

Yamada, M., 2011, *Phys. Plasmas* **18**(11), 111212.

Yamada, M., 2012, *Prog. Theor. Phys. Supp.* **195**, 167.

Yamada, M., L.-J. Chen, J. Yoo, S. Wang, W. Fox, J. Jara-Almonte, H. Ji, W. Daughton, A. Le, J. Burch, et al., 2018, *Nat. Commun.* **9**(1), 1.

Yamada, M., H. Furth, W. Hsu, A. Janos, S. Jardin, M. Okabayashi, J. Sinnis, T. Stix, and K. Yamazaki, 1981, *Phys. Rev. Lett.* **46**, 188.

Yamada, M., H. Ji, S. Hsu, T. Carter, R. Kulsrud, N. Bretz, F. Jobes, Y. Ono, and F. Perkins, 1997a, *Phys. Plasmas* **4**, 1936.

Yamada, M., H. Ji, S. Hsu, T. Carter, R. Kulsrud, Y. Ono, and F. Perkins, 1997b, *Phys. Rev. Lett.* **78**, 3117.

Yamada, M., H. Ji, S. Hsu, T. Carter, R. Kulsrud, and F. Trintchouk, 2000, *Phys. Plasmas* **7**, 1781.

Yamada, M., R. Kulsrud, and H. Ji, 2010, *Rev. Mod. Phys.* **82**, 603.

Yamada, M., F. Levinton, N. Pomphrey, R. Budny, J. Manickam, and Y. Nagayama, 1994, *Phys. Plasmas* **1**, 3269.

Yamada, M., Y. Nagayama, W. Davis, E. Fredrickson, A. Janos, and F. Levinton, 1992, *Rev. Sci. Instru.* **63**, 4623.

Yamada, M., Y. Ono, A. Hayakawa, M. Katsurai, and F. Perkins, 1990, *Phys. Rev. Lett.* **65**, 721.

Yamada, M., F. Perkins, A. MacAulay, Y. Ono, and M. Katsurai, 1991, *Phys. Fluids B* **3**, 2379.

Yamada, M., Y. Ren, H. Ji, J. Breslau, S. Gerhardt, R. Kulsrud, and A. Kuritsyn, 2006, *Phys. Plasmas* **13**, 052119.

Yamada, M., J. Yoo, J. Jara-Almonte, W. Daughton, H. Ji, R. M. Kulsrud, and C. E. Myers, 2015, *Phys. Plasmas* **22**(5), 056501.

Yamada, M., J. Yoo, J. Jara-Almonte, H. Ji, R. M. Kulsrud, and C. E. Myers, 2014, *Nat. Commun.* **5**, 4774.

Yamada, M., J. Yoo, and C. E. Myers, 2016a, *Phys. Plasmas* **23**(5), 055402.

Yamada M. et al., 2020, *Bull. of Am. Phys. Soc.* Div. Plasma Phys.

Yamada, M., J. Yoo, 2021, *Bull. of Am. Phys. Soc.* Div. Plasma Phys. GO09-00006.

Yamada, M., J. Yoo, and S. Zenitani, 2016b, in *Magnetic Reconnection* (Springer), edited by W. Gonzalez and E. Parker. pp. 143–179.

Yokoyama, T., K. Akita, T. Morimoto, K. Inoue, and J. Newmark, 2001, *Astrophys. J. Lett.* **546**, L69.

Yokoyama, T., and K. Shibata, 1995, *Nature* **375**(6526), 42.

Yoo, J., 2013, *Experimental Studies of Particle Acceleration and Heating during Magnetic Reconnection*, Ph.D. thesis, Princeton University.

Yoo, J., J.-Y. Ji, M. V. Ambat, S. Wang, H. Ji, J. Lo, B. Li, Y. Ren, J. Jara-Almonte, L.-J. Chen, et al., 2020, *Geophys. Res. Lett.* **47**(21), e2020GL087192.

Yoo, J., B. Na, J. Jara-Almonte, M. Yamada, H. Ji, V. Roytershteyn, W. Fox, and L.-J. Chen, 2017, *J. Geophys. Res.* **122**, 9264.

Yoo, J., S. Wang, E. Yerger, J. Jara-Almonte, H. Ji, M. Yamada, L.-J. Chen, W. Fox, A. Goodman, and A. Alt, 2019, *Phys. Plasmas* **26**(5), 052902.

Yoo, J., M. Yamada, H. Ji, J. Jara-Almonte, and C. E. Myers, 2014a, *Phys. Plasmas* **21**(5), 055706.

Yoo, J., M. Yamada, H. Ji, J. Jara-Almonte, and C. E. Myers, 2014b, *Phys. Plasmas* **21**(5), 055706.

Yoo, J., M. Yamada, H. Ji, J. Jara-Almonte, C. E. Myers, and L.-J. Chen, 2014c, *Phys. Rev. Lett.* **113**(9), 095002.

Yoo, J., M. Yamada, H. Ji, and C. Myers, 2013, *Phys. Rev. Lett.* **110**, 215007.

Yoon, P., A. Lui, and M. Sitnov, 2002, *Phys. Plasmas* **9**(5), 1526.

Zenitani, S., M. Hesse, A. Klimas, and M. Kuznetsova, 2011, *Phys. Rev. Lett.* **106**(19), 195003.

Zweibel, E., and D. Haber, 1983, *Astrophys. J.* **264**, 648.

Zweibel, E., and M. Yamada, 2009, *Annu. Rev. Astron. Astrophys.* **47**(1), 291.

Zweibel, E. G., and A. Brandenburg, 1997, *Astrophys. J.* **478**, 563.

Zweibel, E. G., and M. Yamada, 2016, *Proc. R. Soc. Lond. A* **472**, 20160479.

Index